# PALEONTOLOGY AND GEOLOGY OF THE WESTERN SALTON TROUGH DETACHMENT, ANZA-BORREGO DESERT STATE PARK, CALIFORNIA

FIELD TRIP GUIDEBOOK AND VOLUME FOR THE 1995 SAN DIEGO ASSOCIATION OF GEOLOGIST'S FIELD TRIP TO ANZA-BORREGO DESERT STATE PARK

## VOLUME I

Edited By
**PAUL REMEIKA**
Park Paleontologist/State Park Ranger I
Anza-Borrego Desert State Park
and
**ANNE STURZ**
Vice-chairperson/San Diego Association of Geologists
University of San Diego/Marine/Environmental Studies

# SAN DIEGO ASSOCIATION OF GEOLOGISTS

## 1995 OFFICERS

CHAIRMAN..... Phil Rosenberg

VICE CHAIRMAN..... Anne Sturz

SECRETARY..... Tissa Munasinghe

TREASURER..... Werner Landry

## 1995 CORPORATE SPONSORS

Allied Geotechnical Engineets, Joe Corones, William J. Elliott, Earth Resource Mapping, F & O Drilling Company Inc., Gallagher Drilling Service, Geotechnical Exploration, Inc., Group Delta Consultants Inc., Hargis + Associates, Barbara Johnston, The JMA Group of Companies, Perry and Nona Crampton, Robertson Geotechnical, Inc., (Gail) Fast Eddie Hammond, Dr. Anne Sturz, Tony Sawyer - Consulting Hydrogeologist, Soil Wash Technologies, Daryl Streiff, TPS Technologies Inc., Transglobal Environmental Geochemistry, Woodward-Clyde Consultants, Zeiser-Kling Consultants, Carole Ziegler.

**Field Trip Leaders:**

Chuck Herzig, Charles Lough, Paul Remeika, Anne Sturz.

**Front cover photograph:** The restored Vallecito Stage Station stands guard against a magnificent backdrop of the Laguna Mountains detachment breakaway margin along the Carrizo Corridor. Courtesy of Paul Remeika.
**Back cover photograph:** Desert ocotillo landscape with the detachment breakaway margin of the Tierra Blanca Mountains, highlighted by Sombrero Peak, in the distance. Courtesy of Paul Remeika.

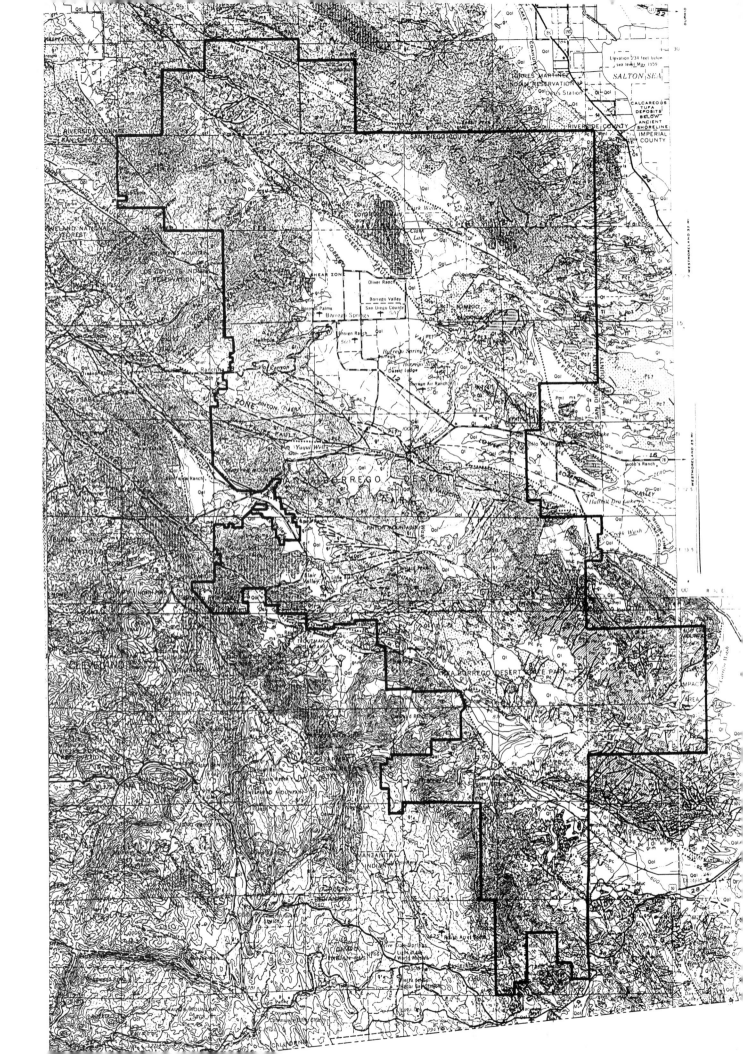

# PREFACE

**The Paleontology and Geology of the Western Salton Trough Detachment** was compiled to accompany the 1995 San Diego Association of Geologist's field trip to Anza-Borrego Desert State Park. Instead of a traditional field trip guide, the itinerary in Volume I provides a preliminary introduction of important geological criteria that allow the recognition of distinct mid-Tertiary to Recent domino-style fault-bounded crustal blocks that evolved as part of the WSTD along the Main Gulf Escarpment. The structural details are spectacularly exposed in the Vallecito-Fish Creek Basin and Borrego-San Felipe Basin, providing a natural scientific laboratory to study detachment-related deformation. We look forward to introducing you to a landscape borne from one of the most seismically-active areas within the state of California. Each participant will see in outcrop active faults of the San Jacinto and Elsinore Fault Zones; invertebrate fossils from the proto-Gulf of California; ancestral Colorado River deltaic sediments; and syn-depositional Neogene vertebrate-bearing sediments of the Vallecito and Borrego Badlands.

The articles in Volume I support the itinerary, present original data provided by on-going research, and reflect a small sampling of endeavors in the state park and beyond its immediate boundaries. The selected bibliography in Volume II stands alone as a comprehensive listing of references pertaining to the tectonic, seismologic and paleontologic evolutionary history of Anza-Borrego, as part of the WSTD.

Compilation of the field trip guidebook and volumes during the past year was accomplished with the enthusiastic assistance of the contributing authors and support from personnel of Anza-Borrego Desert State Park. Additional projects are currently underway. Some of these include research on Split Mountain megabreccia landslide deposits; magnetostratigraphy of the Borrego Badlands; tectonic investigations of Borrego Mountain; petrified wood identification of the late Pleistocene Clark Mountain local flora; study of camelid footprints, trackways and associated ichnofauna (part of the Arroyo Seco local fauna) from Blancan basin-margin deposits of the Vallecito-Fish Creek Basin; and detachment-related interpretation of the Borrego-San Felipe Basin based on detailed stratigraphic mapping, correlation and revision.

This guidebook and volumes serve as a beginning; pages via which interested colleagues may read to further explore byways and backroads, often neglected by casual visitors, in pursuit of answers. Many answers to paleontologic, geologic and seismologic questions, however, remain unresolved. As remarkable fossil discoveries are made, and as our interpretations about the geology are fine-tuned, we find ourselves approaching the brink to understanding the past, present and future tectonic histories of Anza-Borrego. Have a good trip!

Paul Remeika and Anne Sturz, Editors
October 1, 1995

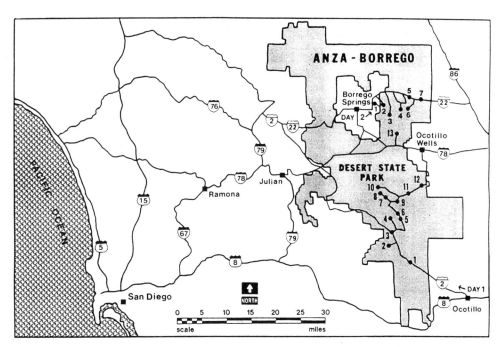

Figure 1. Field trip routes through Anza-Borrego, showing numbered stops and locations as mentioned in text.

# TABLE OF CONTENTS

**Introduction: A Visitor's Perspective -Center-stage at Font's Point-**
Paul Remeika ............................................................................................................................................. 1

**Basin Tectonics, Stratigraphy, and Depositional Environments of the Western Salton Trough Detachment: The 1995 San Diego Association of Geologist's Field Trip Guide to Anza-Borrego Desert State Park, California**
Paul Remeika ............................................................................................................................................. 3

**Preliminary Report on the Echinodermata of the Miocene and Pliocene Imperial Formation in Southern California**
Charles L. Powell, II ................................................................................................................................ 55

**Cretaceous Palynoflora and Neogene Angiosperm Woods from Anza-Borrego Desert State Park, California: Implications for Pliocene Climate of the Colorado Plateau and Age of the Grand Canyon**
Paul Remeika and R. Farley Fleming ...................................................................................................... 64

**Fossil Vertebrate Faunal List for the Vallecito-Fish Creek and Borrego-San Felipe Basins, Anza-Borrego Desert State Park and vicinity, California**
Paul Remeika, George T. Jefferson, and Lyndon K. Murray .................................................................. 82

**An Additional Avian Specimen Referable to *Teratornis incredibilis* from the early Irvingtonian, Vallecito-Fish Creek Basin, Anza-Borrego Desert State Park, California**
George T. Jefferson ................................................................................................................................. 94

**The Borrego Local Fauna: Revised Basin-margin Stratigraphy and Paleontology of the Western Borrego Badlands, Anza-Borrego Desert State Park, California**
Paul Remeika and George T. Jefferson ................................................................................................... 97

**The Mid-Pleistocene Stratigraphic Co-occurrence of *Mammuthus columbi* and *M. imperator* in the Ocotillo Formation, Borrego Badlands, Anza-Borrego Desert State Park, California**
George T. Jefferson and Paul Remeika ................................................................................................. 104

**A Diverse Record of Microfossils and Fossil Plants, Invertebrates, and Small Vertebrates from the late Holocene Lake Cahuilla Beds, Riverside County, California**
David P. Whistler, E. Bruce Lander, and Mark A. Roeder .................................................................. 109

**Stromatolites: Living Fossils in Anza-Borrego Desert State Park**
H. Paul Buchheim ................................................................................................................................. 119

**Interagency Cooperative Agreement between California Park Service, Colorado Desert District, Anza-Borrego Desert State Park and United States Department of the Interior, Bureau of Land Management, California Desert Conservation District**
Paul Remeika ........................................................................................................................................ 125

**Neogene Stratigraphy of the Borrego Mountain Area, Anza-Borrego Desert State Park, California**
C. Herzig, A. Carrasco, T. Schar, G. Murray, D. Rightmer, J. Lawrence, Q. Milton, and T. Wirths ...... 132

# INTRODUCTION: A VISITOR'S PERSPECTIVE
## CENTER-STAGE AT FONT'S POINT

**Paul Remeika**

California Park Service, Colorado Desert District, Anza-Borrego Desert State Park, 200 Palm Canyon Drive, Borrego Springs, California 92004

In one aspect Font's Point is a sudden place. Roads end. Carved by time. Centered in the arid Borrego Badlands, four million seasons of geology and paleontologic history band across a rawboned desertscape in richly-colored reds, browns, greens and grays. Conglomerates, sandstones, claystones and mudstones are the catalyst, chronicling the metamorphosis of differed paleoenvironments laid bare to view like so many pages of an open book. Compressed and hardened, each layer is a page recording a rock-leafed history of landscapes, fossil lifeforms, and climatic changes that are no longer present at Anza-Borrego.

In another aspect, Font's Point is well named. Over 200 years ago history met geology when the Spanish explorer Juan Baustista de Anza passed this high cornice, leading a band of men, women, and mules northward to Monterey. The path he forged followed San Felipe Wash. Father Pedro Font, who served as official chaplain, diarist and observer on Anza's historic marshes in 1775 and 1776, described this vantage point of the Borrego Badlands as "the sweepings of the earth".

Humbled, nongeologists verbalize about the ancient sweepings with a litany of superlatives-- breathtaking, incredible, spacious and bewildering. An open window, the visual drama of the northern half of Anza-Borrego -a 360-degree view- spreads out like a map. On the southern horizon stretch the Fish Creek, Vallecito and Pinyons, guardian ramparts of the Carrizo Corridor. In the middle distance, twin buttes of Borrego Mountain, split asunder by faulting, sit as granitic ribs that came up through the sediments like a basement elevator. In front of them, is the unfenced bleakness of San Felipe Wash, Sleepy Hollow and Borrego Sink. To the east, the desert seems to have no horizon. Where the land steps downward, the placid inland pool of the Salton Sea, 235-feet below sea level, occupies the lowermost portion of the geologically young Salton Trough, down-dropped by faulting that is rafting Baja California away from mainland Mexico. To the west, the granitic bastion of the San Ysidro Mountains, abuts the Anza-Borrego desert. To the north, Clark's Dry Lake lays in contrast to the precipitous paired mountain chains of Coyote Mountain, lesser in size, and the overlwelming 8,000 foot high dusk of the Santa Rosa Mountains. Where stone climbs into sky, monster fault lines control the mood of the land, dynamically breaking its continuity.

Old but not ancient, this upturned weathered edge may be the best place in North America to view sediments of the Pliocene and Pleistocene Epochs. But standing atop this precarious promontory and looking back into the past is no easy task. Many wonder how the sea of soft rock below came about, groping to visualize the past when water (gulf, river, delta and stream) drowned the land. Or what geological forces created this erosional masterpiece. And why? Once a wealth of muds and sands, this silent ordering of the ages is now an orchestrated tableau of ruined beds of dry-packed earth, forming a gutted and forbidding landscape. Any thought of verdant gallery forests and babbling brooks must be imaged -like a desert mirage- by its very absence. The visitor has to stretch his or her mind to visualize a landscape devoid of cacti, ocotillo and creosote bush. Think for a minute of a delta and images of aquatic, wet lowland habitats come to mind. During the Pliocene Epoch, Anza-Borrego was located south of the border as a receiving basin for the ancestral Colorado River while it carved out the Grand Canyon. Earlier, deltaic-marine waters of the northern Gulf of California were here. On a grandiose scale, marine sediments graded into deltaic facies, inaugurated by petrified wood, shellfish heaped in great abundance, and informative vertebrate trackways -such is the outcome where river meets sea. In turn, all become overlain by shallow pond, lacustrine bog and floodplain deposits as our local mountains were uplifted. Piled on top of one another, brown earthscape high, dry and unforgiving. Gone are shelly beaches and reefs, the Colorado River, camels, horses, cheetahs, bears and ground sloths. Many creatures are extinct. But they were here. Their story is part of our inheritance, given to us by diligent earth scientists who patiently probe the badland's history, turning back the paleontologic and geologic clock to yesteryear when this desertland was home to checks and balances of an Ice Age life system.

As the geologic story approaches our own time, acres of sedimentary rock contain enough side canyons and disoriented washes for a lifetime of adventureous exploring. Where the eventful drama of prehistoric life once teemed, it is now, by the very nature of the area, a baked and arid rocky geography of sunken mesas and corrugated hills of dry mud. Daily, forces of erosion gently soften contour lines, evening out highs and lows. Ironically, the area is experiencing weathering, wind, rain and generations of flashflooding that helped in its formation. Nowhere else is this combination exactly the same. And nowhere else of equivalent size can its rocks, histories, landscapes and faunas be told. They are Anza-Borrego.

Figure 1. Below Font's Point the Borrego Badlands expose a Plio-Pleistocene rock-leafed history. From top to bottom, sediments include vertebrate-bearing basin-margin alluvia of the Ocotillo formation interbedded with finer grained basinal claystones of the Brawley and Borrego Formations.

# BASIN TECTONICS, STRATIGRAPHY, AND DEPOSITIONAL ENVIRONMENTS OF THE WESTERN SALTON TROUGH DETACHMENT

## THE 1995 SAN DIEGO ASSOCIATION OF GEOLOGIST'S FIELD TRIP GUIDE TO ANZA-BORREGO DESERT STATE PARK, CALIFORNIA

**Paul Remeika**
California Park Service, Colorado Desert District, Anza-Borrego Desert State Park, 200 Palm Canyon Drive, Borrego Springs, California 92004

## INTRODUCTION

ABDSP was established, in part, to protect and preserve in public trust a unique out-of-doors storehouse of paleontological and geological natural resources of the Vallecito-Fish Creek Basin (VFCB) and the Borrego-San Felipe Basin (BSFB). These small-scale Cenozoic paleobasins contain syn-extensional deposits that evolved in response to strike- and dip-slip fault-induced paleoenvironments of the Western Salton Trough Detachment (WSTD).

ABDSP's rich and diverse paleontological resources, represented by five paleofaunas (four local faunas) (LF) (Downs and White, 1968; Remeika, 1992a; Remeika and Jefferson, 1993; and this volume) list 213 vertebrate taxa that range from Hemphillian to Rancholabrean North American Land Mammal Age (NALMA), including large grazing herbivores, carnivores, aquatic mammals, microtine rodents, amphibians, reptiles and fish, with a dozen holotypic specimens. Add nearly 300 species of marine and deltaic-marine invertebrates and the collected assemblage (500+ taxa) may represent the largest repository of Plio-Pleistocene vertebrate and invertebrate fossils in North America.

The VFCB and BSFB have received renewed attention as an outgrowth of ABDSP's General Plan Inventory of Natural Features (Remeika 1992b, 1993). The purpose of this 2 day introductory field trip is to examine and systematically review, in detail, the unique tectonic, stratigraphic and paleontologic phenomena recorded within Neogene sediments exposed in both post-detachment paleobasins, part of the geodynamic WSTD.

The roadlog for the first day provides a reappraisal of the VFCB, including its tectonic evolution with an informal revision of the stratigraphy from the uppermost (youngest) western exposures of the sequence to the bottom (older) easternmost deposits. The paved portion follows the Imperial Highway north from the town of Ocotillo to the Imperial/San Diego county line, and County Highway S-2 north to Canebrake. It also includes abbreviated portions of the Carrizo Corridor within the state park south of Agua Caliente County Park. For additional coverage of S-2 northward to Scissor's Crossing, the reader is referred to Todd, et al. (1994) and Remeika and Lindsay (1992). The dirt portion is a through-going, four-wheel drive " back through time" field session, that departs from S-2 at Vallecito Creek and travels eastward across the three-dimensional clastic depositional-dip sections of the Vallecito and Fish Creek Badlands to Split Mountain Gorge, via the Diablo Dropoff. Stops along this route update research endeavors conducted since the Society of Vertebrate Paleontology's successful paleogeologic off-road field trip (White, et al. 1991).

The VFCB has been the subject of past field trips (Gastil and Bushee, 1961; Downs, 1966; Hart and Dowlen, 1974; Robinson and Threet, 1974; Norris, et al. 1979, 1993; Crowell and Sylvester, 1979b; Kerr and Kidwell, 1991; White, et al. 1991; Remeika and Lindsay, 1992; Todd, et al. 1994). Most consisted of short excursions to points of local geologic or paleontologic interest (e.g. Split Mountain Gorge, Vallecito Creek, Fish Creek Wash, County Highway S-2).

The roadlog for the second day is based on preliminary stratigraphic mapping of the westernmost

Borrego Badlands. It provides a neotectonic and stratigraphic reassessment of the basin-margin outcrop geology for the northern portion of the BSFB, exposed within the seismogenic San Jacinto Fault Zone (SJFZ) along County Highway S-22 through the Borrego and Santa Rosa Badlands. The paved portion of the route includes several off-road side-trips (four-wheel drive recommended) to local points of geological interest. The BSFB has been the subject of two past field trips (Gath, et al. 1986; Reynolds and Remeika, 1993).

To establish stratigraphic control within the WSTD, this report challenges traditional ideas about the Neogene VFCB and BSFB by providing a non-traditional, non-uniform forum for local stratigraphic schemes, correlations and interpretations of the sedimentary package based on fifteen years of field-related investigations of the distinctive facies associations used herein. It contradicts the basic premise of guidebooks by advocating the informal definition of new formations or groups of formations where necessary, or retaining rock units previously defined locally or regionally.

**PLEASE NOTE: It is a violation of federal and state regulations and Public Resource Codes to disturb, collect, or destroy fossiliferous and geologic resources on public lands under the jurisdictions of the California Park System (state) and the Bureau of Land Management (federal).**

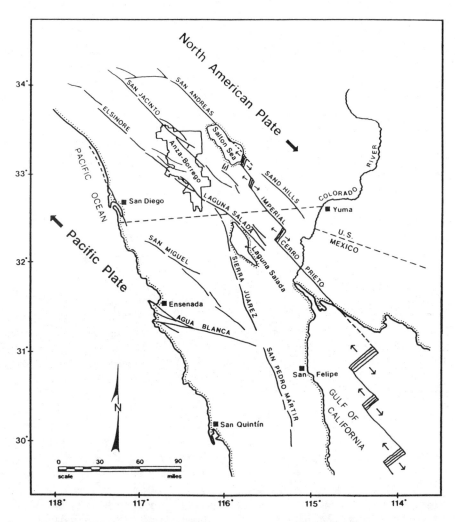

FIGURE 1. Simplified diagrammatic fault map of the Western Salton Trough Detachment, East Gulf Escarpment. Map shows the major NW-trending fault zones and their relationship to Anza-Borrego, the Salton Sea, Laguna Salada, and with the inferred spreading centers of the Salton Trough-Gulf of California structural depression.

# ROAD LOG

## DAY 1: Vallecito-Fish Creek Basin

**0.0 (0.0)** EXIT Interstate 8 at the Ocotillo/Imperial Highway off-ramp. At the junction with the Imperial Highway, make a right turn and meet in the dirt parking lot N of the Desert Kitchen cafe and gas station complex. From here, the road log proceeds N through the small desert community of Ocotillo.

**0.2 (0.2)** Interstate 8 underpass. Proceed N through the small desert community of Ocotillo.

**0.4 (0.2)** During the summers of 1977, 1982 and 1983, catastrophic floodwaters from the Jacumba Mountains roared down Devil's Canyon and In-Ko-Pah Gorge (from the W), covered roads with mud and debris, and damaged many residential and commercial buildings in Ocotillo. Interstate 8 and the nearby San Diego and Arizona Eastern Railroad were also heavily damaged.

**0.8 (0.4)** SLOW DOWN. The Imperial Highway makes a 30 MPH turn to the left.

**1.4 (0.6)** Four-way traffic stop intersection with Shell Canyon Road. Note weathered green- and white-colored sandbags along the side of the road. These are remnants from the last major flood event (1983). Shell Canyon Road provides paved access to the Bureau of Land Management's (BLM) 17,000 acre Coyote Mountains Wilderness Area, including Fossil Canyon, designated as an Area of Critical Environmental Concern (ACEC), part of an interagency cooperative agreement between ABDSP and the BLM (Remeika, this volume). Fossil Canyon is closed to vehicles and fossil collecting is prohibited. CONTINUE STRAIGHT.

**1.9 (0.5)** To the right (3:00) is the complexly-faulted SW range-front of the Coyote Mountains. The Coyote Mountains consist of over 1400 m of metasedimentary marbles, quartz-mica schists and metaquartzites (Miller and Dockum, 1983), subordinate crystalline basement, and Miocene superjacent volcanic and sedimentary strata related to the extensional attenuation of the crust along the major NW-striking Elsinore Fault Zone (EFZ) (W strand of the San Andreas transform system) (Miller and Kato, 1990).

The well-defined EFZ can be traced for over 220 km from the Sierra de los Cucapas (Laguna Salada), Mexico to Corona where it bifurcates into the Whittier-Narrows and Chino Faults. The EFZ is an important Neogene tectonic feature related to major N-NW-striking Late Cenozoic transpeninsular fault zones that delineate the easternmost Peninsular Ranges detachment breakaway margin of the Western Salton Trough Detachment (WSTD) (Frost, et al. 1993) (Figure 1). It is part of the 1500 km-long Main Gulf Escarpment (MGE) (Axen, 1995) that extends from Anza-Borrego southward down the length of Baja and the western Gulf of California. The MGE is interpreted to be the footwall of breakaway margins of E-directed extensional fault systems and/or W-tilted half-grabens above E-dipping listric fault systems (Dokka and Merriam, 1982; Stock and Hodges, 1990). Locally, the MGE governs the structural relief of the Coyote Mountains and the VFCB including, but not limited to, sedimentary accumulation rates and facies development.

**5.3 (1.4)** Beneath power lines the road enters BLM's California Desert Conservation Area, Yuha Desert District.

**6.1 (2.8)** To the left, low-lying mudhills consist of reddish-brown siltstones and claystones of the Diablo formation (informal name of Remeika, 1991a). To the right, the Diablo formation (part of the informal Colorado River group of Reynolds and Remeika, 1993) and the light-green colored sediments of the Burrobend formation (new informal name this volume) of the Imperial group (new informal name this volume) are steeply juxtaposed against the thrust-front of the Coyote Mountains Frontal Fault (CMFF).

**7.2 (1.1)** To the right is dirt access to the Domelands trailhead parking area at the base of the Coyote Mountains, part of the BLM's ACEC. Trails lead to marine fossil-bearing sediments of the Latrania member (informal name of Stump, 1972) which is here raised to informal formational level. Fossil collecting is prohibited within this area.

The controversial Imperial Formation of Hanna (1926) is herein elevated to informal group level as the Imperial group. As used here, rock exposures of the Imperial group were first discovered in Carrizo Wash in the early 1850's (Blake, 1855). Originally named the Carrizo Formation (Kew, 1914) and the Imperial Formation (Hanna, 1926), Woodring (1932), Tarbet and Holman (1944), and Tarbet (1951) redefined the stratigraphic package based on local stratigraphic correlations, but could not reach an agreement on placement of member boundaries.

The Imperial group includes a two-fold lithologic division: **(1)** a lowermost coarse-grained turbidite deposit of the marine Lycium member (informal name of Winker, 1987), which is here raised to informal formational level, the Lycium formation, that grades upward into gypsiferous deposits identical to the Fish Creek Gypsum (?) (Ver Planck, 1952) in Gypsum Canyon S of the Flatiron (Coyote Mountains) and N of Boundary Mountain (Fish Creek Mountains), and discontinuously exposed transgressive bioclastic limestones and a coarse-grained, fossiliferous marine sandstone of the Latrania formation. Based upon diagnostic mollusks, Arnold (1904), Mendenhall (1910), Kew (1914), and Bramkamp (1935) considered the lower Imperial group to be Miocene or middle Miocene in age in the Coachella Valley; and **(2)** upper widely distributed fine-grained deltaic-marine mixed lithologies marginal to the northern Gulf collectively named as the Burrobend formation. The Burrobend formation is coeval with the Bouse Formation (Metzger, 1968) of the Colorado River Valley, and is constrained to the lower Pliocene based on identified foraminifera (Ingle, 1974; Stump, 1972) and magnetostratigraphy (Opdyke, et al. 1977; Johnson, et al. 1983).

The Latrania formation represents the lowermost transgressive bioclastic and bioturbated marine sandstones and conglomerates containing coraliferous sandstones discontinuously exposed only on the western margin of the Salton Trough (dismembered from the Gulf) along inboard strands of the San Andreas Fault Zone (SAFZ). Informal subdivisions include the Lion Sandstone (Vaughan, 1922), or Latrania Sands (Hanna, 1926; Keen and Bentson, 1944; Powell, 1986; Kidwell, 1988; Deméré, 1993), lower marine shales (Woodring, 1932), lower member (Tarbet, 1951), Andrade member (informal name of Christensen, 1957), Lycium member (informal name of Woodard, 1963, 1974), and "northern" Miocene and "southern" Pliocene units (Powell, 1986, 1988).

The Latrania formation is very fossiliferous in the VFCB, dominated by a molluscan fauna that has a strong Panamic affinity to the Caribbean, Surian and Peruvian Provinces (Powell, 1987). Conrad (1855), Kew (1914), Vaughan (1917), Hanna (1926) and Bramkamp (1935) did the first detailed paleontology descriptions, followed by specific reports on the archaeogastropods (Schremp, 1981), bivalves (Powell, 1986), and a rocky shore ichnofacies community (Watkins, 1990). Taxa discovered along the southern flanks of the Vallecito Mountains are characteristic of a euhaline, moderate to high-energy subtidal to inner shelf, rocky to sandy habitat (Tucker, et al. 1994). The megafossils indicate continental shelf water depths of less than 50 m (Powell, 1986).

The marine invertebrates of the Latrania formation are numerous and diverse, including over 100 species of bivalves, 72 species of gastropods, 16 species of echinoids, numerous corals, barnacles, trace fossils, ostracodes, bryozoans, and foraminiferans. Faunal evidence indicates these fossils are closely related to species of the Gulf of Mexico-Caribbean regions.

Calcareous nannoplankton (zones CN9-CN11) and planktonic foraminifers (zones N17-N19) (McDougall, et al. 1994), and marine vertebrate fossils (Mitchell, 1961; Deméré, 1993; Thomas and Barnes, 1993) have also been documented from the Latrania formation. Radiometric age determinations (K/Ar) of about 6 Ma (Matti, et al, 1985), and $6.04 \pm 0.18$ and $5.94 \pm 0.18$ Ma (Smith, 1991; Matti and Morton, 1993) from a basalt flow in the overlying Painted Hill Formation (Allen, 1957) at Super Creek in Riverside County, 7.6 - 8.0 Ma for an ash in the Latrania at Garnet Hill derived on chemical correlation with volcanic ash in the Ventura Basin which was dated using diatom biostratigraphy (Rymer, et al. 1994, 1995), and a maximum K/Ar age of $10.1 \pm 1.2$ Ma from the underlying Coachella Fanglomerate at the Trout Farm in Whitewater Canyon (Krummenacher in Peterson, 1975) constrain the oldest mollusks recovered from the Latrania formation to a late Miocene age.

This report contradicts a plethora of models that imply a south to north marine inundation of the Salton Trough up to San Gorgonio Pass during the Miocene (McDougall, et al. 1994; Rymer, et al. 1994, 1995; Powell, 1995). The Salton Trough represents a geodynamic structural depression, the northernmost equivalent of the proto-Gulf of California, that evolved in response to Pliocene to Recent wrench-fault transtension associated with NW-trending right-slip strands of the SAFZ. There is a lack of field evidence to support the existence of a proto-Salton Trough extension to the northern Gulf of California prior to the Pliocene.

This interpretation is reinforced by Miocene (marine), as well as Pliocene (deltaic-marine and delta-plain) sediments dismembered in the Salton Trough along the inboard side of the SAFZ (Mission Creek strand). Thomas and Barnes (1993) provide an estimate of 240 km of Miocene offset along the SAFZ based on a correlation of fossiliferous localities. Peterson (1975) estimates up to 215 km of offset along the tectonically-active Mission Creek Fault during the Neogene. More recent estimates of dismemberment suggest 180 km between deltaic outcrops at Willis Palms (Indio Hills) and the Bouse Formation of the Colorado River Valley since ~ 5 Ma. Furthermore, Winker and Kidwell (1986) provide palinspastic reconstruction of Pliocene delta-plain wood-bearing sediments dismembered from outboard counterparts (Fortuna Basin and San Luis Basin) farther to the south as part of the Bouse Embayment. Offset clastic lacustrine sediments of the Pliocene Borrego Formation (Tarbet and Holman, 1944) in the Mecca Hills represent the first unequivocal Plio-Pleistocene basinal rock unit deposited after deltaic progradation in the Salton Trough, and constrain its age to the late Neogene.

**7.5 (0.3)** The NW-striking CMFF is responsible for uplifting the Coyote Mountains along active low-angle, strike-slip strands dramatically conveyed by a significant thrust-fault component (with a reverse fault plane that strikes N80°E and dips between 24° to vertical toward the range-front) (Pinault, 1984). Along this portion of the CMFF, extending to the Carrizo Badlands Overlook, bedrock discontinuously overrides Neogene sedimentary rock (Latrania, Burrobend, or Diablo formations). Deformation of Quaternary alluvial fan-fluvial basin-margin deposits along the southward continuation of the EFZ (Rockwell and Pinault, 1986) can be mapped beyond the International Border where it is linked to the NW-striking Laguna Salada Fault in northernmost Baja California (Mueller and Rockwell, 1995).

Primary geomorphic evidence of recent activity along the CMFF is well expressed at the surface. Discrete, though segmented, NW-trending lineaments, well-preserved Holocene fault scarps, shutter ridges, deflected and offset stream channels, and displaced alluvial fans as much as 40 m in a right-lateral sense indicate repeated and frequent events of dominantly dextral strike-slip faulting (Pinault and Rockwell, 1984). There is also a considerable dip-slip component, that heretofore, has not been recognized. Pinault (1984) estimates that the CMFF moved at a rate of about 3.4-5.5 mm/yr. during the Holocene with a recurrence interval of about $350 \pm 125$ years for large events.

**8.3 (0.8)** Leaving BLM's California Desert Conservation Area, Yuha Desert District, and entering San Diego County and Anza-Borrego Desert State Park (ABDSP) (Bow Willow Sector). This magnificent parkland was established in 1933 as California's first desert state park. It has grown to about 620,000 acres, and includes 1/5 of San Diego County. It is the largest state park in the contiguous United States (with over 800 km of dirt roads, and 12 wilderness areas) protecting animal, plant, geologic, paleontologic, and archeologic resources for the benefit of current and future park visitors. ABDSP is about 48 km wide E to W, and ranges from sea level along its easternmost boundary to over 2,000 m atop Combs Peak in the Bucksnort Mountains. Lengthwise, the park stretches 96 km long N to S, from just beneath Toro Peak (elevation 2902 m) in the Santa Rosa Mountains to just shy of the International Border near Jacumba.

**8.4 (0.1)** Milepost marker 56.

**8.6 (0.2)** Mortero Wash on the left with access to the Volcanic Hills, Indian Hill and Dos Cabezas. Four-wheel drive is strongly recommended. North Mortero Wash on the right with access to the Dolomite Mine.

**9.0 (0.4)** Volcanic Hills on the left. The Jacumba Basalt (Miller, 1935) is composed of dark-colored olivine basalts and lighter andesites associated with massive flows, breccia, air-fall cinder deposits, and minor volcaniclastic sandstones. Good exposures are south of Red Hill (elevation 561 m) along Lava Flow Wash. The volcanics have also been informally assigned to the Alverson Canyon formation (Tarbet and Holman, 1944), Alverson andesite (Dibblee, 1954), and Alverson formation (Ruisaard, 1979). Radiometric age determinations (K/Ar) range from $14.9 \pm 0.5$ Ma, $16.9 \pm 0.5$ Ma, $18.5 \pm 0.9$-$18.7 \pm 1.3$ Ma to $21.5 \pm 3.9$ Ma (Hawkins, 1970; Minch and Abbott, 1973; Eberly and Stanley, 1978; Hoggatt, 1979; Ruisaard, 1979; Gastil, et al. 1981; Mace, 1981) and constrain the initial regional-scale synrift events within the Salton Trough.

Contrary to most views, the structurally and lithologically similar volcanic exposures of the Jacumba Basalt in the Coyote Mountains, Volcanic Hills, and Table Mountain are dismembered by a composite of post-Miocene strike- and dip-slip separation along the WSTD.

**9.5 (0.5)** Milepost marker 55. Locally, silicified wood-bearing sandstones of the Diablo formation (Pliocene) are baked by a resurgence of volcanic activity (flows) associated with the Jacumba Basalt in the Volcanic Hills.

**10.1 (0.6)** Pull-out parking area on the left.

**10.5 (0.4)** Milepost marker 54.

**11.1 (0.6)** Jojoba Wash on the left and West Dolomite Mine Trail on the right. Enter Carrizo Canyon State Wilderness Area of ABDSP.

**11.4 (0.3)** S-2 climbs the low gravelly terrace held up by fine-grained redbeds of the Diablo formation. The terrace is capped by light-brown, unconsolidated Quaternary sandstones and gravels that form desert pavement.

**11.5 (0.1)** Milepost marker 53.

**12.5 (1.0)** Milepost marker 52.

**12.7 (0.2) STOP 1 Carrizo Badlands Overlook** (Figure 2). TURN RIGHT onto the dirt roadway and PARK at the display panel for a dramatic view of the VFCB.

The complex VFCB is a 373 km² syndetachment sedimentary basin, stepped-down basinward along the easternmost Peninsular Ranges detachment breakaway margin in the southern portion of ABDSP, 43 km south of Borrego Springs, California. This structural domain is interpreted to have evolved during post-Miocene transtensional unrest between the evolving EFZ on the west (Allison, 1974a, b; Gibson, 1983; Gibson, et al. 1984) and the San Jacinto Fault Zone (SJFZ) on the east (Sharp, 1967). Lithospheric attenuation between these young master faults is relative to transform-related motion, crustal extension, and deformation along the westernmost aggrading Pacific-North American plate-tectonic boundary of the Salton Trough-Gulf of California (STGC) structural rift depression (Elders, et al. 1972; Gastil, et al. 1975; Crowell and Sylvester, 1979a, b; Elders, 1979; Hutton, et al. 1991; Stock, et al. 1991; Winker and Kidwell, 1986; Lucchitta, 1987), part of the WSTD. Locally, movement direction (shear sense) of the detachment faults is generally top to the east, indicating asymmetric extension, with exhumation of the Peninsular Ranges crystalline complex.

Viewed from the Carrizo Badlands Overlook, the VFCB varies from 19 to 24 km in width and has an asymmetric half-graben configuration. This paleobasin is fault-bounded on four sides, delineated by the bedrock Fish Creek, Vallecito-Pinyon, Tierra Blanca and Coyote hinterland uplifts. Structural relief is

Figure 2. Looking east over the Vallecito and Carrizo Badlands from the Carrizo Badlands Overlook. In the distance is the Fish Creek Mountains.

between 300-1000 m. Extension and differential deformation is accomodated by dilational movements along NW-striking, high-angle, dip-slip fault strands within the EFZ, westward rotation of the Peninsular Ranges (Gastil, et al. 1975) along strike-slip transfer faults, and more specifically, clockwise rotation of the VFCB (Woodard, 1963; Johnson, et al. 1983).

Typical Cordilleran source rock lithologies include late Mesozoic mid-crustal eastern-zone tonalitic plutons, associated high-grade migmatitic fabrics, Paleozoic metasedimentary, and undated prebatholithic miogeoclinal screens (Miller and Dockum, 1983; Gastil and Miller, 1984; Todd, et al. 1994) that collectively yield a local Peninsular Range suite (Peterson, et al. 1968; Peterson and Nordstrom, 1970). Recent interdisciplinary studies of this core complex continue to add new data (Todd, et al. 1994; Grove, 1994; Lough, 1993).

Along the Tierra Blanca Mountains, the dynamic EFZ is delineated by a linear, though segmented, aggregate of NW-SE trending basin-bounding fault segments, forming a 2 km-wide zone, associated with late Quaternary fault scarps that form significant strike- and dip-slip components along the abrupt, crystalline range-front footwall. The EFZ consists of three NW-striking conjugate segments, each up on the southwest. The segments, *en echelon* to the SE, are the Vallecito Valley Fault (VVF) (west [main] strike-slip segment); a zone which includes two sub-parallel branches of the Tierra Blanca Mountains Frontal Fault (TBMFF) (central segment); and the Vallecito Creek Fault (VCF) (east segment). The latter two segments make up a zone of composite dip-slip faults that trend northwest parallel to the range-front with subordinate dextral shear that displace late Pleistocene and Holocene alluvial fans. RETURN to the pavement.

**12.9 (0.2)** TURN RIGHT (north) on S-2.

**13.2 (0.4)** Entrance to boulder-choked Canyon Sin Nombre on the right with access to the Carrizo Badlands. The sandy drainage is cut longitudinally by the CMFF. Four-wheel drive is recommended for visitors interested in exploring this remote jeep trail.

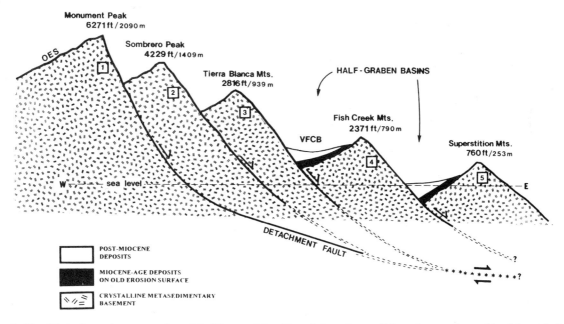

Figure 3. Simplified schematic cross-section of the Western Salton Trough Detachment (Mount Laguna-Superstition Mountains), Anza-Borrego Desert State Park and vicinity. Note how the once coherent old erosion surface (OES) has fractured into discrete bedrock blocks. Each block continues to slip down (east) along master fault planes (hanging wall normal faults), illustrating successively lower elevation profiles, as they subside domino-style into the Salton Trough. At depth, incipient brecciation, chloritic alteration and mylonization documents the detachment. On the surface, half-graben basins, such as the Vallecito-Fish Creek Basin (VFCB) develop in-between, with Miocene sedimentary detritus preserved on footwall bedrock exposures. Rock units exhibit steeply-dipping bedding planes due to block tilt. Post-Miocene deposits are progressively younger up-section and reveal gentler dips. Not to scale. Master faults include: 1. Laguna Mountains Escarpment; 2. Vallecito Valley Fault (west branch of Elsinore Fault Zone) (EFZ): 3. Tierra Blanca Mountains Frontal Fault (central branch of EFZ); 4. Coyote Creek-Superstition Mountains Fault (west branch of San Jacinto Fault Zone) (SJFZ); 5. Superstition Hills Fault (east branch of SJFZ).

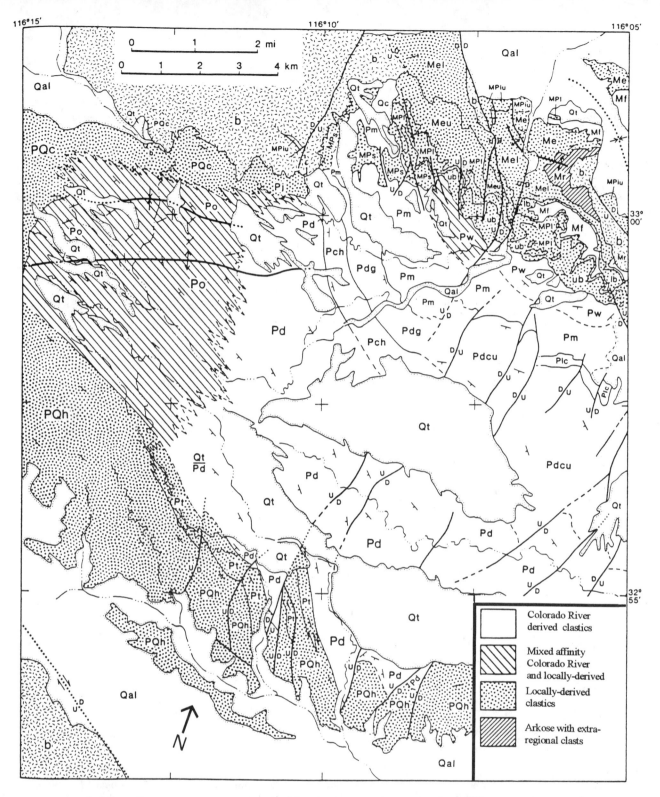

Figure 4. Simplified geologic map of the Vallecito-Fish Creek Basin, illustrating distribution of provenance suites. b= crystalline basement, undifferentiated; Mr= Red Rock Canyon member; Me= Elephant Trees member; Mel= lower fanglomerate; lb= lower boulder bed; Meu= upper fanglomerate sandstone; Mf= Fish Creek Gypsum; Mpiu= Imperial group, undifferentiated; Mpi= Lycium formation; ub= upper boulder bed; Mps= Stone Wash member; Pw= Wind Caves member; Pm= Mudhills member; Plc= Lavender Canyon member; Pdg= Deguynos member; Pch= Camels Head member; Pdcu= Deguynos and Camels Head members, undifferentiated; Pd= Diablo formation; Pj= Jackson Fork member; Pqc= Canebrake Conglomerate, undifferentiated; Po= Olla member; Pt= Tapiado member; Pqh= Huesos member. Modified after Kerr (1982), Remeika (unpublished mapping, 1981-1984), Winker (1987).

**13.6 (0.3)** Milepost marker 51. SLOW DOWN. The modern roadway (built in 1952; paved in 1960) curves downward through Sweeney Pass. Near the bottom remnant scars of old Sweeney Grade (1930's-1940's) are visible in the road cut on the left. Road cuts to the right reveal good cliffside exposures of Canebrake Conglomerate (Dibblee, 1954). This rock unit is composed of gray, crudely-bedded, massive (@2,333 m) pebble to boulder gravels, sandstones and micaceous clays of predominantly granitoid (redeposited La Posta Tonalite of Todd, 1994; and Chappell and White, 1974) composition, exposed along the southeast flank of the Pinyon Mountains in the vicinity of the Mud Palisades, and at Sweeney Pass. It is restricted by short transport distances into the subsiding basin axis, representing proximal deposition of a vertically-stacked W-NW thickening of coarse-grained alluvial fan sediments. Basinward, it grades into a variety of fossiliferous finer-grained fluvial, lacustrine and distal alluvial fan settings typical of the Palm Spring Formation, exhibiting very complex intertonguing relationships between the proximal and distal fan deposits. In this vicinity, the Canebrake Conglomerate is complexly deformed within the EFZ.

This volume informally recognizes four (I-IV) distinct depositional episodes for the Canebrake Conglomerate. These reflect episodic responses to dynamic climatic, hydrologic, and tectonic controls on the overall fluvial regime, as expressed in basinward discharge, sediment type, and net rate of sediment accumulation (including facies patterns, low-angle unconformities, and kinematic indicators) during episodes of uplift documented in the VFCB. These episodic events are categorized as Canebrake I for deposition of the Pliocene Olla member (informal name of Winker, 1987); Canebrake II for the lower half of the upper Pliocene Palm Spring Formation lacustrine lakebeds (informal Tapiado member of Woodard, 1963) and floodplain deposits (informal Huesos member of Woodard, 1963); Canebrake III for Plio-Pleistocene distal alluvial fans (informal Vallecito member of Woodard, 1963) of the Palm Spring Formation; and Canebrake IV for Quaternary fan and fluvial deposits exposed at Agua Caliente, Canyon Sin Nombre, and Bow Willow Creek (Bow Willow formation) (new informal name this volume).

**14.6 (1.0)** Milepost marker 50 at Sweeney Canyon. This canyon is on strike with the EFZ and a significant segment of the EFZ (VVF) can be traced up the canyon, separating undifferentiated Canebrake Conglomerate on the right from desert varnished metasedimentary bedrock of the Jacumba Mountains on the left.

**15.5 (0.9)** Ahead (on the right), the Canebrake Conglomerate is unconformably overlain by Quaternary coarse-grained pebbly, arkosic sandstones. Sedimentary structures indicate the sediments are fluvially derived and may represent deposits from the ancestral Carrizo Creek. The Bow Willow formation occupies the distal end to Woodard's (1963) informal Mesa conglomerate of the northernmost VFCB. One *Mammuthus* sp. molar fragment was retrieved from these sediments in the early 1980's along Vallecito Creek, and more recently, Aves, small- and medium-sized Mammalia, *Camelops* sp., and petrified wood were recovered from finer-grained exposures along Bow Willow Creek near the historic site of Las Palmitas.

**15.6 (0.1)** Milepost marker 49.

**15.9 (0.3)** Entrance road to South Carrizo Creek on the right.

**16.0 (0.1)** CROSS the main drainage of Carrizo Creek. Note sand berms that line the highway and patchwork repairs to the pavement. This low-lying dip of S-2 annually floods from rainfall runoff funneled down Carrizo Gorge. Large floods during 1982-1983, caused massive amounts of flood waters to flow across this portion of S-2 for 8 months before the roadway was repaired. The Bow Willow formation makes up the sediments on either side of the wash. On the right across Carrizo Creek, the TBMFF (central segment of the EFZ) has deformed recent terrace gravels juxtaposed against the Quaternary Bow Willow formation.

Carrizo Gorge (to the left) is one of only two known occurrences of modern algal stromatolites from ephemeral stream environments N of the Mexican border (Buchheim, 1990). The other site is San Felipe Creek. Buchheim (this volume) documents species of *Schisothrix* encrusting granitic cobbles and boulders throughout Carrizo Gorge. The stromatolites build a knobby surface of digitate heads by the rapid and abundant growth of filamentous blue-green algae (mucilaginous cyanobacteria) which induces the precipitation of, and biotic cementation by, calcite, lamination, and laterally linked hemispheroids.

**16.3 (0.3)** On the left is the entrance to the Bow Willow Primitive Campground, with access to Bow

Willow Canyon and Sombrero Peak (elevation 1,403 m).

**16.4 (0.1)** In the left foreground (8:00) is the low hillock of Egg Mountain, held up by the Bow Willow formation. The steep-sided ridgeline is horizontally off-set by post-Pleistocene movement along the TBMFF. Its latter half extends E to Bow Willow Creek.

**16.6 (0.2)** Milepost marker 48. The TBMFF bifurcates into two branches (left and right) where S-2 crosses the Bow Willow alluvial fan.

**17.5 (0.9) STOP 2 MOUNTAIN PALM SPRINGS.** MAKE A LEFT TURN and DRIVE up the dirt roadway to the primitive campground/trailhead parking lot adjacent to light-colored exposures of brecciated leucotonalite. Directly to the W is Southwest Grove, the only large palm grove in ABDSP visible from a paved highway. Along the roadway are excellent stands of teddybear cholla and taller Hoffman's cholla. Elephant trees abound throughout the rocky terrain and along hiking trails between here and Bow Willow. On the right is the impressive E-facing footwall escarpment of the structurally-elevated (relative to the down-dropped hanging wall block to the E), highly fractured Tierra Blanca Mountains, made up of "S-type" La Posta Tonalite. It is controlled by the dominantly dip-slip right strand of the TBMFF which may serve as a detachment fault to the main Laguna Mountains Escarpment (listric) accomodation fault. The magnitude of vertical separation denoted by the TBMFF is considerable in the western VFCB, where 4 km of Neogene W-dipping basin fill is present (White, et al. 1991), much of which represents the main sedimentary response to extensional attenuation of the WSTD.

Based on Landsat Thematic Mapper and 1:24,000 B&W aerial imagery of Mountain Palm Springs, the elongate W block of the range-front is controlled by the left branch of the TBMFF, and displays up to 1 km of subordinate horizontal offset. The left branch is accessible via a rocky hiking trail to North Grove. Additional NW-trending splinter faults marked by Mary's Grove, Surprise Canyon Grove and Palm Bowl Grove, extend northward to Torote Canyon. These lineaments intersect with the VVF S of Inner Pasture.

Overall, the present-day sense of slip along the TBMFF is dominantly dip-slip. This supports Lamar and Merifield (1975), Todd and Hoggatt (1976, 1979), and Lowman (1969, 1972, 1976, 1977) assertions that there is no field evidence of appreciable lateral offset along the southern segment of the EFZ between here and the unbroken bedrock of the Sawtooth Range which crosses the EFZ at Campbell Grade on S-2 at the southeastern end of Mason Valley (Lowman, 1969, 1976). Gravity and geodetic analysis of the TBMFF indicates that there is minute deep dextral shear across the EFZ. The basin geometry between Carrizo Creek and Agua Caliente is broken on the W by listric normal faulting, with syndepositional down to the E dip-slip faulting of the desert floor controlled along a vertical 4 km subsurface scarp (Gibson, 1983; Gibson, et al. 1984). This evidence strongly suggests that brecciated crystalline bedrock (La Posta Tonalite) of the Tierra Blanca Mountains is down-dropped as gravitational hanging wall slide masses displaced from footwall counterparts in the Laguna Mountains (Todd and Hoggatt, 1979), typical of the breakaway zone of domino-style mountain blocks associated with the WSTD (Figure 3). RETURN to S-2.

**18.7 (1.2)** MAKE A LEFT TURN (north). Enter the Sombrero Peak State Wilderness Area of ABDSP.

**18.8 (0.1)** Milepost marker 47.

**19.6 (0.8)** Junction of Sweeney Pass Road (S-2) with Bow Willow Creek Road (on the right). Street sign is incorrectly marked by "Great South of 1849", an indirect reference to the Southern Overland Mail Route of 1849. Bow Willow Creek Road does not represent the precise route of the Butterfield Overland Mail, nor did the company itself exist in 1849.

**19.7 (0.1)** Entrance to Indian Canyon (on the left) across its cone-shaped alluvial fan, extending outward from the canyon mouth onto the desert floor. Indian Canyon and Canebrake Canyon are antecedent gaps (wineglass canyons), developed in dip-slip environments. Their base is defined by an active alluvial fan built up along the foot of the mountain range, the stem is the narrow defile cut perpendicular across the range-front, and the bowl is distinguished by open intermontane uplands (N and S forks of Indian Valley, and Inner Pasture of Canebrake Canyon).

**19.9 (0.2)** Milepost marker 46.

**20.2 (0.3)** Well of Eight Echoes on the right. This is a

good location to see evidence of recent seismicity within the EFZ. Of particular interest is the TBMFF (on the left) and the presence of a noticeable 11.5 km-long fault scarp along the base of the mountain front. This surface rupture is from the historic M7.0-M7.5 earthquake of February 23, 1892 (Strand, 1980). Slickensides indicate 5-6 m of dip-slip and 1-2 m of dextral slip (Rockwell, 1989). It is listric at depth and mimics the MGE scarp along the Sierra San Pedro Mártir of Baja California, Mexico. This neotectonic activity suggests that the westernmost VFCB continues to subside, underlain by an E-dipping detachment fault. The TBMFF can be traced along the brecciated range-front to SE of Agua Caliente County Park where it terminates perpendicular to the NW-striking VCF.

The VVF is regarded by Buttram (1962) and Moyle (1968) as the main segment of the EFZ. It can be traced, with certainty, from the SW edge of the Coyote Mountains (CMFF) N through the Tierra Blanca Mountains, across Vallecito Valley to the Sawtooth Range at Campbell Grade. The VVF, at least in the Tierra Blanca Mountains, is characterized by strike-slip faulting. At the kilometer scale, major lineaments of the EFZ continue N through Mason Valley, Banner Canyon up to Temecula and the southernmost Los Angeles Basin (Whittier-Narrows). Gray (1961), Lamar (1961), Weber (1975), Kennedy (1977), Crowell and Ramirez (1979), Lamar and Sage (1973), Hull (1991), and Morton and Miller (1987), report that the total dextral slip across the northern EFZ ranges between 2-32 km. Contrary, Lampe, et al. (1988) suggest the southern EFZ may have small (~ 2-3 km) or nonexistent displacement.

**20.9 (0.7)** Milepost marker 45. Note that modern-day alluvial fan surfaces are truncated along the base of the Tierra Blanca Mountains by the aforementioned seismicity. Also, on the right, note deformed cuesta surface profiles (WSW-dip) of the Quaternary Canebrake Canyon alluvial fan to the N and NE.

**21.8 (0.9)** On the left is the S entrance to the small desert community of Canebrake, with access to Canebrake Canyon and the Canebrake Ranger Station. Recent seismicity within the EFZ includes the M4.5 Canebrake Earthquake of 1969 and the M4.8 Agua Caliente Earthquake on September 13, 1973 (Allison, 1974a,b, 1978). The epicenter of the latter main shock is where the confluence of Inner Pasture and Canebrake Canyon intersect on the VVF. Focal mechanism solutions for aftershocks indicate right-lateral strike-slip motion (Allison, 1974a). Between 1900-1974, at least 10 M4.0-M4.9 earthquakes have occurred on this portion of the EFZ (Pinault, 1984). Studies by Langenkamp and Combs (1974) and Allen, et al. (1965) confirming a marked S increase in microseismicity along the EFZ from Corona to the Mexican border.

**21.9 (0.1)** Milepost marker 44.

**22.8 (0.9)** On the left is the middle entrance road to Canebrake.

**22.9 (0.1) STOP 3 VALLECITO CREEK** (prior to Milepost marker 43). TURN RIGHT (east) on the established dirt roadway maintained along Vallecito Creek. Four-wheel drive vehicles are strongly recommended beyond this point. Nearby, sediments of the Vallecito Badlands are juxtaposed against sheared granitic rocks along the TBMFF.

The thick sedimentary and volcanic deposits exposed throughout the Carrizo, Fish Creek, and Vallecito Badlands of the VFCB provide excellent Mio-Pleistocene stratigraphic completeness (deposition of >5000 m of vertically-stacked, superimposed upward-thickening and coarsening syntectonic sediments that dip gently westward), structural intactness, best preserved and most accessible Neogene stratigraphic and paleontologic package in North America (Downs and White, 1968; White, et al. 1991; Remeika and Lindsay, 1992). They are ideally situated both spatially and temporally for well-defined lithostratigraphy (Woodard, 1963, 1974; White, et al. 1991), magnetostratigraphy (Opdyke, et al. 1977; Johnson, et al. 1983) and the development of a refined biostratigraphy.

Neogene sedimentary rocks of the western Imperial Valley have been subjected to intensive scrutiny and study. They were first described in detail by Kew (1914) and Hanna (1926). The first systematic approach to stratigraphic nomenclature was by Tarbet and Holman (1944), then later redefined and mapped by Tarbet (1951), Ver Planck (1952), Dibblee (1954), and Durham and Allison (1961). Although Woodard (1963, 1974), and Winker (1987) combined the results of previous investigators to redefine the stratigraphic column in the Split Mountain Gorge area of the VFCB, no study prior to the present one has attempted to establish correct stratigraphic refinement, and aid in the recognition of syntectonic strata exposed throughout the VFCB.

Based on unpublished research and field mapping (Figure 4), this report recognizes five well-preserved mid-Miocene to Pleistocene (Hemphillian-Irvingtonian) epicontinental lithostratigraphic non-marine, marine, deltaic and terrestrial depositional suites. The latter documents allocyclic sedimentary development in the western VFCB. Facies distributions coarsen, become less consolidated upward, contain progressively younger, and less steeply dipping strata, and systematically record in ascending stratigraphic order, from E to W, much of the large-magnitude tectono-stratigraphic package and paleontologic history of ABDSP relative to the WSTD (Figure 8).

The detailed sedimentary paleofacies is differentiated basinwide into useful genetic-stratigraphic affinities. From oldest to youngest, they are sub-divided petrographically, stratigraphically and paleontologically into: **(1)** steeply to moderately tilted Miocene pre-rift arkosic sandstones and fanglomerates of the Split Mountain Formation (Tarbet and Holman, 1944) exposed in the walls of Split Mountain Gorge. These intercalate with Miocene synrift volcanic flows (Eberly and Stanley, 1978; Mace, 1981; Ruisaard, 1979; Hawkins, 1970) and volcaniclastic deposits of the Jacumba Basalt exposed in the Fish Creek Mountains (Ruisaard, 1979), Volcanic Hills (Fourt, 1979) and Coyote Mountains (Eberly and Stanley, 1978; Mace, 1981; Gjerde, et al, 1982); **(2)** moderately tilted rift-related macroinvertebrate-rich coralgal sediments (Stump, 1972) of the Latrania formation of the lower Imperial group that mark the northern proto-Gulf of California termination of late Miocene marine transgression into the southernmost developing-subsident Salton Trough region. They are exposed in the Carrizo Badlands (Foster, 1979), Coyote Mountains (Bell-Countryman, 1984), and Fish Creek Badlands (Winker, 1987). These carbonate sandstones discontinuously overlie turbidite sandstones (Lycium fomation) and Fish Creek Gypsum (Ver Planck, 1952; Dean, 1988) in the vicinity of Split Mountain Gorge; **(3)** represents a moderately tilted Pliocene lithostratigraphic fluviomarine (Gulf of California)/deltaic (Colorado River) mixed affinity package composed of a progradational sequence of fine-grained sandstones, silts and clays of the Burrobend formation of the upper Imperial group. It includes the informally named Wind Caves, Mudhills, Deguynos and Camels Head members of Woodard (1963), and Winker (1987); **(4)** non-marine, wood-bearing delta-plain arenites and overbank splays of the Diablo formation (Remeika, 1991a; Remeika and Lindsay, 1992; Reynolds and Remeika, 1993) provisionally assigned to the lower half of the Colorado River Group (Reynolds and Remeika, 1993). Includes fine to very fine-grained Colorado River-like terrigenous quartz-rich sediments classically exposed in the Fish Creek Badlands (White, et al. 1991) and Carrizo Badlands (Smith, 1962); **(5)** strictly moderately to gently tilted locally-derived (Peninsular Range suite) Plio-Pleistocene synrift marginal coarse-grained facies distributions shed from adjacent mountains. Accumulation of siliciclastics is governed by autocyclic and allocyclic mechanisms between the Canebrake Conglomerate (Dibblee, 1954) and coeval distal to axial finer-grained fluvial-lacustrine sediments of the Palm Spring Formation (Woodring, 1931), provisionally assigned to the informal Anza-Borrego group (Reynolds and Remeika, 1993), named after the state park where excellent exposures are present.

**23.4 (0.5)** Intersection with View of the Badlands Wash. TURN LEFT and PROCEED N up the wash. Locally-derived epiclastic facies distribution between the Canebrake Conglomerate and Palm Spring Formation emphasize an allocyclic response of synrift clastics to late Pliocene extensional deformation, uplift of sediment provenance areas (ancestral Vallecito-Pinyon and Tierra Blanca Mountains), and basin subsidence of the westernmost VFCB. On the basis of stratigraphic study, subsidence of the alluvial basin along a prominent high-angle normal fault commenced between 3.5-2.5 Ma ( e.g. Hull, 1991), and caused a significant rearrangement of stratigraphic formations (Burrobend, Diablo, Canebrake and Palm Spring) and coarsening upward sedimentologic architecture. This is dramatically conveyed by the basin asymmetry that controls the distribution pattern and accumulation of synrift terrigenous coarse clastics of the proximal Plio-Pleistocene Canebrake Conglomerate, and the distal basin-margin Plio-Pleistocene Palm Spring Formation.

**23.5 (0.1)** The siliciclastic sediments exposed along the wash belong to the Vallecito member of the Palm Spring Formation. The non-marine Palm Spring Formation was named by Woodring (1932) for fossiliferous terrestrial sediments that overlie the fine-grained deltaic sequence on the north side of Carrizo Wash. It represents a finer-grained basinward facies that is laterally transitional and coeval with the proximal Canebrake Conglomerate. The type section is measured between Arroyo Tapiado to Arroyo Hueso in the basin-center. Locally-derived siliciclastics consist of about 225 m of interbedded siltstones and claystones, alternating with poorly-indurated arkosic sandstones, lithic arkoses, intraformational pebbly conglomerates, and subordinate fresh-water marls that are internally complex, deposited in three superposed facies that exhibit an overall coarsening-upward trend. These include, the lowermost Tapiado (basin-center lacustrine lakebeds), intermediate Huesos (floodplain), and uppermost Vallecito (alluvial plain) members (Woodard, 1963).

The Palm Spring Formation exhibits a continuous lithostratigraphic sequence (Woodard, 1963), displaying a documented Blancan-Irvingtonian LMA fossil vertebrate record (Downs and White, 1968; White, et al. 1991) that provides a clear record of evolution, immigration and extinction throughout the section, dated through magnetostratigraphy (Opdyke, et al. 1977; Johnson, et al. 1983), and vertebrate zonation (Vallecito Creek LF) spanning an interval from 2.6 Ma to less than .9 Ma within the VFCB.

The Vallecito member is recognized as an alluvial-braidplain deposit of coarse sandstone and conglomerates with a significant amount of siltstones and sandstones that represent a variety of moderate scale depositional systems which were active on distal basin-margin bajada fans. This rock unit is coeval, in large part, with the similarly-deposited Ocotillo Conglomerate (Dibblee, 1954) of the Borrego Badlands. It is exposed from View of the Badlands Wash to June Wash, is clearly related to the Canebrake III fluvial episode, is stratigraphically higher than the Huesos member, and straddles the Plio-Pleistocene boundary. It is also very fossiliferous, yielding a comparable Irvingtonian-age vertebrate local fauna (Vallecito Creek LF) to the Borrego LF (Remeika and Jefferson, 1993), as well as to the El Golfo LF (Shaw, 1981) of the northern Gulf of California.

**23.7 (0.2)** The faint pathway of the historic Butterfield Overland Mail Route (1858-1861) crosses View of the Badlands Wash here. This untouched segment of the route extends from the Palm Spring way-station at Mesquite Oasis to the Vallecito stationhouse on S-2. It is one of the best preserved stretches on the entire route. Over 137 years ago, the Butterfield Overland Mail Route carried mail and passengers from Missouri to San Francisco in twenty-five days or less. From Yuma, the overland route coursed into Mexico then recrossed the border and slanted NW across the uninviting desert floor of Anza-Borrego (Wright and Bynum, 1954).

**24.2 (0.5)** Note that the strata strike to the NW and dip to the SW, suggesting deposition in an asymmetric half-graben. Basin geometry is inferred from block tilting and prograding of Pliocene to recent alluvial and fluvial sedimentary facies. Faulting was active during sediment deposition, as shown by intraformational and angular unconformities, abrupt lateral and temporal changes in facies, provenance, accumulation rates, bedding and paleocurrent orientations, and by an upsection decrease in dip angle. This typical basin-fill architecture is manifested throughout the Palm Spring Formation from View of the Badlands Wash eastward to Rainbow Basin N of the mouth of Arroyo Tapiado.

**25.3 (1.1) STOP 4 BADLANDS RIDGE.** (Anza-Borrego Zone 56). This hogback defines one faunal zone (reference point) within the VFCB biostratigraphic sequence (based on local geographic-stratigraphic collecting levels of Downs and White, 1968), and is approximately 457 m below the top of the modified measured section of Opdyke, et al. (1977). This site corresponds to the Olduvai event in the Matuyama Magnetochron, dated 1.7-1.9 Ma, marking the important Pliocene-Pleistocene boundary. The presence of *Pewelagus* sp. from a locality just below the hogback, marks the last occurrence of Archaeolaginae in the latest Blancan of North America.

Above the hogback, there is a typical Irvingtonian fauna which includes several species of microtines (Zakrzewski, 1972); pocket gophers (White and Downs, 1961; Becker and White, 1981); modern leporids (*Sylvilagus* sp. and *Lepus* sp.) (White, 1984; White, 1991); equids (Downs and Miller, 1994), and amphibians and reptiles representing the Vallecito Creek LF (Downs and White, 1968). The fauna in this part of the section occurs in the Vallecito member and suggests an environment of a temperate savannah of rolling plains adjacent to meandering streams and shallow playa lakes. Riverine deposits are widespread as episodic floods resulted in the accumulation a many cyclic overbank deposits. RETURN to Vallecito Creek.

**27.2 (1.9)** TURN LEFT (downstream) to Arroyo Hueso.

**27.8 (0.6)** On the left hillsides, *Mammut*-bearing siliciclastic sediments are locally folded along a narrow belt of deformation extending from Mesquite Oasis (Palm Spring) to the mouth of Arroyo Hueso. Here, the large Vallecito Creek splinter fault (easternmost strand of the EFZ) splays off the main branch of the TBMFF (Woodard, 1963), subparallel to the dry streambed of Vallecito Creek. In this vicinity it cuts through the Vallecito member, extending its surface expression southeastward into the Carrizo Badlands below Carrizo Peak (elevation 802 m) in the Coyote

Figure 5. West-dipping sediments of the Palm Spring Formation exposed in Rainbow Basin along Arroyo Tapiado. Note Sombrero Peak in the distance.

Mountains.

**28.2 (0.4)** Palm Spring turnoff. CONTINUE STRAIGHT. Historic Palm Spring, 0.9 km up this spur road at Mesquite Oasis, is located where a line of mesquite trees and several young *Washingtonia filifera* palm trees mark the former site of the Butterfield Overland Mail way-station, conveniently situated between the Carrizo and Vallecito stage stations. The year-round seep that provides water for the local vegetation is produced by the small yet structurally complex syncline developed S of Palm Spring (W of the Arroyo Hueso entrance). The short wavelength fold in the Vallecito member sediments reflects shallow thrusting and compression driven by the rift-border VCF.

The Vallecito Creek Fault (VCF) indicates commencement of a progressive eastward fault migration of the EFZ into the basin architecture with time. As constrained by stratigraphic data, the geometry of faults related to the hanging wall of the Tierra Blanca mountain front implies a significant synrift increase in basin width, with a Pliocene-Quaternary composite throw of ~ 4 km. The VCF is the major NW-striking feature in the area bounding the Tierra Blanca Mountains-Vallecito Badlands. It can be traced, as a straight lineament, from the mouth of Arroyo Seco del Diablo up the right fork (NW) of Squaw Canyon (Agua Caliente County Park) to Vallecito County Park. The fault is expressed as a series of aligned benches, deformed sediments, NE-facing fault scarps with 5-7 m of vertical down to the E offset that abruptly truncate Holocene alluvial fans, and a wide zone of intensely fractured and hydrothermally altered rock, hot springs, and a prominent linear valley through Agua Caliente County Park (Buttram, 1962). A minor strand also extends up Carrizo Valley, vertically offsetting the Canebrake IV sediments in road cuts to either side of S-2.

**29.6 (1.4)** Arroyo Hueso. CONTINUE STRAIGHT. On the mudhills skyline ahead and to the left there is a wedge of green-colored mudstones. This marks the passage of the Middle Mesa Fault NE through the

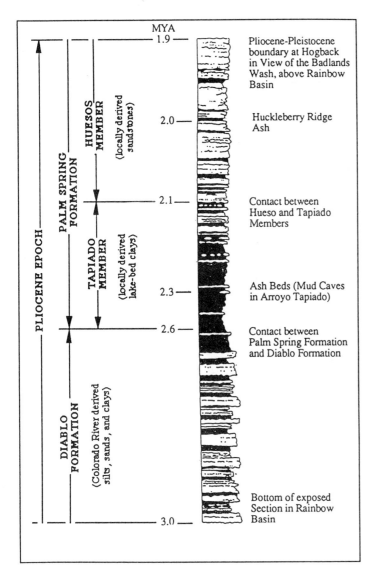

Figure 6. Simplified stratigraphic section of Rainbow Basin, Arroyo Tapiado. After Remeika and Lindsay (1992).

Huesos member. This is the first in a series of cross-basin orthogonal left-lateral transcurrent faults that form several intrabasinal fault blocks of compression-folding and small-scale displacement of sediments on the NE side of the VCF. These fault block sediments are very fossiliferous.

The Huesos member is identified as light-colored cross-bedded channel sandstones that cap the higher ridges of the badlands skyline. These are underlain/overlain by poorly bedded siltstones and carbonaceous claystones that display a characteristic layer-cake sequence that is so prevalent within the Vallecito Badlands. The sediments are an interbedded sequence of tan-brown siltstones, buff feldspathic sandstones, and greenish-colored claystones of braided-stream, lacustrine and floodplain origin. Together they form the distal margin of a vast sedimentary wedge of alluvium, fed by rivers and streams that flowed from crystalline-rich bedrock exposures, stretching S from the rising and dismembered Vallecito-Pinyon Mountains. Small ponds and lacustrine waterways are suggested for this transition zone by the presence of freshwater ostracodes, gastropods, minnows and sticklebacks, amphibians, reptiles (including *Clemmys marmorata* and *Eumeces* sp.), anseriform birds, *Gerrhonotus multicalinatus,* and *Pliopotamys minor.*

**29.8 (0.2)** Hollywood and Vine. Prior to the entryway vertical-standing SW-dipping sandstones with matrix-supported granitic clasts are typical of the Huesos member. They are folded and upthrown along the VCF. The Huesos Member consists of alternating tabular sandstone bodies and finer-grained cyclic overbank sediments that represent basinwide cycles of channel and floodplain deposition. The tall, rusty signpost is an original Jasper trail marker, installed by former County Supervisor James Jasper in the 1890's along the Butterfield Overland Mail Route.

**30.3 (0.5)** Approaching Arroyo Tapiado, sediments on the right are about 150 m thick and produce an Irvingtonian fauna. The sediments form a structural fault block with the same apparent dip as the down-thrown block in the section N of the Middle Mesa Fault, but the strike is deflected 10° to the S. Field observations and seismological evidence indicates that these sediments resemble those of the Huesos member opposite Palm Spring and the uppermost Tapiado member in this vicinity. They are exhumed, uplifted and dissected along the VCF which has migrated and tilted sediments basinward. This tectono-sedimentary behavior of progressive basinward stepping-down of bounding faults from the W is consistent with active extension accomodated in the geometry of the half-graben.

**31.8 (1.5)** Arroyo Tapiado turnoff. MAKE A LEFT TURN in Arroyo Tapiado, and FOLLOW the established roadway northbound into the heart of the Vallecito Badlands. This main arroyo drains the southern flanks of the Pinyon Mountains where its headwaters occur in the massive boulder and cobbled sediments of the Canebrake Conglomerate. Arroyo Tapiado is generally steep-walled and entrenched below the surface of West Mesa throughout its lower half, and steep and V-shaped in its upper half.

**32.4 (0.6)** Ahead, a sub-parallel strand of the VCF cuts across the wash in a NW-SE direction. Note that sandstones of the Huesos member (on the left) are tilted steeply to the southwest, juxtaposed against non-deformed sediments on the right.

**32.9 (0.5)** The open, highly-eroded amphitheater of Rainbow Basin displays in cross-section an orderly yet complex progression of terrestrial sedimentation defining the Palm Spring Formation (Figure 6). The deposits dip 15° to 25° to the west. This is the stratotype for the Palm Spring Formation. From E to W, the lowermost greenish lacustrine claystones of ancient Lake Tapiado are the Ten Caves submember (informal name of White, et al. 1991) of the Tapiado member. They are overlain by the channel sandstones and playa-margin sediments of the Four Mile submember (informal name of White, et al. 1991), which are, in turn, overlain by the Huesos member. From playa-margin sediments of Rainbow Basin many microtines, plus *Borophagus* sp., *Canis* sp., *Urocyon* sp., *Arctodus* sp., *Mustela* sp., *Satherium* sp., *Trigonictis* sp., *Acinonyx* sp., *Felis* sp., *Smilodon* sp., *Equus* sp., *Platygonus* sp., *Blancocamelus* sp., and *Hemiauchenia* sp. of the Vallecito Creek LF have been recovered.

**33.2 (0.3)** Ahead, on the eastern skyline, distinctive reddish-colored sandstones come into view and appear to conformably underlay the Tapiado member. These sediments represent uppermost exposures of the Diablo formation and yield an abundance of ferruginous sandstone concretions and silicified wood. The contact is 2.6 Ma. This date is consistent with an abrupt increase in the rate of subsidence within the Pliocene Loreto Basin of Baja California Sur (Umhoefer, et al, 1994), and is closely synchronous with Lonsdale's (1989) estimate for the beginning of gulf spreading and related wrench tectonics of the WSTD along the MGE. It is entirely plausible that the kinematic evolution of the EFZ and SJFZ commenced between 2.6 - 2.5 Ma (Hull, 1991: Harbert, 1991).

The Colorado River group, including the Diablo formation, represents a strictly Plio-Pleistocenelithostratigraphic time-transgressive deltaic stratigraphic package composed of a non-marine progradational sequence of fine-grained, quartzitic delta-plain and lacustrine deposits independently derived from the ancestral Colorado River. It also includes the lacustrine Borrego Formation (Tarbet and Holman, 1944) and Brawley Formation (Dibblee, 1954) exposed in the Borrego-San Felipe Basin (BSFB) to the N. This separate designation is straightforeward and precise, permitting future investigators to clearly differentiate and convey specific intent regarding the strata, fauna, and facies relations of this lithologically unique series of rock units.

**34.2 (1.0) STOP 5 CANYON OF TEN CAVES.** (Anza-Borrego Zone 45). On the left side of the arroyo is the dry tributary of Canyon of Ten Caves. This is the type section of the Tapiado member. The Tapiado member represents a subordinate green lacustrine claystone facies (Carey, 1976; Cunningham, 1984) of a perennial lake (basin depocenter) deposited on the Diablo formation (Figure 6). This contact has a paleomagnetic signature of 2.6 Ma (Opdyke, et al. 1977).

The Ten Caves submember is the basin-center lakebed sequence of blue-gray and olive-green gypsiferous claystones, mudstones and calcareous siltstones, replaced by the Four-Mile submember that displays a very fossiliferous mixed/alternating fluvial/deltaic facies transition into the overlying Huesos member. Both record a progressively upward coarsening semiaquatic shoreline sequence in marginal lacustrine sediments of Pliocene Lake Tapiado, deposited concordantly on the Diablo formation along the basin-margins. This thick fluviatile sequence replacement of depositional environments characterizes the modern Valle de San Jose (e.g. Lake Henshaw), which occupies a similar shallow footwall-uplift basin developed within the EFZ.

On the right side, the Diablo formation is one of the most extensively exposed sedimentary unit in the VFCB. The massive sandstones yield many fusiform, spheroid and tabular concretions that litter surface exposures. In marked contrast, the overlying Palm Spring Formation sediments (restricted in outcrops to the western VFCB) are tan to olive gray, coarse to fine-grained, poorly sorted, feldspar-rich, arkosic sandstones with intraformational gravel conglomerates.

The Diablo formation grades upward from the Imperial group, and represents a prominent non-marine, high-energy delta-floodplain arenitic deposit

Figure 7. The Middle Mesa Fault exposed in Arroyo Tapiado.

debouched by the ancestral Colorado River. Originally, Woodard (1963) informally designated the basal 300 m of fine reddish-brown, massive crossbedded arenaceous and argillaceous sandstones of Woodring's (1932) Palm Spring Formation as the Diablo member. Winker (1987) informally included Woodard's (1963) upper submember of the Diablo member as part of the initial basin-margin Olla member of the Canebrake Conglomerate.

Remeika (1991) recognized Diablo-like lithologies as distinctive in appearance, are extensively exposed in ABDSP, and are not genetically nor petrographically synonymous with the locally-derived Palm Spring Formation. Throughout the WSTD, many mapped "Palm Spring Formation" outcrops are exclusively delta-plain sediments of the Diablo Formation or lacustrine lakebed sediments of the Borrego Formation and do not realistically represent true Palm Spring Formation lithologies. The continued misuse of stratigraphic context, limits, depositional relationships, and relative stratigraphic positions of Colorado River-derived rock units has lead to confusion and ambiguity within the literature.

**34.3 (0.1)** Ahead, the NE-trending Middle Mesa Fault is dramatically exposed in the left cliffside (Figure 7). This left-lateral transcurrent fault has dismembered the Four Mile submember (left block), juxtaposed against the Ten Caves submember (right block). This fault extends from the VCF to Split Mountain Gorge, and may continue beyond to the Coyote Creek Fault (CCF). A complex network of NE-trending sinstral faults occur in the VFCB and are similar in orientation and sense of slip to cross-faults of the CCF that produced the M6.5 Borrego Mountain Earthquake of 1968, to the left-lateral (?) Inspiration Point Fault of the BSFB, and the initial M6.2 event on the Elmore Ranch Fault in the 1987 Superstition Hills Earthquake sequence. NE-trending faults field mapped and suggested by aerial imagery between the EFZ, SJFZ, and the SAFZ have been interpreted to accomodate distributed dextral shear between the Pacific and North American Plates.

**35.0 (0.7) STOP 6 CAVE CANYON.** (Anza-Borrego Zones 45 and 46). Cave Canyon is among the longest and largest natural mud cave formations reported in North America. Characteristic pseudokarst features preserved within the Ten Caves submember include recessed subterranean stream channel outlets and caves that have internally eroded into halls, galleries, aisles, chambers, and crawlways. Locally, Cave Canyon exhibits a channel length of over 500 m.

The Ten Caves submember represents about 116 m of massive gray to greenish gray suspension-deposited gypsiferous claystones and mudstones with minor calcareous siltstone interbeds. Generally devoid of easily recognizable fossils, the presence of gastropods, ostracodes, amphibian bones, chelonian fragments, and fish spines, supports a paleoenvironmental interpretation for a predominant lacustrine habitat. In the E-facing upper canyon walls to the W, in marked contrast, is the very fossiliferous coarse clastic lacustrine-fluvial, playa-margin transitional deposits of the Four Mile submember which grade upward into the Huesos member.

Crossing the arroyo here, are two locally-derived airfall tuff layers. The oldest (lowermost) ash, exposed to view at the mouth of Cave Canyon has a fission-track age of $2.3 \pm 0.4$ Ma (N.M. Johnson, pers. comm., 1983). The upper ash is buried by a recent rock fall. Stratigraphically separated by 26 m, each is a unique datum marker. Both are exposed, up canyon, above Four Mile Wash.

**35.4 (0.4)** Ahead, the sheer cliffside is made up of the buff sandstones, calcareous siltstones, and claystones of the Four Mile submember.

**36.8 (1.4)** On the left is Four Mile Wash (closed to vehicles). This interbedded sequence (archtype) is indicative of the gradual progradation of a fluvial sandstone facies over Lake Tapiado (Ten Caves submember). Interfingering sandstones represent stringers of the Huesos member. Across from the entrance to Four Mile Wash, remains of rodents, birds, *Borophagus* sp., *Canis* sp., *Felis* sp., *Mylohyus* sp., and *Platygonus vetus* have recently been found within these layers. On the right, the bluish-gray mudstone hill is the type locality of *Coendou stirtoni*. Turtle fragments, *Sylvilagus hibbardi*, and several bird bones were collected from this screen wash site. From other nearby localities between Arroyo Tapiado and Arroyo Hueso reptiles, amphibians, fish, turtles, anseriform birds (Howard, 1963), *Dipodomys* sp. (Cunningham, 1984), and five genera of fresh water molluscs (Taylor, 1966) have been recovered by crews from the Los Angeles County Museum (LACM), Imperial Valley College Museum (IVCM) and ABDSP.

**38.0 (1.2)** Note the presence of two airfall ash beds. These ashes (examined at Cave Canyon) indicate incipient syntectonic volcanism occurred sporadically nearby, possibly a reactivation from the Volcanic Hills during the Pliocene (2.3 Ma). Rare vertebrate footprints of equids and camelids, plus the remains of the megaherbivore camel *Blancocamelus* sp. have been preserved between the ash beds.

**38.1 (0.1)** Arroyo Tapiado fork. The main branch (left fork) continues straight. MAKE A RIGHT TURN (east) and FOLLOW the established roadway (right fork). Note sediments in the natural wash cuts are composed of coarse-grained gray sandstones, thin to medium-bedded, with sporadic limonitic-staining, and abundant ripplemarks. This section represents the shoreline of Lake Tapiado. Preserved root casts and mudcracks indicate subaqueous to subaerial conditions along the shoreline margin.

**38.5 (0.4)** Arroyo Seco del Diablo turnoff. CLIMB the short hillside on the right (Cut Across Trail) onto West Mesa. The smooth-appearing upland surface of West Mesa represents the remnant of an extensive pediment that briefly covered much of the area.

**40.8 (2.3)** Arroyo Seco del Diablo junction. TURN LEFT. The drainage of Arroyo Seco del Diablo cuts through a nearly uninterrupted sequence of trough cross-stratified paleochannel arenites of the Diablo formation, notable for a remarkable diagnostic collection of silicified wood and ephemeral ichnofossils.

**42.3 (1.5) STOP 7 ARROYO SECO DEL DIABLO.** The Diablo formation is a particularly rich source of silicified wood material. Remeika, et al. (1988) was the first to document the existence of ancestral *Umbellularia salicifolia* (Lauraceae), *Populus* sp., and *Salix* sp. (Salicaeae), and *Juglans pseudomorpha* (Juglandaceae). Recently, sabal palm fronds have also been discovered in Olla Wash. The palm fronds mark the first occurrence of fossil leaf material from ABDSP.

**44.7 (2.4) STOP 8 CAMEL RIDGE.** (Anza-Borrego Zone 35). Camel Ridge consists of a mixed affinity sequence of mudflats and channel sandstones derived from the ancestral Colorado River interbedded by locally-derived sandstones (Figure 9). Nearly 100 vertebrate tracks of camelids, felids and shorebirds were discovered by P. Remeika in the soft, cohesively-

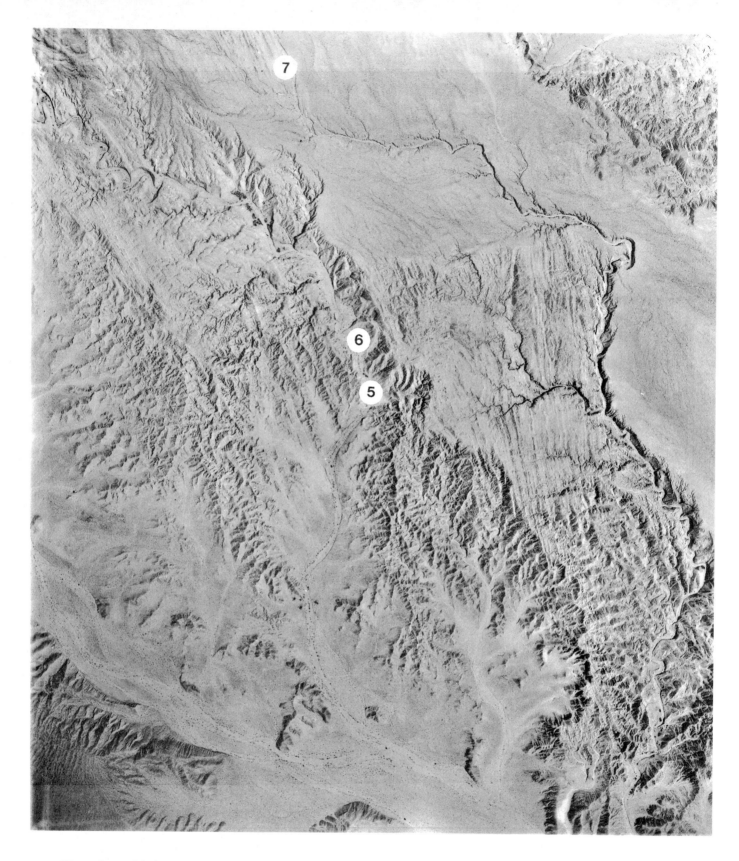

Figure 8. Aerial photograph of the Vallecito Badlands. Field trip stop 5 is located at Canyon of Ten Caves in Arroyo Tapiado. Stop 6 is Cave Canyon. Stop 7 is petrified wood locality in Arroyo Seco del Diablo.

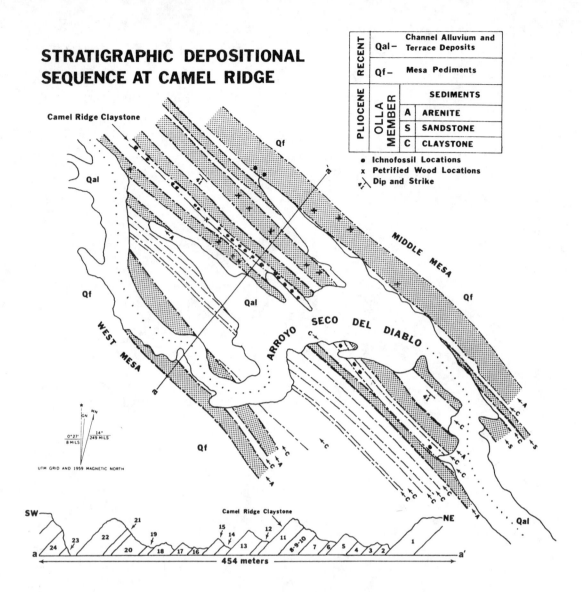

Figure 9. Generalized geologic cross-section of the Camel Ridge area, illustrating interbedded relationship (mixed affinity) between west-dipping basin-margin depositional sequence and fossil resource localities. Arenitic sandstones are derived from a Colorado River source area. Coarse-grained sandstone and micaceous claystone beds are locally-derived.

banded, micaceous claystones and arenites. Of the many tetrapod footprints studied, two ichnogenera of *Pecoripeda* compare favorably with *Camelops* sp. and *Hemiauchenia* sp. (Stout and Remeika, 1991). Also present are rare footprints of *Felipeda* sp. cf. *Felis* sp. or immature *Smilodon* sp., and shore bird tracks (*Avipeda*) comparable to *Calidris* sp. and *Tringa* sp. (Jorgensen, pers. comm., 1994). Preservation of animal tracks probably resulted from seasonal intermittent flooding and deposition of sediments into a shallow freshwater lake. The discovery of Camel Ridge marks the first occurrence of fossil tetrapod tracks in the Colorado Desert. The taxa identified so far, are referable to the Arroyo Seco LF. At Anza-Borrego, the Arroyo Seco LF (2.6 to 3.3 Ma) is comparable, in one stratigraphic sense, to the combined Hagerman (of Idaho), and the Benson (of Arizona) paleofaunas. The flora and fauna of the Diablo formation depicts a more equable climate than exists today. RETURN to the junction with the Cut-Across Trail.

**48.6 (3.9)** Junction with Cut-Across Trail. MAKE A LEFT TURN, southbound.

**48.8 (0.2)** Fish Creek turnoff via the Diablo Dropoff. TURN LEFT onto the established roadway crossing Middle Mesa.

**50.1 (1.3) STOP 9 DIABLO DROPOFF.** (elevation 359 m). Lunch stop with a spectacular view of the Fish Creek Badlands. Rates of basin subsidence and tilting of upper-plate rocks is confirmed by the presence of Plio-Pleistocene syntectonic granitic and gneissic detrital fan sedimentary systems (Canebrake Conglomerate I-IV), preserved to the NW on the SE flank of the Pinyon Mountains, that are directly associated with the WSTD. They episodically prograded into the least-tilted western basin-margin setting as an upward-thickening and coarsening megasequence. Fanning dips and clast imbrications show a basinward paleoflow, suggesting substantial proximal stratigraphic depositions from the Tierra Blanca Mountains hanging wall (on the W), and from the side wall of the NE-striking Pinyon Mountains transfer fault (on the NW).

Syndeposition is significant because it is closely synchronous to post-Miocene regional uplift, progressive unroofing, and shedding of rock units from the ancestral Peninsular Ranges, inferred detachment breakaway faulting in the domical Vallecito-Pinyon mountain-complex (Lough and Stinson, 1991; Lough, 1993), and to a strike-slip wrench-tectonic transformation along the WSTD plate margin of the STGC (Lonsdale, 1989).

**50.2 (0.1)** The narrow roadway enters the Carrizo Badlands State Wilderness Area (Fish Creek Sector). Four-wheel drive is strongly recommended. SAFELY NEGOTIATE the roadway from Middle Mesa down several hundred feet into Dropoff Wash, and intersect with Fish Creek Wash after about one mile of rough travel.

**51.2 (1.0)** T-intersection with Fish Creek Wash, which drains the northern half of the VFCB. Downstream, Fish Creek is channeled through the deep, sinuous, and straight-walled Split Mountain Gorge, ultimately draining into San Sebastian Marsh. MAKE A LEFT TURN onto the established roadway, destination Sandstone Canyon. The route enters the Fish Creek Badlands, part of the Vallecito Mountains State Wilderness Area.

**51.5 (0.3)** Beneath the perforated mesa tops, the W-tilted strata expose a long series of meander belts including paleochannel sandstones, point bars, natural levees and overbank splays. Sands and silts of the Diablo formation are superposed vertically thorughout the canyoned walls in contrasting bands of massive sandstone and thin reddish-brown claystone.

The type section of the Diablo formation is between Loop Wash and Dropoff Wash along Fish Creek Wash. Basin-center Diablo sediments include extra-regional pale pink to cream-colored, fine to very fine-grained, massive, concretionary, quartzitic paleochannel Colorado River-suite affinity arenites and gravelly sandstones, interbedded with subordinate Borrego Formation-like brown ripple-laminated, gypsiferous overbank claystones and siltstones. The distinctive arenites form prominent strike ridges, and are locally fossiliferous, yielding wood-bearing temperate hardwood species of the Carrizo Local Flora (Remeika, et al. 1988; Remeika, 1994); extralimital palynofloras (Fleming, 1994); a Blancan-age vertebrate succession (Arroyo Seco LF of Downs and White, 1968); brackish-freshwater invertebrates, fish bones, reworked delta-front shellhash, and an increasing megavertebrate ichnogenera (Stout and Remeika, 1991).

**51.7 (0.2)** Evidence of the West Mesa Fault can be seen across the wash. Left-lateral strike-slip movements have locally displaced the reddish-brown claystone against a weathered exposure of arenitic sandstone.

**52.0 (0.3)** The prominent Layer Cake (from which the Layer Cake LF gets its name) erosional knoll stands as a mute remnant stack of locally-derived fine-grained, thin-bedded micaceous silts and clays interbedded within delta-plain deposits like those of the Olla member. It has a paleomagnetic signature of 3.32 Ma (Opdyke, et al. 1977).

The Layer Cake LF is one of the most nearly complete successions of Blancan taxa known, and although incomplete, represents the oldest Blancan assemblage in California. Vertebrates include *Teratornis incredibilis* (see Jefferson, this volume), *Hypolagus regalis*, *Nekrolagus* sp., *Perognathus hispidus*, *Prodipodomys* sp., *Sigmodon lindsayi*, *Dinohippus* sp., and *Megatylopus* sp. Host rocks extend from Sandstone Canyon to Blackwood Basin.

**53.3 (1.3)** As the established roadway veers to the right, note the increasing presence of thin olive-green Ollalike claystones and thicker grayish fluvial sandstones interfingering into the Diablo formation. Travel up-canyon approaches the basin-margin.

**54.2 (0.9) STOP 10 SANDSTONE CANYON.** (Anza-Borrego Zone 29). In this vicinity, sediments are clearly fluviatile. Minute amounts of iron oxides and manganese have stained sandstones in subdued tones of pale orange and reddish-brown. The only known complete skull of *Nekrolagus* sp. was found in exposures N of the canyon entrance, along with two skulls of *Hypolagus vetus*. In the lower part of the section, rodent and lagomorph skulls are usually complete, although rare. There is a skull (with the tympanic bullae in place) of *Dipodomys* sp. from a locality a few feet stratigraphically above the entrance to Sandstone Canyon (Cunningham, 1984). Nearby, petrified sabal palm fronds mark the first occurrence of leaf material in ABDSP. The wood was discovered, in situ, from trough cross-stratified paleochannel sandstone interbeds. In addition, ephemeral tetrapod footprints of *Canipeda* sp. cf. *Borophagus* sp., *Hippipeda* sp. cf. *Dinohippus* sp., *Pecoripeda* spp. cf. *Megatylopus* sp., *Blancocamelus* sp., and *Hemiauchenia* sp. have been found recently near Sandstone Canyon and Olla Wash.

**54.6 (0.2)** TURN AROUND and RETURN to Fish Creek Wash. MAKE A RIGHT TURN and follow the established roadway back to Dropoff Wash.

**57.8 (3.2)** Dropoff Wash. CONTINUE STRAIGHT, destination Split Mountain Gorge. Down canyon, the established roadway winds through the lower section of the Diablo formation. The Diablo is interpreted as representing a delta-plain habitat, the uppermost subaerial facies of a thick deltaic sequence within the VFCB.

**58.8 (1.0)** Loop Wash junction (west end). The open geologic setting to the N is part of Blackwood Basin. Deposited to a thickness of more than 500 m, the Diablo formation separates the brackish-marine Burrobend formation from the overlying terrestrial Palm Spring Formation.

The Diablo paralic sequence was also inundated by marine tidal currents and probably formed strandline beaches, arranged perpendicular to the old coast line as shown by the presence of mudcracks, ripplemarks, and shellhash. There is an impoverished deltaic fauna (about 3.5 Ma), including *Dendostrea? vespertina*, *Anomia subcostata*, and *Argopecten deserti*, ascribed to the Burrobend formation. In the sediments in question, seven species of formaminifera from the Cretaceous Mancos Shale are documented (Merriam and Bandy, 1965), indicating a Colorado River source. The Mancos Shale is a Cretaceous deposit located in Utah, New Mexico and southwestern Colorado.

**59.7 (0.9)** The massive brown and soft green claystones on either side of Fish Creek Wash mark the gradational boundary of the Diablo formation with the uppermost Camels Head member (Winker, 1987) of the Burrobend formation, and has a paleomagnetic age of 3.72 Ma (Opdyke, et al. 1977). Briefly, these deposits represent a richly fossiliferous sequence of fine-grained claystones and calcareous sandstones deposited into the proto-VFCB during the early Pliocene.

The thick, deltaic-bedded uppermost half of the Imperial group is present from Wind Caves to Middle Mesa with reference to Fish Creek Wash, and across into the Coyote Mountains (Stump, 1972; Foster, 1979; and Powell, 1986). This rock unit records a major shift in the sedimentary fabric of the Imperial group, as noted by Merriam and Bandy (1965), Muffler and Doe (1968), Winker and Kidwell (1986), Winker (1987), and White, et al. (1991). It represents an early-stage terregenious fluviomarine deposit controlled largely by ancestral Colorado River detritus transported into the extensional northern Gulf of California embayment (proto-VFCB) across the NW-striking, right-lateral, strike-slip SAFZ.

The recognition of, and distinction between, several informal depositional schemes for the Burrobend formation has been previously noted in the literature. This subdivision includes: **(1)** a clay-dominated progradational pro-delta deposit variously named the Coyote Mountain Clays (Hanna, 1926; Bell-Countryman, 1984; Deméré, 1993), or Mudhills member (Winker, 1987); and **(2)** finely-interstratified silt and mud tidal rhythmites with infrequent *Dendostrea?*-dominated coquina accumulations representing climatically sensitive delta-front and transitional tidal flat lithologies of the Yuha Reefs (Hanna, 1926), or further refined into the informal Capote and Descanso members in the Coyote Mountains (Christensen, 1957), and the Deguynos and Camels Head members in the Fish Creek and Carrizo Badlands (Woodard, 1963). Furthermore, Winker (1987) summarized the Burrobend formation exposures along Fish Creek Wash into four informal deltaic sequences, deposited in a very shallow subtidal to lowest intertidal paleoenvironment of the northern Gulf of California. From oldest to youngest, they include initial turbidites (Wind Caves member), pro-delta claystones (Mudhills member), delta-front rhythmites (Deguynos member) and transitional

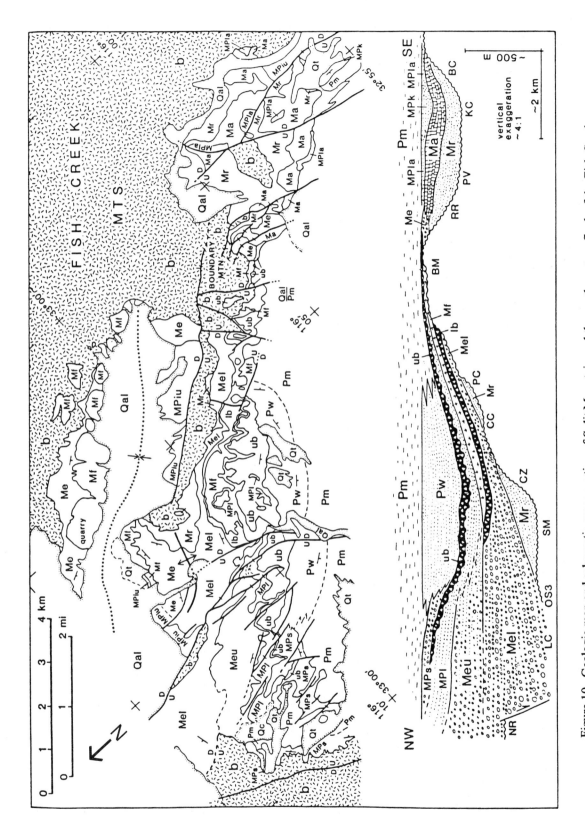

Figure 10. Geologic map and schematic cross-section of Split Mountain and the southwestern flank of the Fish Creek Mountains, illustrating lower nonmarine units and lower marine units. Cross-section thicknesses are approximate. b= crystalline basement, undifferentiated; Mr= Red Rock Canyon member; Me= Elephant Trees member; Mel= lower fanglomerate; lb= lower boulder bed; Meu= upper fanglomerate sandstone; Mf= Fish Creek Gypsum; Mpl= Lycium formation; Mpla= Latrania formation; Mpiu= Imperial group, undifferentiated; Mps= Stone Wash member; ub= upper boulder bed; Pw= Wind Caves member; Pm= Mudhills member.

mudflats (Camels Head member). A fifth facies, the Lavender Canyon member, includes a delta-front distributary bar sandstone deposit exposed in Lavender Wash within the Carrizo Impact Area. These latter informal members are adopted herein for the Burrobend formation.

Age of these deltaic deposits is more precisely known, as a result of well-calibrated paleomagnetostratigraphy of the VFCB (Opdyke, et al. 1977; Johnson, et al. 1983). Paleomagnetic chrons identified in the Fish Creek Wash section include the Gilbert (reversed-polarity) Magnetochron (5.40 to 3.32 Ma) and Gauss (normal-polarity) Magnetochron (3.32 to 2.43 Ma). Based on ostracodes (Quinn and Cronin, 1984), benthic foraminiferal assemblages (Ingle, 1974), and planktonic foraminifera (Stump, 1972) collected from the Mudhills and Deguynos members, the rift-related deltaic package documents an early Pliocene age (zones N19-N21). The stratigraphy and sedimentology is further confirmed by siliciclastic-carbonate distribution of epifaunal oyster-anomiid lateral accretionary coquina beds (Kidwell, 1987; Watkins, 1990a) characteristic of the Deguynos member, and reworked Late Cretaceous calcite-filled foraminifera (Merriam and Bandy, 1965) and diagnostic palynological assemblages derived from the Colorado Plateau (Fleming, 1994; Fleming and Remeika, 1994; and Remeika and Fleming, this volume), beginning about 4.5 Ma (Fleming, 1994).

Up-section, distal paleofacies of the Camels Head member grade into the major wood-bearing Diablo formation (White, et al. 1991; Remeika and Lindsay, 1992). Studied in outcrop, both genetically-related rock units reveal a complex lateral and vertical juxtaposition of basinal lithofacies, grading from subaqueous delta-front deposits into tidal flat to subaerial delta-plain basin-center and basin-margin depositional relationships.

**60.1 (0.4)** Camels Head Wash (Anza-Borrego Zone 7). In this area, the presence of petrified wood, fossil vertebrate, and abundant shellfish mark the location where Colorado River met the sea. Many large camelid footprints of *Pecoripeda* sp. cf. *Megatylopus* sp. have been found above and below deposits yielding fossil oysters. A *Dinohippus* sp. maxilla was found in this geological interval, and represents Anza-Borrego's earliest fossil horse.

**60.3 (0.2)** The low green mudhills to the left mark the transitional boundary between the Camels Head-Deguynos members. These beds are fossiliferous with brackish-water forms similar to those that today inhabit the Gulf of California.

**60.5 (0.2) STOP 11  LOOP WASH** (east end). Coquina beds of oysters, pectinids, barnacles, bits of urchin tests, and other intertidal shellhash represent deltaic life that coexisted about 4.0 Ma.

**60.7 (0.2)** CONTINUE STRAIGHT following the established roadway of Fish Creek Wash. For the next 2.4 km, the wash cuts down through the Deguynos member of the Burrobend formation. Note that, at one moment, high energy waters built up immense lenticular longshore accumulations of shellhash, while during another, the water was sufficiently still for the tidal deposition of predominantly laminated silts and clays (rhythmites) with high rates of sedimentation. This distinctive ribbed appearance is laterally-persistent, representing seaward-sloping fine-grained green, yellow, gray and tan clayey siltstones, silty claystones, and fine sandstones. Piled one on top of the other, sediments form intricate profiles of banded rock sheets. The laminae are generally wafer-thin, defined by horizontal partings of silty claystone. Locally, exposures of rhythmite-bedding in upward-coarsening cycles are topped by sandstone or coquina (forming "elephant knee" hogbacks. The delta-front beds of the Burrobend formation are about 500 m thick in this area.

**61.5 (0.8)** Ahead is the prominent Elephant Knees hogback. Weathered a dark brown, shell material and wave-rippled sedimentary structures reveal the high energy nature of these shoal deposits. The presence of *Dendostrea? vespertina*, *Anomia subcostata*, and *Argopecten deserti* implies these sediments represent a shallow embayment with intertidal, brackish-water conditions.

**62.4 (0.9)** Ahead, note that claystone beds of the Deguynos member dip steeply to the SE near the Little Devil Fault trace.

**62.9 (0.5)** Mudhills Wash. The low mudhills consist of deeply weathered claystones of the Mudhills member of the Burrobend formation. It is a massive claystone facies, about 600 m thick, suggesting an inner shelf-intertidal pro-deltaic paleoenvironment during the early Pliocene, about 5.0 Ma. The Mudhills member grades upward into the Deguynos member that forms the prominent Elephant Knees hogback, clearly visible S above Mudhills Wash.

Figure 11. The entrance of Split Mountain is spectacularly guarded by cliffsides composed of the Elephant Trees member of the Split Mountain Formation.

**63.1 (0.2)** Junction with North Fork of Fish Creek Wash. CONTINUE STRAIGHT. North Fork offers access to the deeply dissected Fish Creek Badlands on the left below the southern flank of the Vallecito Mountains. Side canyons of Oyster Shell Wash, Lycium Wash, and Mollusk Wash intersect North Fork with access to remnant exposures of the Latrania formation.

**63.3 (0.2)** ENTER Split Mountain Gorge. This rocky chasm displays many features characteristic of desert watercourses. These include undercut banks, vertical canyon walls, and boulder-choked tributaries shaped by rampaging flashfloods. Seasonally, Split Mountain Gorge funnels torrents of floodwater, filling Fish Creek Wash from bank to bank.

**63.4 (0.1)** The upper boulder bed (informal name of Kerr, 1982; Winker, 1987) is a massive, chaotic landslide breccia that forms the prominent resistant knob on the right and caps much of the slope on the left of the wash.

**63.5 (0.1)** The bend in the canyon marks the narrowest confines of Split Mountain Gorge. The alternating, ledge-forming sandstone beds exposed in the canyon walls are part of the Lycium formation. Due to the abrupt steepness of the cliffsides, rockfalls are common along this stretch of Fish Creek Wash. The Lycium formation consists of about 75 m of medium-bedded, coarse-grained olive green marine sandstones that have yielded forams (Tarbet and Holman, 1944) and fragmented megavertebrates of late Miocene age. Exposed in the east-facing cliffs of the gorge, the Lycium sandstones interfinger with a tongue of the Elephant Trees member (informal name of Winker, 1987) of the Split Mountain Formation. This is strong evidence of a gulf shoreline where rift-related shallow-marine turbidite sandstones were deposited coeval with proximal alluvial fan sediments along the western gulf margins.

**63.8 (0.3)** The anticline and Split Mountain Fault. On the left, at the level of the wash, is the anticlinal fold involving bedded marine turbidite sandstones (Lycium formation) squeezed tightly and deformed against the down-thrown lower boulder bed (informal name of Kerr, 1982; Winker, 1987) landslide megabreccia. Around the corner, the Split Mountain Fault cuts the upper flank of the Split Mountain anticline with a left-lateral displacement of over 333 m.

**63.9 (0.1)** Beyond the entrenched meander is a straight segment of Fish Creek Wash which follows the trace of the Split Mountain Fault. Down-canyon, for the next 2.0 km, note the sharp contrast in stratigraphic units on either side of Split Mountain Gorge.

**64.1 (0.2) STOP 12 LOWER BOULDER BED.** The W-facing cliff on the right is capped by a thick olive-gray, ledge-forming rock unit extending downslope to the canyon floor. This locally-significant lower boulder Bed, approximately 50 m thick, represents a second landslide/avalanche megabreccia deposit exposed in Split Mountain Gorge. This limited outcrop records the aftermath from a mass of dislodged basement rock that bulldozed its way down across the proximal alluvial fan surface. It was triggered by seismic activity and occurred in a terrestrial environment, whereas the upper boulder Bed occurred in a marine environment.

Note the immense blocks of earth and smaller boulders that now litter the canyon floor. Many of these came crashing down during the Borrego Mountain Earthquake of 1968.

**64.3 (0.2)** The stratigraphically lowest sedimentary rocks of the VFCB have been referred to as the Split Mountain Formation (Tarbet and Holman, 1944; Tarbet, 1951; Dibblee, 1954; Stump, 1972; Kerr, et al. 1979), and Anza formation ( informal name of Woodard, 1963, 1974; Kerr, 1982; Cassiliano, 1994), exposed in the vertical canyoned walls on both sides of the wash.

The Split Mountain Formation rests nonconformably upon the progressively west-tilted metasedimentary and crystalline (La Posta Tonalite, in part) basement hanging wall fault block of the Fish Creek Mountains. This rock unit, nearly 833 m thick, unit is composed of a sequence of land-laid braided-stream sandstones, fanglomerates, and mud flow deposits that chronicle the initial deroofing of the ancestral Vallecito-Pinyon mountain highlands. It is coeval with Miocene detachment faulting and volcanism, and includes all non-marine Miocene rock units directly below the Imperial group, but does not include Eocene-age Ballena Gravels exposed in the Vallecito Mountains (Kerr, 1982; Lough, 1994). It is coeval with the Diligencia Formation (Bohannon, 1975) and, in part, with the Coachella Fanglomerate (Vaughan, 1922) of the northern Salton Trough.

Winker (1987) (Figure 10) subdivided the chronology by refining the stratigraphy depositionally, rearranging the Split Mountain Formation into two informal members, the lowermost Red Rock Canyon member and the upper Elephant Trees member. In the latter member, there are various bedding types throughout the sequence, ranging from massive, matrix-supported boulder to cobble conglomerates to lesser debris flows and sheetflood deposits suggesting alluvial fan deposition (Kerr, 1982). Note that exposures of the Elephant Trees member are composed of distinctive non-marine red to red-brown monomictic matrix and framework-supported alluvial fanglomerates, boulder conglomerates, breccias, debris flows, and coarse-grained pebbly feldspathic sandstones (Kerr, 1982) (Figure 11). This member is coeval with the Coachella Fanglomerate that has been dated at 10 Ma (Krummenacher in Peterson, 1975).

**65.2 (0.9)** On the left-hand side of the roadway, the reddish-colored sandstones at the base of the cliff represent the Red Rock Canyon member. It is interpreted as a braided stream facies within the Split Mountain Formation. In the southern VFCB, the Red Rock Canyon member is a locally-derived non-marine arkosic series of pebbly sandstones intercalated with tholeiitic to alkalic volcanic (basaltic) flows and tuffs of the Jacumba Basalt (Mendenhall,1910; Woodard, 1974).

**65.8 (0.6)** EXIT Split Mountain Gorge. On both sides of the strike valley ahead, light-colored, frosty-looking exposures of Fish Creek Gypsum can be seen. The gypsum beds were brought to the surface by local down-faulting and folding associated with the WSTD. Detachment faulting has dismembered the Split Mountain Formation and younger rock units along the SE end of the Vallecito Mountains, as well as here along the entrance to Split Mountain Gorge. Contrary to published reports, the Fish Creek Gypsum may not constitute the oldest and lowermost lithologic member of the Imperial group.

**66.3 (0.5)** To the right is the United State Gypsum mining operation which began in 1946. At present, this multi-bench operation is the largest gypsum mine in the United States, with over one million tons of gypsum (96-99 per cent purity) mined annually (Sharpe and Cork, 1995). This glass-like equigranular evaporite is actively quarried for wallboard, soil binder, chalk, window panes, and various casting plaster. It is hauled by a narrow gauge railroad to the calcining and wallboard plant at Plaster City, 41.6 km to the south. The large, undeveloped reserves are over

50 km thick here, but thin rapidly westward.

The Fish Creek Gypsum (Ver Planck, 1952; Dean, 1988; Sharpe and Cork, 1995) is composed of marginal marine mudstones and anhydrite/gypsum evaporites of the Imperial group. These deposits represent the relics of a desiccated gulf basin model, precipitated from saline waters in a shallow synrift depression of the Gulf of California. As discontinuously exposed in Split Mountain Gorge, there is no clear consensus as to whether these evaporites may be nonmarine (Winker, 1987), the first record of marine waters into the northern Gulf of California (Kidwell and Kerr, 1991), or the basal unit of the Imperial group (Stump, 1972; Dibblee, 1954).

S of the gypsum mining operation (N of Boundary Mountain) and in Gypsum Wash in the Coyote Mountains, gypsum-like deposits interfinger with uppermost shallow-water bioclastic (?) turbidite (Lycium formation) sandstones and grade upward into very fossiliferous transgressive skeletal limestones and sandstones of the Latrania formation. Above Split Mountain Gorge, the gypsum appears to sit atop the upper boulder bed. More field work is required before the Fish Creek Gypsum's stratigraphic position is fully understood. Regionally, the Fish Creek Gypsum may correlate to a global lowering of sea level and the formation of anhydrites during the late Miocene. For example, gypsum deposits are documented by the Deep Sea Drilling Project Cruise Leg 13 for this interval of desiccation responsible for the deposition of the Rosetta Evaporites (gypsum) throughout the drained basin of the Mediterranean Sea (Hsu, et al. 1973; Hsu and Cita, 1973).

**66.7 (0.4)** Intersection with the paved Split Mountain Road. TURN LEFT, destination Ocotillo Wells. The low hill on the left marks the site of an old celestite (strontium sulfate) mining operation. Locally known as the Roberts and Peeler Deposit, this exposure of intercrystallized gypsum and celestite was one of only three deposits in California mined for strontium minerals.

**69.2 (2.5)** Leaving Anza-Borrego Desert State Park. CONTINUE STRAIGHT.

**70.1 (0.9)** Halfhill Lake (dry playa) on the right represents a sag pond developed along the southerly alignment extension of the CCF.

**70.7 (0.6)** During the M6.5 Borrego Mountain Earthquake (1968), the electrical power sub-station's large transformers were shifted about, shearing anchor bolts and breaking the X-bracing (Sharp and Clark, 1972). Roadway makes a sharp turn to the left.

**71.0 (0.3)** The sandy hills on the right are the backside of the Ocotillo Badlands. The badlands are bounded on two sides by left-stepping *en echelon* segments of the CCF that have up-warped and deformed Plio-Pleistocene sediments of the Borrego Formation and younger Ocotillo Conglomerate (Brown and Sibson, 1989).

**71.6 (0.6)** Roadway makes a sharp turn to the right.

**74.8 (3.2)** T-intersection with Highway 78 at Ocotillo Wells. Benson Dry Lake and the Ocotillo Wells County Airport are straight ahead. This small desert community is a year-round haven for off-highway vehicle enthusiasts. MAKE A LEFT TURN onto State Highway 78, destination Borrego Springs. Ocotillo Wells is known for the appreciable ground breakage sustained during the 1968 Borrego Mountain Earthquake. Several active NW-SE trending strands of the CCF cut through the area. The main strand crosses State Highway 78 at 0.2 miles E of the Burro Bend Cafe, trending SE along the western margin of the Ocotillo Badlands. Overall, 38 cm of right-lateral strike-slip occurred along 49.6 km of the CCF (Sharp and Clark, 1972). Post earthquake aftershocks and microseismicity nearly doubled the original ground displacement. In this vicinity, up to 23 cm of vertical slippage was also recorded. Several residential homes and commercial buildings were severely damaged.

**75.3 (0.5)** On the right, Borrego Mountain (East Butte) represents a fractured sliver of crystalline basement (late Cretaceous) uplifted above the San Felipe drainage plain by the CCF.

**75.4 (0.1)** Entrance to the 40,000-acre Ocotillo Wells State Vehicular Recreation Area at Main Street. CONTINUE STRAIGHT.

**76.4 (1.0)** Ocotillo Wells ranger station entrance on the right. CONTINUE STRAIGHT.

**77.9 (1.5)** Buttes Pass. MAKE A RIGHT TURN onto the dirt roadway. FOLLOW the established roadway N towards Buttes Pass. Although four-wheel drive is recommended, two-wheel drive vehicles can usually negotiate this portion of the route up to the bulletin board parking area.

**78.9 (1.0)** Intersection of Buttes Pass and Borrego Mountain Wash. MAKE A RIGHT TURN.

**79.4 (0.5) STOP 13 BORREGO MOUNTAIN.** PARK at the Bulletin board parking area. Hike to ridgeline above the entrance to scenic Hawk Canyon for dramatic overview of the Borrego Mountain area. E-directed Block-faulting has dismembered East and West Butte. Miocene sediments exposed along Blow Sand Canyon and in Borrego Mountain Wash correlate to rock units of the Calcite Scenic Mine Area along the SE end of the Santa Rosa Mountains. These sediments are capped by Mio-Pliocene (?) conglomeratic sandstones that extend westward along the N flank of Yaqui Ridge (Ship Rock). These are classically uplifted (normal fault) and dismembered from equivalent deposits exposed along State Highway 78 W of Tamarisk Grove. RETURN to State Highway 78.

**80.9 (1.5)** MAKE A RIGHT TURN on State Highway 78, destination Borrego Springs.

**82.4 (1.5)** Junction with Borrego Springs Road. MAKE A RIGHT TURN onto Borrego Springs Road.

**83.2 (0.8)** San Felipe Wash has entrenched itself in 1.6 km-wide Texas Dip below a thick wedge of late Quaternary terrestrial sediments laid down atop an older pediment surface of Cactus Valley. Texas Dip was formed during the Quaternary when rushing flood waters breached a natural dam at The Narrows (along State Highway 78) and catastrophically bull-dozed out Texas Dip. The temporary dam was caused by seismic activity associated with the San Felipe Fault and had pooled up waters from San Felipe Creek along the strike valley of the Mescal Bajada half-graben.

**88.7 (5.5)** Two-way traffic stop intersection of Yaqui Pass Road and Borrego Springs Road. La Casa del Zorro is on the left. Gas, oil and minor camping supplies are available at the Borrego Springs Chevron gas station (diagonally across the road) on the right. CONTINUE STRAIGHT.

**94.1 (5.4)** Christmas Circle. PROCEED RIGHT (around the rotary) for access to downtown Borrego Springs.

**END OF DAY 1**

# DAY 2: Borrego-San Felipe Basin

**0.0 (0.0)** START at the Christmas Circle rotary. DRIVE EAST on Palm Canyon Road (S-22).

**0.5 (0.5)** Junction with Di Gorgio Road.

**1.5 (1.0)** Four-way traffic stop intersection at Borrego Valley Road.

**2.7 (1.2)** Borrego Springs Airport on the left. CONTINUE STRAIGHT (east), approaching the Borrego Badlands (Figure 12), dominated by the southernmost cuesta promontory of Font's Point (elev. 431m above sea level).

**3.0 (0.3)** Milepost marker 22.

**4.0 (1.0)** Milepost marker 23.

**4.1 (0.1)** End of Palm Canyon Road at Old Springs Road (right). CONTINUE on S-22 as it makes a sharp left turn heading north on Pegleg Road (S-22).

**4.7 (0.6)** Milepost marker 24. Profile of the badlands to the right reveals that late Quaternary faulting associated with the Coyote Creek Fault (CCF) has uplifted and deformed the northern flank (Ocotillo Rim) of the Borrego Badlands.

**5.0 (0.3)** To the right (in the foreground), behind the white storage tank and shed is an elongated NW-SE sand dune feature possibly uplifted along the CCF. The co-seismic CCF extends NW along the W side of Coyote Mountain to Desert Gardens, the Coyote Badlands, and through Box Canyon. In this vicinity, the CCF, is dominantly a dip-slip fault with a minor (?) strike-slip component, the master western branch of the SJFZ. On April 9, 1968, the M6.5 Borrego Mountain Earthquake hit Anza-Borrego. It was a classic example of seismic triggering of a main shock, with an epicenter in the Borrego Badlands near Third Wash, and epicentral distribution of aftershocks covering a broad area around the Ocotillo Badlands-Borrego Mountain area. Ground rupture occurred along 31 miles on the central segment of the CCF near Ocotillo Wells, terminating SW of Vista del Malpais. Desert Gardens marks the epicenter of the 1969 M5.9 Coyote Mountain Earthquake. Aftershocks of this event were restricted to depths of between 10-14 km (Thatcher and Hamilton, 1973). Focal mechanism

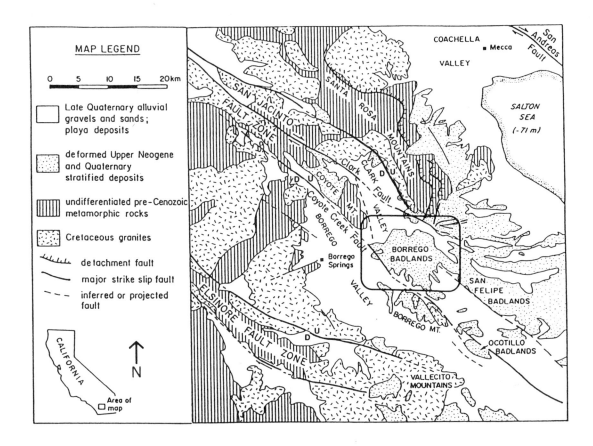

Figure 12. Simplified regional geology of the Borrego Badlands area, Anza-Borrego Desert State Park, California. Modified after Pettinga (unpublished data, 1992).

define an elongated high-angle left-slip cross-fault whose major axis was oriented NE (Peterson, et al. 1987).

**5.5  (0.5)** At this location (marked on the right by a young palo verde tree), a slight rise in the roadway may mark the approximate trace of the active CCF. *In situ* stress measurements along the seismogenic fault zone across the central rupture break have been made by Keller, et al. (1978). Using theodolite alignment arrays, they found a slip rate of about 10 mm/yr., calculating a recurrence interval of about 200 years for major events on the fault. Sharp (1981) has estimated minimum horizontal displacements along the fault zone of between 5.7 and 8.6 km during the late Pleistocene. This contradicts recent field evidence which strongly suggests dip-slip displacement along the northern segment of the CCF.

**5.7  (0.2)** Milepost marker 25. CONTINUE STRAIGHT approaching Coyote Mountain. On the right, beyond the row of tamarisk trees (wind-break), unconsolidated, light-colored mudhills are remnants of a playa that occupied the eastern half of Borrego Valley during the late Quaternary. Coeval fan-delta deposits in the vicinity of Dump Wash yield tufa fragments and the freshwater snail *Physella* sp.. Further S, low-lying shoreline relic features are associated with an enormous tufa bed near the Borrego Sink.

**6.5  (0.8)** End of Pegleg Road (S-22) at Henderson Canyon Road (left). CONTINUE on S-22 as it veers to the right. This segment of S-22, known as the Borrego-Salton Seaway, was built in 1968 and more-or-less follows the approximate trace of the old Truckhaven Trail across the Borrego Badlands/Santa Rosa Badlands.

**6.7  (0.2)** Milepost marker 26.

**6.9  (0.2)** Junction with Rockhouse Canyon Road on the left.

**7.4  (0.5)** Junction with unmarked Beckman Wash

on the right. TURN RIGHT onto dirt roadway leading SE (opposite small, gray-colored microwave tower. PROCEED S along this dirt track. Four-wheel drive is recommended beyond this point.

**7.7 (0.3)** The dirt roadway is built on late Pleistocene Font's Point sandstone (informal name of Remeika and Pettinga, 1991) that has been slightly deformed. The surface is uplifted on the S forming the Ocotillo Rim, and gently dips N-NW.

**8.7 (1.0)** ENTER ABDSP.

**9.3 (0.6)** To the left (east) bluff exposures preserve angular unconformity near top of the Ocotillo Conglomerate (Dibblee, 1954). The Ocotillo Conglomerate has been redefined as the Ocotillo formation (informal name of Remeika, 1992a). The vertebrate-bearing Ocotillo formation represents a locally-deposited middle to early late Pleistocene continental fluvial-floodplain succession of gravelly sandstones, silts and claystones exposed throughout the northernmost half of the western Borrego Badlands (informally referred to as Fault Block A) and Coyote Badlands (?) (Remeika, 1992a). The stratotype is located west of Inspiration Wash in the Arroyo Otro-Mammoth Cove section, above and below the Ocotillo Rim NW of the Inspiration Point Fault. To establish stratigraphic control and age determinations, Remeika and Pettinga (1991) and Remeika (1992a) recognized three previously undefined lithofacies for the Ocotillo Conglomerate of Dibblee (1954). In ascending order they include a 74 m-thick subaqueously-deposited distal alluvial fan sequence (informal Mammoth Cove sandstone member), an 87 m-thick series of freshwater lacustrine claystones (informal Las Playas member) and a 216 m-thick mixed composition fluvio-lacustrine interbeds (informal Inspiration Wash member to the west and Short Wash Member to the east) that laterally grade into the Las Playas member.

The resistant caprock of Font's Point sandstone may be younger than 0.37 Ma. To the E, in Arroyo Otro, a pronounced angular unconformity near the top of the Inspiration Wash section is demonstrated by uplift, syndepositional erosion and its close proximity to the active CCF before deposition of the overlying Font's Point sandstone. The upper contact is defined by an erosion surface having a paleomagnetic signature of 0.37 Ma (D.F. Scheuing, written comm., 1991). Font's Point sandstone can be differentiated from underlying beds on the basis of stratigraphic position, presence of caliche-mottled horizons, pluvial-shoreline (tufa) deposits and Rancholabrean (?) vertebrate fossils. Additional evidence for syndepositional deformation includes clastic dikes, intraformational unconformities, discordance in bedding dip across bedding planes, thrust faulting in fold hinges (Mammoth Cove) and chaotic folding suggesting liquefaction.

**9.4 (0.1)** Possible Bishop Tuff bed in saddle to the left. BEAR RIGHT. White beds to the right may be reworked tuff.

**9.7 (0.3)** Saddle. PROCEED S over lip of saddle. Fossiliferous overbank sediments include finer-grained siltstones and claystones of the Inspiration Wash member.

The Inspiration Wash member in the westernmost Borrego Badlands (Fault Block A) overlies the distal tongue of the Mammoth Cove sandstone member. Between Arroyo Otro (to the east) and Beckman Wash, it consists of about 200 m of fluvially-deposited terrestrial floodplain facies of interbedded sheetflood gravelly sandstones and fine-grained overbank siltstones/claystones. Striking E, the section laterally grades into purplish-gray playa claystones of the Las Playas member. Within the claystones, ripplemarks and mudcracks are common.

Diagnostic fossil vertebrates include *Smilodon gracilis, Artodus* sp., *Mammuthus columbi, Equus bautistensis* and *Camelops huerfanensis* of the Borrego LF. A late Irvingtonian LMA is indicated. The occurrence of *Microtus californicus, Lepus* sp. cf. *L. californicus*, and *?Eucheratherium* sp. appear only in the youngest levels of Valle Escondido (possible early Rancholabrean LMA) above a recently discovered tephra which may correlate to the Bishop Tuff at .74 Ma (Rymer, 1991) marking the Matuyama-Bruhnes geomagnetic boundary. Depending on the time scale used, the base of the ABDSP Irvingtonian is between 1.6-1.9 Ma with the transition between Irvingtonian-Rancholabrean at .35-.45 Ma.

At present interdisciplinary studies continue on the Borrego LF and its host sediments (especially material recovered from the Font's Point sandstone). Continued mapping, including stratigraphic revision and correlation between Fault Blocks A and B, vertebrate biostratigraphy, magnetostratigraphy and tephrochronology of the local Borrego Badlands section, will greatly enchance its importance in the Pleistocene history of western North America.

**9.8 (0.1)** S side of road is a resistant cemented arkosic sandstone marker bed that holds up the

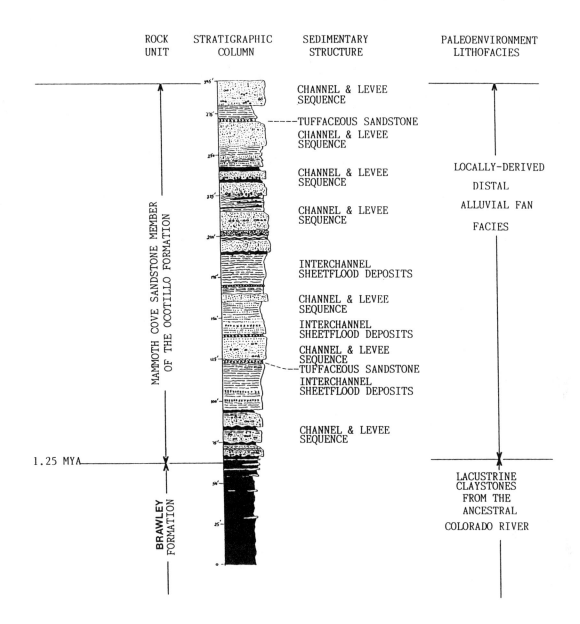

Figure 13. Simplified stratigraphic column of the Mammoth Cove Sandstone member of the Ocotillo Formation exposed in Mammoth Cove below the Ocotillo Rim, Borrego Badlands.

Ocotillo Rim above Mammoth Cove. DRIVE ACROSS N-dipping resistant sandstone and park near canyons running S to Mammoth Cove Overview.

### 9.9 (0.1) STOP 1 MAMMOTH COVE OVERVIEW.
HIKE S for view of the Mammoth Cove sedimentary sequence (Fault Block A) and deformation fabric beneath the Ocotillo Rim. Landscape of the small drainage intrabasin of Mammoth Cove is characteristic of arid regions undergoing crustal seismogenic neotectonics and extreme denudationary processes.

Highly-eroded tan-gray crossbedded and massive sandstones are Mammoth Cove sandstone member (Figure 13). Closely coupled to the strike-slip CCF, several high-angle cross-faults indicate contractional Quaternary tectonism in the deformation process of strain rotation. Folding is gentle to isoclinal.

Fault-parallel shortening, discretely shown by NE-trending tightly-folded lacustrine evaporites of the Brawley Formation may demonstrate the presence of a detachment surface at shallow depth, suggesting a conjugate relationship that allows fault blocks to decouple and rotate on cross-faults. Previous published paleomagnetic investigations confirm that a clockwise rotation has occurred in the Borrego Badlands (Bogen and Seeber, 1986; Scheuing, et al. 1988). D.F. Scheuing (unpublished data, 1988) has documented block rotation of $23 \pm 13$ degrees and 37-40 degrees for Mammoth Cove, depending on the relative age of the sediments; older Brawley Formation (Dibblee, 1954) sediments show twice the rotation observed than for the Mammoth Cove sandstone member. RETURN to vehicles and proceed W.

**10.1  (0.2)**  ABDSP boundary signs. Temporarily EXIT the state park.

**10.2  (0.1)**  TURN RIGHT (east) up unnamed tributary wash and RE-ENTER ABDSP. STAY LEFT as the wash forks.

**10.4  (0.2) STOP 2 BISHOP TUFF.** PARK at end of wash. Inspiration Wash member exposures to the N contain 6-inch thick bed of white ash which may represent the middle Pleistocene Bishop Tuff derived from the Long Valley-Glass Mountain Center in eastern California (Izett and Naeser, 1976). Within the Salton Trough, the Bishop Tuff is reported from the Mecca Hills (Babcock, 1974; Rymer, 1991) and Coyote Badlands. Contemporaneous Bishop Tuff exposures (Fryant ash member) in the Coyote Badlands are defined by five closely-spaced airfall ash deposits fingerprinted in Ash Wash (G.J. Miller, pers. comm., 1985). Identification of the ash was based on its chemcial composition (A.M. Sarna-Wojcicki, pers. comm., 1984). The Coyote Badlands is inferred to be right-laterally displaced from the western Borrego Badlands (fault block A) along the strike-slip CCF. Beckman Wash samples are in the process of being geochemically examined. A similar ash bed has recently been discovered in fault block B. RETRACE the route back to Beckman Wash and RETURN to S-22.

**13.3  (2.9)**  MAKE A RIGHT TURN (east) on S-22 (directly opposite the small microwave tower).

**13.5  (0.2)**  Milepost marker 27. At 8:00 Coyote Mountain (elevation 1064 m above sea level) represents a fractured, elongated block of lower-plate cataclastic, mylonitic and metasedimentary rocks (Theodore, 1967a,b, 1968, 1970) that have been dramatically down-faulted (W-directed) from the footwall Santa Rosa Mountains and uplifted on the W along the dynamic CCF on its western side. The E face may also be bounded by the Coyote Mountain Frontal Fault (?). Note triangular facets developed in the bedrock. These are very prominently developed along its W edge. These tectono-geomorphic features are reliable criterion of active range-front dip-slip faulting. Also note steep profiles of immature alluvial fans. Their steepness strongly reflects basin dropping relative to mountain uplift. Older fans are covered with a modest coating of desert varnish.

**13.8  (0.3)**  To the left is the closed triangular-shaped tectonic graben of Clark Valley (elevation 133 m above sea level), sandwiched between Coyote Mountain on the W and the Santa Rosa Mountains on the N and E. It represents a dynamic extensional wrench basin, structurally controlled by a composite of master dip-slip faults [Coyote Creek, Buck Ridge, and Coyote Mountain Frontal (?)] and a strike-slip fault (Clark) in a transtensional regime adjacent to fault stepovers that produce localized extension within the SJFZ. The Buck Ridge Fault may serve as a listric accomodation fault. The margin between Clark Valley and the Santa Rosa Mountains is discretely marked by the Clark Fault (CF) which lies beneath Holocene alluvium of the Clark's Dry Lake depocenter. The basin presumably formed during block-fault detachment with related wrench tectonics between Coyote Mountain and the Santa Rosa Mountains and the clockwise rotation of the Borrego Badlands away from the irrotational Coyote Mountain block. In 1986, geodetic measurements in Clark Valley included a resurvey of benchmarks installed prior to 1969. Hudnut and Seeber (1986) report astroazimuth investigations produced evidence for clockwise rotation of the basin ($5 \pm 3$ microradians per year) and for right-lateral deformation on the CF. Also note upper-plate crystalline megalandslide deposits N of Clark Valley. These may reflect a complex zone of detachment and related seismicity along the Buck Ridge Fault.

Reflecting the recency of crustal movements, the desert here is actively undergoing extension, rapid subsidence, rotation, and tilting (Nicholson and Seeber, 1989), part of a widening oblique transform motion within this portion of the WSTD. There are no foothills. Fault scarps are everywhere. Sesimic activity, always persistent, continues to occur. Note the

Figure 14. County Highway S-22 provides an unobstructed view of the structurally-uplifted footwall segment of the Santa Rosa Mountains.

abandoned building in the middle of Clark's Dry Lake. This structure once served as the University of Maryland's Clark Lake Radio Observatory.

14.3 (0.5) S-22 is built on the featureless, creosote-covered paleosols of the Font's Point sandstone, a late Pleistocene alluvial deposit that is locally deformed. The remnant surface is incised by many N-flowing ephemeral drainages along this portion of the NW Borrego Badlands.

14.5 (0.2) Milepost marker 28. Entering ABDSP (Borrego Badlands Sector).

14.8 (0.3) Truckhaven Trail on the left.

14.9 (0.1) CROSS unsigned Inspiration Wash.

15.4 (0.5) Milepost marker 29. To the NE (left) is an excellent unobstructed panorama of the structurally-uplifted footwall segment of the Santa Rosa Mountains (Figure 14). Its titanic lower-plate ramparts are due to Plio-Recent footwall uplift and guard northern Clark Valley on such a grandiose scale that they attain the greatest vertical rise of any mountain range in the state park. Local mountain heights include Villager Peak at 1918 m and Rabbit Peak at 2222 m, with clear lines of sight following the ridgeline up to Toro Peak at 2902 m above sea level.

15.9 (0.5) Entrance to Font's Wash on the right. MAKE A RIGHT TURN, destination Font's Point. Four-wheel drive is recommended to negotiate loose sand. Font's Wash drains much of the NW Borrego Badlands N of the Ocotillo Rim. Road conditions along this major watercourse are subject to change without notice. Seasonally, rampaging flashfloods fill the arroyo from bank to bank, funneled N into the topographically lower Clark's Dry Lake playa. Font's Wash displays many ephemeral components characteristic of desert arroyos, including choked tributaries, undercut banks, eroded hillslopes, incised channels and pediments with narrow interfluves.

16.1 (0.2) To the right bluff exposures of green lacustrine claystones and red calichified sandstones belong to the late Pleistocene Font's Point

Figure 15. Mudhills at the entrance to Font's Wash display fossil-bearing sediments of the Font's Point Sandstone underlain by the Las Playas member.

sandstone. The presence of the Font's Point sandstone documents neotectonic uplift of the Santa Rosa Mountains.

**16.6 (0.5)** To the left (Figure 15), the first series of varicolored mudhills include N-dipping conglomerates (basin-margin interbeds of the Lute Ridge member of Canebrake Conglomerate IV). These coarse sediments yield articulated tufa fragments that may correlate to tufa horizons preserved at Stop 6. These overlay deeply-weathered unconsolidated reddish-brown sediments of the Font's Point sandstone (distal equivalent of the Lute Ridge member) and lacustrine claystones of the Las Playas member of the Ocotillo formation (Remeika and Jefferson, 1993). The mixed salmon-brown and greenish-colored micaceous claystones and siltstones with sandy interbeds represent a lacustrine fluvial-playa sequence analagous to the modern Clark's Dry Lake playa.

A bone-bearing sandstone interbed in the Font's Point sandstone yields a late Pleistocene (late Irvingtonian-early Rancholabrean LMA) megafauna (Borrego LF). Over the past several years, paleontologists (under permit) have unearthed the fossil remains of now-extinct *Nothrotheriops* sp. cf. *N. shastensis*, *Canis latrans*, *Mammuthus columbi*, *Equus bautistensis*, *Camelops huerfanensis*, and *Odocoileus* sp. Work initiated in the 1970's by Ted Downs and Harland Garbani (LACM), in the 1980's by the late George Miller (IVCM), and continued in the 1990's by Paul Remeika and George Jefferson (ABDSP) have all led to cyclic prospecting and recovery of a rich vertebrate megafauna (Irvingtonian-Rancholabrean) represented by 50 mammalian genera. This Borrego LF (Remeika, 1992a; and Remeika and Jeffeson, this gudebook) consists of a fossil assemblage of at least 5,000 cataloged specimens, collected from more than 400 paleontologic localities in the Ocotillo formation. Most of the composite fauna is based upon fragmentary remains and only a few articulated specimens have been found.

**17.1 (0.5)** Roadway veers left across the obscured trace of the Inspiration Point Fault. Within Mammoth Cove (Fault Block A), this prominent left-lateral cross-basin fault is nearly orthogonal to the main right-lateral strike-slip CCF. It trends NE across the entire breadth of the SJFZ. Recent mapping confirms a considerable dip-slip component to the fault. Hudnut and Clark (1989) suggest that slip on related cross-faults locally decrease normal stress on the main faults. Hudnut, et al. (1989) hypothesize that surface rupture of the Inspiration Point Fault may trigger rupture on either the CF or CCF by a mechanism similar to that which occurred during the November 23 and 24, 1987 Superstition Hills Earthquake (M6.2 and M6.6) sequence.

**17.2 (0.1)** To the left, bluff exposures of Fault Block B are E-dipping Font's Point sandstone capped by Quaternary alluvium.

**17.5 (0.3)** Junction with Short Wash Trail on the left. CONTINUE STRAIGHT. The roadway soon veers right out of the main wash into a tributary arroyo. Proceed SW, destination Font's Point.

**18.0 (0.5)** N-dipping unconsolidated brownish claystones and arkosic sandstones of the Font's Point sandstone, and siltstones and claystones of the Las Playas member (P. Remeika, unpublished data, 1995) make up the low cohesion slopes of badland relief along the roadway.

**18.2 (0.2)** On the right, steeply-dipping salmon-brown colored siltstones of the Short Wash member are complexly-folded approaching Cottontail Fault. Strata include a mixed affinity package of finer-grained siltstone and claystone interbeds derived from the Colorado River (?) and coarser clastics from local source areas. They form subtle topographic highs, indicating a fault-propagated NW-plunging, anticlinal fold. Note the greenish claystone marker bed that defines the core of the fold. This marker bed can be followed from the Inspiration Wash area (S-curves) E to Palo Verde Wash. Ongoing deformation is suspected by a systematic decrease in the dip of bedding towards the N, away from the anticlinal axial surface. This section (Fault Block B) is coeval to the section in Inspriation Wash Fault (S-curves) (Fault Block A), indicating left- lateral displacement along the Inspiration Point Fault.

Various-sized cameloid footprints, comparable to the coeval ichnogenera *Pecoripeda* spp. cf. *Camelops huerfanensis* (Remeika, 1992a) and more recently, *Titanotylopus* sp., have been discovered in the Short Wash member.

In 1994, P. Remeika discovered *Camelops* sp. fragments in the Short Wash member in Font's Wash. This is the first documented occurrence of vertebrate fossils from this member. G.T. Jefferson, assisted by paleo-volunteers, is currently excavating the site.

**18.4 (0.2)** CROSS trace of the high-angle, E-trending Cottontail Fault. This secondary cross-fault is also transverse to the nearby major NW-trending CCF. Its presence implies that crustal deformation is being accommodated by local folding, dip-slip faulting and block rotation. Activity on this short fault has juxtaposed Font's Point sandstone (south) down against the Short Wash member (north).

**18.7 (0.3)** DRIVE ACROSS semi-resistant calichified surface of Font's Point sandstone.

**19.4 (0.7)** Y-intersection. STAY LEFT on one-way loop drive. Sedimentary exposures are entirely of Font's Point sandstone.

**19.7 (0.3) STOP 3 FONT'S POINT.** The approach to Font's Point is distinguished by a rather flat, subdued landscape of ocotillo and creosote bush. Now, at the brink, is one of the most sublime spectacles in all of Anza-Borrego. HIKE to the Ocotillo Rim at Font's Point, elevation 431 m above sea level. This is the climactic overlook of the Borrego Badlands (Figure 16). Unparalleled for scenic grandeur, the view is of one of the most seismically active regions in California: a transition region from the continental transform regime of the SJFZ/SAFZ to the Gulf of California oceanic rifting regime (Hudnut, et al. 1989).

On the southern horizon are the Fish Creek, Vallecito and Pinyon Mountains. The low mound of hills E of the Fish Creek Mountains is Superstition Mountain. On November 23 and 24, 1987, major twin earthquakes, measuring M6.2 and M6.6 respectively, hit along the Superstition Hills Fault (part of the SJFZ). Weathered desiccation cracks atop Font's Point were noticeably enlarged. During the Borrego Mountain Earthquake of 1968 (M6.5), over six feet of rock at the tip of Font's Point tumbled down the cliffside. An additional five feet of cliffside collapsed along the SE tip during the Landers Earthquake of June, 1992 (M7.6).

To the S, the laterally-offset twin buttes of Borrego Mountain form a natural barrier along the San Felipe Wash drainage. San Felipe Wash follows the approximate trace of the Coyote Creek Fault. Related Mio-Pliocene strata in the vicinity of Hawk Canyon and Borrego Mountain Wash have been thrust over and folded up against the crystalline bedrock, indicating NW-directed shortening of the crust. Nearby, the tectonically subsiding Borrego Sink (sag pond) is the lowest structural feature in the area at 156 m. As a result, the sink receives the greatest percentage of surface water streamflow and groundwater inflow entering the Borrego Valley aquifer.

To the E are the shimmering waters of the Salton Sea. To the N stand the impressive crystalline, mylonitic and metasedimentary basement terranes of the Santa Rosa Mountains and Coyote Mountain, separated by the enclosed half-graben of Clark's Dry Lake. To the W is the great mountain barrier protecting Borrego Valley. The San Ysidro Mountain Range rises abruptly nearly a vertical mile from the valley floor. San Ysidro Peak at 2049 m above sea level is the high point punctuating the skyline.

The late Pleistocene Font's Point sandstone forms the stratum capping sediments exposed in the receding cliffside of Font's Point. The fossiliferous Font's Point sandstone represents the distal portion of the Canebrake IV interval. It forms the thick, blocky reddish-brown caprock (arkosic sandstones) and underlying coarse-grained braided stream sediments of the promontory. Here it is crudely bedded and locally conglomeratic (pebbly). Distinctive mottled pedogenic zones of caliche indicate a semi-arid transitional environment of deposition during the late Pleistocene. Nearby, early Rancholabrean (?) vertebrate fossils recovered from this rock unit include *Accipiter* sp., *Buteo* sp., *Neotoma lepida*, *Mammuthus columbi*, *Equus* spp., *Equus (?hemionus)* sp., *Camelops* sp., and *Cervus* sp. Study of this faunal element will greatly aid in our understanding of, and interpretation of, the Borregan scene during its transition (desertification) to an arid Colorado Desert environment.

The gray-buff base of the cliff is the fossiliferous Mammoth Cove sandstone member. It represents a locally-derived granule to bouldery conglomeratic distal alluvial fan facies. The coarseness and thickness of this sandstone unit generally increase proximally NE while grain size and fan gradient decrease down fan SW, ultimately pinching out basinward east of Vista del Malpais. The Mammoth Cove sandstone alluvial wedge is stratigraphically and petrologically correlative with the Canebrake Conglomerate III interval of the Truckhaven Rocks. Lenticular bandings of sediment with parallel bedding, climbing ripple lamination, thin overbank splays of fine-grained biotite-rich silts and clays, and torrential bedding indicate rapid deposition. Basinward, the Mammoth Cove sandstone member mantles the uppermost fine-grained riverborne redbeds of the Brawley Formation and interfingers with the Short Wash member. The basal contact (with the Brawley Formation has yielded a paleomagnetic signature of 1.25 Ma (D.F. Scheuing, written comm., 1989). Extinct vertebrates, such as *Arctodus* sp., *Mammuthus imperator*, *Equus (Dolichohippus) enormuus*, and *Camelops* sp. index the Irvingtonian LMA.

The Brawley Formation (Dibblee, 1954) is composed of distinctive basinal interbedded light red-brown, nonmarine, fine-grained lacustral Borrego-like claystones and siltstones from the ancestral Colorado River (Colorado River group). Contains subordinate locally-derived sandstone tongue marker beds of mixed affinity (informal Ghost Palms member of the Ocotillo formation) (P. Remeika, unpublished data, 1994) that pinch out basinward (to the E). It also contains a distinctive, laterally-continuous dark-green gypsiferous claystone marker bed. Typically shallow-water, low-energy lacustrine environment of deposition that represents a continuation of relict estuarine lakebed conditions deposited above the Borrego Formation. Along the basin-margin, mixed affinity sediments are recognized as the Short Wash member of the Ocotillo formation. Fossiliferous with rare foraminifers and diagnostic brackish-freshwater invertebrates and plant macrodetritus (Taylor, 1981; Stearns, 1901), and reworked petrified wood. Microfaunal hydrobiids *Tryonia* sp., *Fontelicella* sp. and Planorbidae occur with the aphaeriid *Pisidium* sp. cf. *P. compressum* and unionid *Anodonta* sp. cf. *A. californiensis*, restricted to the lacustral claystones of the Brawley and Borrego Formations.

The Borrego Formation (Tarbet and Holman, 1944; and Tarbet, 1951) represents a thick (900 m-1900 m) basinward lacustrine to brackish-water, shallow low-energy nearshore (distal) interdistributary facies of the underlying Diablo formation. Exposures only occur N of the active delta cone in the Salton Trough and constrain the age of the "true" Salton Trough to the Pliocene. Basinal sediments exposed in Rainbow Wash and Hills of the Moon Wash directly below Font's Point are fine to very fine-grained light-gray to reddish-brown interbedded claystone and siltstone deposited by the ancestral Colorado River (Colorado River group). Close examination of clay exposures reveals that many are thinly laminated while others appear structureless. All have been complexly deformed. This rock unit straddles the Plio-Pleistocene boundary. Contains distinctive high-energy Diablo formation-like concretionary arenitic interbeds that actively infilled the area by crevasse-splay sedimentation. This unit is fossiliferous with diagnostic

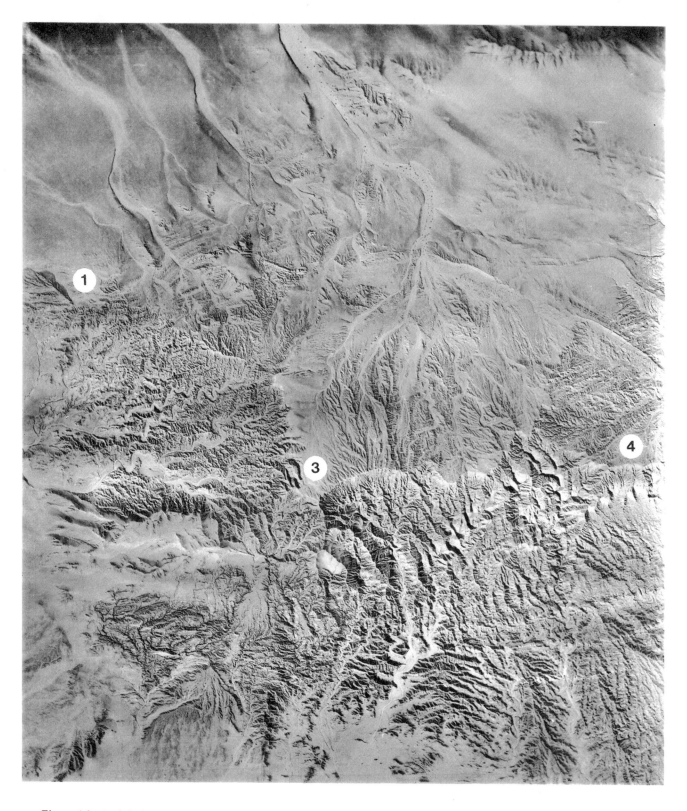

Figure 16. Aerial photograph of the western Borrego Badlands, including Mammoth Cove (stop 1) west of Font's Point (stop 3). Stop 4 is located at the Vista del Malpais overlook.

brackish-freshwater invertebrates and plant macrodetritus. No vertebrate fossils are reported to have been discovered from this unit yet.

The Diablo formation is a widespread delta-plain deposit that is easily recognized throughout the W side of the Salton Trough. In the Borrego Badlands and Truckhaven Rocks (Santa Rosa Badlands), this unit interfingers with undifferentiated distal, coarse-grained arkosic sandstones similar to the upper Miocene (?) Split Mountain Formation. Fossiliferous with diagnositic silicified woods (Carrizo local flora), brackish-freshwater invertebrates (similar to taxa recovered from the Borrego and Brawley Formations, and reworked delta-front shellhash from the upper brackish-marine Burrobend formation. RETURN to vehicles and RETRACE the 3.8-mile route N (downwash) to the paved Borrego-Salton Seaway S-22.

**23.5 (3.8)** MAKE A RIGHT TURN (east) on S-22.

**24.0 (0.5)** Milepost marker 30.

**24.1 (0.1)** Santa Rosa Mountains roadside pullout on the right. Geologically the Santa Rosa Mountains mark the E boundary of the geomorphic Peninsular Range Province. The range is composed of cordilleran reddish-brown Paleozoic metasedimentary rocks (marble, biotite schist, quartzite, banded augen gneiss and amphibolite and younger light-colored Mesozoic grantitic (meta-quartz diorite, granodiorite and pegmatite) core complexes overprinted by the late Cretaceous to early Tertiary Eastern Peninsular Range Mylonite Zone (EPRMZ). This 100 km-long deep-seated ductile shear zone affected the pre-late Cretaceous crystalline/metasedimentary rocks of both the Santa Rosa and San Jacinto Mountains and is part of a much larger mylonite zone that extends southward from Palm Springs to the tip of Baja California (Sharp, 1967; 1979).

According to Simpson (1984), mylonization mostly affected mid-Cretaceous granodiorites, quartz diorites and quartz monzonites of the Peninsular Range Batholith, with minor involvement of older quartzites, marbles and mica-schists. K-Ar and fission-track studies have yielded an upper age limit of ca. 63 Ma for mylonite development (Dokka and Frost, 1978: Wallace and English, 1982). Sharp (1967, 1979) suggested that the EPRMZ represents post-mid Cretaceous unroofing of the batholith along a deep-seated shear zone. On the basis of E-plunging mineral elongation lineations, the EPRMZ originated from W-directed thrusting in response to the closure of a back-arc basin some 100 km inboard from a main E-dipping subduction zone along the ancestral California coast. SW tectonic transport of the structurally high E Peninsular Range Batholith over the lower W batholith has also been postulated by Todd and Shaw (1979).

Post-Pliocene offset within the SJFZ has divided the NNW-SSE trending and E-dipping EPRMZ into three regionally significant fault block ranges: the Santa Rosa, Coyote and San Ysidro Mountains.

**24.2 (0.1)** Note the twin upthrown hillocks of Canebrake Conglomerate IV to the right (1:00). These late Quaternary proximal gravels represent deposits situated on Fault Block B. They are left-laterally displaced from Fault Block A gravels at 3:00 by the Inspiration Point Fault that trends NE-SW across the SJFZ from Lute Ridge (CF) to the Ant Hill (CCF).

**24.9 (0.7)** Blowsand Turnout (unmarked) is on the left. Located between the highway and the Santa Rosa Mountains, is a series of low-lying rolling gravelly hills known collectively as Lute Ridge. Although not obvious from the highway vantage point, the NE edge of Lute Ridge is a classic, textbook strike-slip normal fault scarp, extending 3.2 km in length on the CF. It represents the largest known fault scarp on the North American continent existing in unconsolidated sediments. Lute Ridge is a pressure ridge, continuing to be domed up, resulting in obvious and not-so-obvious unconformities throughout the area. The hummocky topography is deeply dissected. Six additional scarplets are visible on the S-facing slopes of Lute Ridge.

**25.0 (0.1)** Milepost marker 31. At 1:00 the highway crosses lengthwise an elongated NW-SE striking structural graben. It is bounded on both sides by prominent NW-trending fault scarps developed in unconsolidated Quaternary proximal gravel fans capped by blowsand sediments.

On the left, hidden from view on the SW flank of Lute Ridge, is an underground seismograph recording station. Located throughout the SJFZ, a battery of instruments--accelerometers, creepmeters, strainmeters, tiltmeters, leveling lines, geochemical sensors, magnetometers, and a dense network of seismometers--measure and record fault movements (Remeika and Lindsay, 1992).

**25.3 (0.3)** On the left (north) at 10:00, the backside of Lute Ridge is riveted with prominent right-lateral, sinistral shear and oblique normal faults. Fault scarps are also evident on the right, cutting across the base of unconsolidated gravels representing the Canebrake Conglomerate IV. At 11:00 note the significant number of large-scale late Quaternary offset landform features directly east of Lute Ridge, along southern fan-frayed base of the Santa Rosa Mountains. Multiple fault scarps, laterally offset stream channels and fans, banks and bars, and deflected drainage lines apparent at the surface show geologic recency of movement. Horst and graben structures appear between fault scarps.

**25.6 (0.3)** To the N on the E edge of Sierra Ridge is Rattlesnake Canyon. There are two well-developed Pleistocene-Holocene alluvial fans adjacent to it on the lower right, developed along the S base of the Santa Rosa Mountains. Both fans are laterally offset by multiple fault scarps along the CF.

**25.8 (0.2)** Junction with Thimble Trail on the right. TURN RIGHT onto the established dirt roadway of Thimble Trail and FOLLOW it S up the sand hill. Four-wheel drive is recommended beyond this point. Thimble Trail traverses fault-controlled topography produced by Quaternary activity along the master CF. Immediately N is the old Truckhaven Trail. Built in 1929 by A. A. "Doc" Beaty, this historic route connected Borrego Springs with the Coachella Valley. Construction was accomplished by volunteer labor, hand tools, horses, and mule-drawn scrapers.

**26.0 (0.2)** As the roadway climbs from the desert floor across a prominent fault scarp, it enters a relatively uneroded closed graben. To the left, the NW-trending steep-walled ridgeline is one of a population of *en echelon* shutter ridges developed adjacent to the CF. These ridges are made up of the proximal Canebrake Conglomerate IV. Multi-stranded fault strands inidcate that this portion of the badlands is presently undergoing fault-parallel shortening accommodated by folding and faulting in response to strike-slip and irrotational-slip movements on the master fault. These strands terminate to the NW against the Inspiration Point cross-fault.

**26.2 (0.2)** Enclosed structural graben between shutter ridges.
**26.6 (0.4)** CROSS broad syncline. Ridgeline exposures are proximal conglomerates of the Lute Ridge member of Canebrake Conglomerate IV.

**27.0 (0.4)** CROSS a scarp and shutter ridge of Font's Point sandstone. View SE reveals complexly-folded Hidden Canyon strata of undifferentiated Ocotillo formation with reddish-brown clays of the Brawley Formation exposed in the middle of the anticlinal fold, detached at a very shallow depth. Surface faulting/folding throughout this portion of the Borrego Badlands, driven by fault propagation at depth, may account for the majority of displacement along the CF beneath the Palo Verde-Arroyo Salado area.

**27.6 (0.6)** CROSS second scarp and associated shutter ridge.

**27.8 (0.2)** TURN LEFT at junction with Short Wash Trail. A rigth turn leads W to Font's Wash. Roadway is on surface of Font's Point sandstone deflated to the depth of the pedogenic carbonate.

**27.9 (0.1)** SAFELY NEGOTIATE the dropoff.

**28.0 (0.1)** For the next 0.4 miles, wash exposures consist of red-brown Font's Point sandstone. It is fluvial in aspect, subaerially-deposited, braided stream and coarse-grained pluvial lake strata of distal alluvial fans. Note overbank deposits, carbonates, soil horizons with disseminated carbonate nodules, and erosion surfaces suggesting a variety of depositional environments. Unidentified petrified wood fragments (possibly of the Clark Mountain local flora) occur throughout this rock unit.

**28.6 (0.6)** Roadway cuts through SE-plunging anticlinal structure. NW-dipping flanks consist of Font's Point sandstone. Core of the anticline is made up of salmon-brown sediments of the Short Wash member (distal equivalent of Canebrake III).

**29.0 (0.4)** TURN RIGHT (west) onto Short Wash.

**29.1 (0.1)** Prior to the tamarisk tree make a left turn following strip sign directions to Vista del Malpais. DRIVE SW up the tributary arroyo (an established roadway) to Vista del Malpais overview. NW-dipping sediments along this route include the Font's Point sandstone, Short Wash member, and Mammoth Cove sandstone member.

**30.0 (0.9) STOP 4 VISTA DEL MALPAIS.**
PARK at the parking lot at road's end. HIKE 100 m to the scenic overlook. Atop the Ocotillo Rim, Vista del Malpais affords an excellent view of the state park and

Figure 17. Vista del Malpais offers good views of the north-dipping Borrego Badlands. Sediments exposed include the Brawley and Borrego Formations derived from the ancestral Colorado River.

the eastern Borrego Badlands sedimentary section (Figure 17). To the S are the Vallecito-Pinyon, Fish Creek and Superstition Mountains. In the middle distance are the laterally offset twin East and West Buttes of Borrego Mountain.

NW, the overall physiographic feature is anticlinal with the fold hinge plunging NE in the vicinity of Valle Escondido. The SW-NE steeply-dipping Valle Escondido cross-fault, perpendicular to the strike of the main bounding CCF, has thrust and dramatically folded Plio-Pleistocene strata along the E-dipping limb of the anticline. The SW edge of this anticline coincides with the surface break termination of the major 1968 Borrego Mountain seismogenic rupture. At present, the kinematic relationship of the cross-fault remains unresolved.

Directly below, the vista takes in much of the Brawley Formation (Dibblee, 1954), and the Borrego Formation (Tarbet and Holman,1944). The mudhill exposures consist of a thick reddish-brown lacustral/brackish water basinal sequence of interbedded light-gray to red claystones with typically light yellow-gray to pale orange siltstones and sandstones. Also note the presence of a green playal claystone marker bed that overlies the Ghost Palms member of the Ocotillo formation. The Borrego beds locally grade downward into the proximal delta-plain facies of the Diablo formation. For the most part, the lacustral terrigenous sediments are compositionally distinctive (Merriam and Bandy, 1965; Muffler and Doe, 1968; Van De Kamp, 1973; Winker, 1987; Remeika, 1991) and are easily recognized in the nonmarine stratigraphic package of the Borrego Badlands. RETRACE route to junction with Short Wash Trail.

**31.0 (1.0)** Junction of Short Wash Trail on the left. At the signpost, CONTINUE STRAIGHT on Short Wash. Sediments along the wash are undifferentiated Font's Point sandstone.

**31.5 (0.5)** CROSS unnamed fault trace. Sediments on the E are Short Wash member juxtaposed against Font's Point sandstone (on the W).

**31.7 (0.2)** CROSS intrabasinal deformation zone marked by two transcurrent (left-lateral) faults. This oblique slip structure connects two anticlinal folds on opposite sides of Short Wash: Fault Wash Anticline on the S and Hidden Canyon Anticline on the N. Note that Font's Point sandstone is caught up in the fault zone with E-dipping Short Wash member sediments dragged down on the W.

**32.2 (0.5)** Hidden Canyon Wash on the left. Along Short Wash, salmon-brown sediments of the Short Wash member overlay the distal sandstone tongue of the Mammoth Cove sandstone member. Good exposures occur between Inspiration Wash and Font's Wash, striking E through Short Wash and lower Palo Verde Wash. Here, sediments consist of a mixed composition sequence of alternating light gray to salmon-brown sandstones and siltstones, well indurated buff-colored sandstones with pebbly gravels, and an occasional greenish lacustrine claystone interbed. Within the claystones, rippledrift cross lamination and mudcracks are common. The majority of sandstones are fine to medium-grained, well to moderately well sorted, with occasional coarse-grained interbeds. Granule, pebble, and torrentially-bedded conglomeratic sandstones are common, reflecting a locally-derived alluvial source area.

**32.5 (0.3)** Junction with Palo Verde Wash. A right turn will take you south to Cut-Across Trail and the Ocotillo Wells Off-road Vehicular Recreation Area. MAKE A LEFT TURN and PROCEED N to S-22.

**34.9 (2.4)** Junction of Palo Verde Wash and Highway S-22. MAKE A RIGHT TURN (east) on S-22. To the left, across the highway at 8:00, is an elongated shutter ridge squeezed up along the diffuse CF. The unconsolidated sediments include clays and cobbled sandstones of the Pleistocene Short Wash member. W of Smoketree Wash, stream channels, banks and bars of Pleistocene and Holocene alluvial fan surfaces have not only been deformed but laterally offset along the CF. In 1988 a soil chronosequence study of the CF where it traverses across recently abandoned stream channels was undertaken by W. S. Bull's Geomorphology 650 field class from the University of Arizona (Tucson). Assuming that the displacement has not been constant since the late Pleistocene, Jackson and Calderon (1988) obtained a recent slip rate of about 4 m/ka. At 3 ka and 5 ka, slip rates were between 2.6 and 2.2 m/ka, respectively. At 55 ka, a lower rate of 1.2 m/ka was determined, and at 120 ka, a lower rate of 0.9 m/ka was obtained. The increase in slip rate suggests that the CF has been very active in late Holocene time.

**35.0 (0.1)** Milepost marker 33. The flat-lying desert floor is characterized by a xerophytic community typical of the Lower Sonoran Life Zone. Common plants include the spindly-branched ocotillo, hardy creosote bush, cholla cacti and burroweed. Gray conglomeratic beds on the left belong to undifferentiated Canebrake Conglomerate. These upper plate rocks are thrust-faulted over lower plate cataclastics of the EPRMZ by a reactivated detachment fault (J.R. Pettinga, unpublished data, 1991).

**35.3 (0.3)** To the right, note bluffs that line the S side of Palo Verde Wash are composed of distictive conglomeratic sandstones of the proximal Lute Ridge member.

**35.6 (0.3)** As the roadway starts to climb atop the beheaded Quaternary alluvial fan, note that the conglomeratic sandstones, gravels, and interbedded red-brown clays are deformed, steeply-dipping to the SW and W.

**36.0 (0.4)** Milepost marker 34. Prepare to turn right in 0.4 mile.

**36.4 (0.4) STOP 5 BORREGO BADLANDS OVERLOOK.** TURN RIGHT into unmarked paved parking loop and park. This scenic viewpoint overlooks Palo Verde Wash and the distant Vista del Malpais area of the eastern Borrego Badlands. The sedimentary fill of the Borrego Badlands includes a complex succession of marginal alluvial and more interior lacustrine lithostratigraphic facies (Ocotillo formation). Distal alluvial sheetflood deposits graphically intertongue with an impinging Colorado River lacustral environment (Borrego and Brawley Formations). Together, the sequence has been subsequently deformed, folded and cut by transcurrent cross-faults (e.g. Inspiration Point Fault) that show appreciable left- lateral separation.

This location marks the epicenter of the 1954 Santa Rosa Mountains Earthquake which measured M6.2 on the Richter Scale. Leave vehicles and look at the fault scarp on the N side of S-22. Here, the CF side-steps left from the Truckhaven Rocks (Santa Rosa Badlands), crossing the highway as three normal fault-scarp remnant features cutting the gravelly terrace of the alluvial fan surface in a NW-SE direction. S, the termination of the CF in surface trace is caused by a complex basement-cover *decollement*. The over-riding cover of Quaternary sediments is being progressively deformed into a series of complex asymmetric *en *echelon* folds. Aftershock data indicates that a 15-17 km section of the buried fault strand ruptured to the S (J.R. Pettinga, unpublished data, 1991). The NW-trending CF represents the easternmost branch of the SJFZ.

Beginning on the N side of the San Gabriel Mountains, the SJFZ (Given, 1981; Hill, et al. 1975; Sharp, 1967) represents a well-defined master break in the earth's crust, part of the San Andreas Fault system which has long been recognized as the major transform lithospheric plate boundary between the North American and Pacific plates. The SJFZ has a well-documented recent history of seismicity (Allen, et al. 1972; Thatcher, et al. 1975). It has a measured length of approximately 288 km. In Anza-Borrego this through-going fault zone consists of many smaller strands and *en echelon* lineaments that network between major zones of displacement. Some of the most spectacular examples of fault-related landforms along the Santa Rosa Mountains include numerous well-defined graben, pull-apart, pop-up and pressure ridges. Primary topographic expressions are sharp and clear, accompanied by crushed zones of crystalline basement rock, aligned canyons and arroyos, beheaded, deflected and offset drainages, linear fault-line escarpments that can be traced for miles on the ground or via aerial imagery, Holocene quake swarms, slumping and landslides, alignment of hot springs and sag ponds, and numerous recorded earthquakes.

As a rule, large-scale movements have been right-lateral, strike-slip. Cumulative right separation is about 24 km (Sharp, 1967). There is also a vertical component. Large earthquakes, registering M6.0 or greater on the Richter Scale (Sanders, et al. 1986; Rasmussen, 1982) occur along the SJFZ on an average every 8 years. The last recorded event was the major twin earthquakes, measuring M6.2 and M6.6 respectively on November 23 and 24, 1987, along the Superstition Hills Fault (Kahle, et al. 1988). RETURN to vehicles.

**36.5 (0.1)** TURN RIGHT (east) onto S-22 entering the Santa Rosa Badlands (Fault Block C). In this vicinity the Borrego-Salton Seaway S-22 crosses the CF, the SE branch of the SJFZ.

**36.6 (0.1)** PASS Smoketree Wash on the left.

**36.7 (0.1)** Junction with Coachwhip Canyon on the left and Ella Wash on the right. PROCEED SW down Ella Wash. Four-wheel drive is recommended beyond this point. This sandy and bouldery arroyo has been incised into a relic alluvial geomorphic surface washed basinward from the fan-frayed Santa Rosa Mountains. The alluvial fan ranges in age from approximately 7-13 ka (Onken and Rathburn, 1988).

**37.4 (0.7) STOP 6 SHUTTER RIDGE.** PARK and WALK to E end of ridge. Faulting associated with the nearby Clark Fault has raised, rotated and deformed older alluvial sediments, dipping 36 degrees to the SW. Exposed here, ca. 266 m above sea level, are two strata of calcareous tufa-covered boulder shoreline deposits similar to those seen at the base of the section at Manix Lake.

The coarse alluvial sands and gravels are proximal Lute Ridge member and record evidence of a pluvial lake occupying the Borrego Valley area prior to Lake Cahuilla within the WSTD. The two distinctly layered cobble beds occur, *in situ*, with thin imbricating laminae of microcrystalline calcium carbonate encrusting alluvial clasts. The presence of calcareous tufa strandlines is an interesting feature and may not only document the existence and desiccation of a vanished lake but shows a combination of tectonism, Quaternary glacial-interglacial climatically-induced warming and increased aridity were fundamental variables in the desertification of Anza-Borrego prior to the terminal Pleistocene (Remeika, 1991b). RETURN to vehicles. Ella Wash continues S downwash, intersecting with Palo Verde Wash in 1.4 km. Otherwise, RETRACE route to S-22.

**38.1 (0.7)** MAKE A RIGHT TURN (east) on S-22.

**38.2 (0.1)** Junction of S-22 and Arroyo Salado Primitive Campground road on the right. CONTINUE STRAIGHT on S-22. Arroyo Salado marks the route of the old Truckhaven Trail. The primitive campground is just that: primitive. There is no water, no trash cans, and no toilet facilities. Campfires are allowed only in metal containers.

**38.4 (0.2)** Milepost marker 35. Coeval sedimentation and deformation fabrics within the W limb of the Santa Rosa Anticline are characterized by increased steeply-dipping, NW-striking Ocotillo formation (undifferentiated). It has been dextrally dragged from E to W along the S end of the Santa Rosa Mountains. In the past, these sediments have been variously referred to as undifferentiated Canebrake Conglomerate and/or Palm Spring Formation (Dibblee, 1954). However, they represent the proximal coarse-grained *Mammuthus*-bearing middle

Pleistocene Ocotillo formation.

**39.3 (0.9)** Milepost marker 36.

**39.5 (0.2) STOP 7 SANTA ROSA ANTICLINE AT TRUCKHAVEN ROCKS.** TURN LEFT onto paved turnout of unmarked scenic overlook and park. The southernmost flank of the structurally-elevated Santa Rosa Mountains exposes a clastic susccession of Mio-Pliocene nonmarine sedimentary fill (Pyramid Peak), consisting of syntectonic sedimentary facies reflecting a sensitive record of complex surface deformation along the BSFB margin. The highly distended landscape is creased by a labyrinth of vertically-walled gorges and slot canyons. The rock sequence, part of the locally-derived Anza-Borrego group, is open folded into the S-plunging Santa Rosa Anticline.

Pettinga (1991) presented a detailed review of the structural geology of the Truckhaven Rocks, and the following discussion draws heavily on his work. The E limb of the Santa Rosa Anticline is cut by E-W trending dextral strike-slip faults. As described by Pettinga (1991), these faults form an anastomosing system, with offsets up to 100 m. Constraining and releasing bands and steps add considerable complexity to the deformation of strata. The S-dipping sequence is locally folded adjacent to the wrench faults, and strain is accommodated by considerable bending shear and stratal transposition within the stratigraphic pile. Fabric analysis indicates that thin bedding-controlled sheets have sheared and rotated relative to underlying and/or overlying strata.

At 11:00 the Truckhaven Fault is exposed in the core of the anticline. It represents a sinistral wrench tear, antithetic to the nearby SJFZ (CF), between the upthrown, S-plunging E limb and the structurally lower W limb (Pettinga, 1991).

The genetic-stratigraphic cover section of the Truckhaven Rocks/Santa Rosa Badlands is ideally suited to discussion because of its unusual completeness, structural intactness and quality of exposures within the state park. Pyramid Peak consists of a Mio-Pliocene syntectonic interstratified nonmarine package (Anza-Borrego group), including a basal, coarse clastic fanglomerate assigned to the Miocene (?) Split Mountain Formation below, and postextensional, torrentially-bedded proximal sandstones of the Plio-Pleistocene Canebrake Conglomerate II-III above, that accumulated marginal to the BSFB. Basinward, the sandstone are laterally equivalent with fine-grained sediments of the Colorado River group.

The basal Split Mountain Formation (Tarbet and Holman, 1944; Woodard, 1974; Winker, 1987) chronicles the initial deroofing of the ancestral highlands. It is approximately coeval with lower to middle Miocene detachment faults (Pettinga, 1991), and volcanism (Jacumba Basalt) recorded from the VFCB (Kerr, 1982, 1984; Kerr and Kidwell, 1991), ranging from 14.9 to 24.8 Ma (Gastil, et al. 1979). Based upon lateral and vertical distribution, J.R. Pettinga (unpublished data, 1991) informally subdivides the stratification into two lithofacies that may correlate with the informal Red Rock Canyon and Elephant Trees members of the Split Mountain Formation (Winker, 1987) at Split Mountain Gorge and undifferentiated at Borrego Mountain. It is unconformably overlain by Canebrake Conglomerate I.

As defined by Dibblee (1984), the Canebrake Conglomerate is a locally-derived light-gray Plio-Pleistocene alluvial fan-floodplain sequence. Deposits consist of a thick 1000 m series of coarse unsorted arkosic conglomerate and lesser coarse sandstones containing a mixed-provenance granitic and gneissic clast suite from the Santa Rosa crystalline basement. Clasts range from subangular boulders to subrounded cobbles and pebbles. J.R. Pettinga (unpublished data, 1991) recognizes several subfacies, ranging from proximal fanglomerates to distal gradation of indurated gravel and mixed-load pebbly sandstones, extending basinward to nonpebbly, finer-grained braided channel sandstones that interfinger with ancestral Colorado River sequences (Olla member). RETURN to S-22.

**39.6 (0.1)** TURN LEFT (east) onto S-22.

**39.9 (0.3)** North Fork of Arroyo Salado overpass.

**40.0 (0.1)** Cannonball Wash overpass. Resistant Diablo formation-like redbeds form low-lying strike ridges littered with ferruginous sandstone concretions. They take on a myriad of bizarre shapes, including cylinders, ameboids and spheroids such as the well-known examples found at the Pumpkin Patch along Tule Wash. It is believed that concretions developed by diagenetic compaction of the deltaic sediments as a result of anomalies from localized concentrations of carbonate and/or hydrous calcium sulfate. Desert rose gypsum crystals (hydrated calcium sulfate) also occur in rosette-shaped aggregates throughout Diablo exposures.

**40.3 (0.3)** Milepost marker 37. N of the highway, light-colored sandstones of the Diablo formation basinward to nonpebbly, finer-grained braided channel sandstones that interfinger with ancestral Colorado River sequences (Olla member). RETURN to S-22.

**39.6 (0.1)** TURN LEFT (east) onto S-22.

**39.9 (0.3)** North Fork of Arroyo Salado overpass.

**40.0 (0.1)** Cannonball Wash overpass. Resistant Diablo formation-like redbeds form low-lying strike ridges littered with ferruginous sandstone concretions. They take on a myriad of bizarre shapes, including cylinders, ameboids and spheroids such as the interfinger with locally-derived, tan-gray continental sandstones of the Pliocene Canebrake Conglomerate. Similar lithologies in the VFCB are represented by the Olla member of the Canebrake Conglomerate I (Winker, 1987). Within the upper reaches of Palm Wash, cameloid ichnofossils (footprints) have been discovered.

**40.4 (1.0)** Milepost marker 38. The unmarked Salton View Scenic Overlook is on the left. View N at 8:00 reveals E-dipping Mio-Pliocene clastic sediments of the Santa Rosa Badlands, overshadowed on the W by Pyramid Peak (aka Traveler's Peak) at an elevation of 1166 m above sea level.

**40.6 (0.2)** Junction with Truckhaven Trail on the right and the Calcite Mine Scenic Area on the left. The Calcite Mine jeep route is a narrow, rough trail, extending 3.0 km off the highway. Four-wheel drive is recommended. Heyday of the district was during World War II. Optical-grade calcite crystals were trench-mined for use in bombsights and anti-aircraft weaponry because of its double-refraction properties. Calcite veining fills cross shears in the indurated sandstones along E-W trending dextral strike-slip faults. PROCEED east on S-22.

**41.3 (0.7)** The tall microwave tower stands on the San Diego/Imperial county line. At this location, S-22 exits ABDSP (E boundary) and enters the BLM's Desert Conservation Area, El Centro District.

**END OF DAY 2**

**ACKNOWLEDGMENTS**

I am grateful to John Quirk, State Park Ranger II, for his support in scheduling additional time in order to allow me to improve the clarity and presentation of this work. An early version of the manuscript benefited from scutiny by George T. Jefferson.

# REFERENCES CITED

Allen, C.R. 1957. San Andreas Fault Zone in San Gorgonio Pass, southern California. Geological Society of America Bulletin 68:315-349.

Allen, C.R., P.St. Amand, C.F. Richter, and J.M. Nordquist 1965. Relationship between seismicity and geologic structure in southern California. Seismological Society of America Bulletin 55:753-797.

Allison, M. L. 1974a. Geologic and geophysical reconnaissance of the Elsinore-Chariot Canyon Fault systems. In Recent Geologic and Hydrologic Studies, Eastern San Diego County and Adjacent Areas, edited by M. W. Hart and R. W. Dowlen. San Diego Association of Geologists Guidebook, California. p. 21-35.

----- 1974b. Geophysical studies along the southern portion of the Elsinore Fault. Master of Science Thesis, San Diego State University, California. 229p.

Allison, M.L., J.H. Whitcomb, C.E. Cheatum, and R.B. McEuen 1978. Elsinore Fault seismicity: the September 13, 1973 Agua Caliente Springs, California, earthquake series. Seismological Society of America Bulletin 68(2):429-440.

Arnold, R.E. 1904. The faunal relations of the Carrizo Creek beds of California. Science 19:1-503.

Axen, G. 1995. Extensional segmentation of the Main Gulf Escaprment, Mexico and the United States. Geology 23(6):515-518.

Bell-Countryman, P.J. 1984. Environments of deposition, Pliocene Imperial Formation, Coyote Mountains, southwest Salton Trough. In The Imperial Basin, Tectonics, Sedimentation and Thermal Aspects, edited by C.A. Rigsby, Society of Economic Paleontologists and Mineralogists 40:45-70.

Blake, W.P. 1855. Preliminary geological report of the expedition under command of Lieutenant R.S. Williamson, United States Topographical Engineers. U.S. 33rd Congress, Ist Session, House Executive Document 129:1-80.

Bogen, N.L., and L. Seeber 1986. Neotectonics of rotating blocks within the San Jacinto Fault Zone, southern California. EOS, Transaction of the American Geophysical Union 67:1200.

Bohannon, R.G. 1975. Mid-Tertiary conglomerates and their bearing on Transverse Range tectonics, southern California. In San Andreas Fault in Southern California, edited by J.C. Crowell. California Division of Mines and Geology Special Report 118:75-82.

Brown, N.N., and R.H. Sibson 1989. Structural geology of the Ocotillo Badlands antidilational fault jog, southern California. U.S. Geological Survey Open-file Report 89-315:94-109.

Buchheim, H. P. 1990. Discovery of fresh water stromatolites in Carrizo Creek, California. Geological Society of America, Abstract with Programs, 22(7):A358-A359.

Cassiliano, M.L. 1994. Paleoecology and taphonomy of vertebrate faunas from the Anza-Borrego Desert of California. Doctoral Dissertation, Department of Geosciences, University of Arizona, Tucson 421p.

Chappell, B.W., and A.J.R. White 1974. Two contrasting granite types. Pacific Geology 8:173-174.

Christensen, A.D. 1957. Part of the geology of the Coyote Mountain area, Imperial County, California. Master of Arts Thesis, Univesity of California, Los Angeles 188p.

Conrad, T.A. 1855. Report on the fossil shells collected by W.P. Blake, geologist to the expedition under the command of Lieutenant R.S. Williamson, United States Topographical Engineers, 1852. In W.P. Blake, Preliminary Geological Report, U.S. Pacific Railroad Exploration, U.S. 33rd Congress, 1st Session, House Executive Document 129:5-21.

Crowell, J. C., and A. G. Sylvester 1979a. Introduction to the San Andreas Fault-Salton Trough juncture. In Tectonics of the Juncture Between the San Andreas Fault System and the Salton Trough, southeastern California, editedy by J. C. Crowell and A. G. Sylvester. Department of Geological Sciences, University of California, Santa Barbara. p. 1-13.

----- 1979b. Excursion Guide. In Tectonics of the Juncture between the San Andreas Fault system and the Salton Trough, southeastern California, edited by J. C. Crowell and A. G. Sylvester. Department of Geological Sciences, University of California, Santa Barbara, p. 148-168.

Dean, M.A. 1988. Genesis, mineralogy and stratigraphy of the Neogene Fish Creek gypsum, southwestern Salton Trough, California. Master of Science Thesis, San Diego State University, California

150p.

Demére, T.A. 1993. Fossil mammals from the Imperial Formation (upper Miocene-lower Pliocene), Coyote Mountains, Imperial County, California. In Ashes, Faults and Basins, edited by R.E. Reynolds and J. Reynolds, San Bernardino County Museum Association Special Publication 93-1:82-85.

Dibblee, T.W., Jr. 1954. Geology of the Imperial Valley region, California. In Geology of Southern California, edited by R.H. Jahns, California Division of Mines and Geology Bulletin 170(2,2):21-81.

Dokka, R.K., and R.H. Merriam 1982. Late Cenozoic extension of northeastern Baja California, Mexico. Geological Society of America Bulletin 93:371-378.

Downs, T. 1966. Southern California field trip, Anza-Borrego Desert and Barstow areas. Society of Vertebrate Paleontology Field Trip. Manuscript on file, Anza-Borrego Desert State Park, California 12p.

Downs, T., and J.A. White 1968. A vertebrate faunal succession in superposed sediments from late Pliocene to middle Pleistocene in California. In Tertiary/Quaternary Boundary, International Geological Congress 23, Prague 10:41-47.

Durham, J.W., and E.C. Allison 1961. Stratigraphic position of the Fish Creek gypsum at Split Mountain Gorge, Imperial County, California. Geological Society of America Special Paper 68:32.

Eberly, L.D., and T.B. Stanley 1978. Cenozoic stratigraphy and geologic history of southwestern Arizona. Geological Society of America Bulletin 89:921-940.

Elders, W.A. 1979. The Geological Background of the geothermal fields of the Salton Trough. In Geology and Geothermics of the Salton Trough, edited by W.S. Elders. University of California, Riverside, Campus Museum Contributions 5:1-19.

Elders, W.A., R.W. Rex, H.T. Meidav, P.T. Robinson, and S. Biehler 1972. Crustal spreading in southern California. Science 178(4056):15-23.

Fleming, R.F. 1994. Cretaceous pollen in Pliocene rocks: implications for Pliocene climate in the southwestern United States. Geology 22:787-790.

Fleming, R.F., and P. Remeika 1994. Pliocene climate of the Colorado Plateau and age of the Grand Canyon: evidence from Anza-Borrego Desert State Park, California. 4th NPS Conference on Fossil Resources, Colorado Springs, Colorado.

Foster, A.B. 1979. Environmental variation in a fossil scleractinian coral. Lethaia 12:245-264.

Fourt, R. 1979. Post-batholithic geology of the Volcanic Hills and vicinity, San Diego County, California. Master of Science Thesis, San Diego State University, California. 66p.

Frost, E.G., L.A. Heizer, R. Blom, and R. Crippen 1993. Western Salton Trough detachment system: its geometric role in localizing the San Andreas system. In The San Andreas Fault System, edited by F.V. Corona, F.F. Sabins, Jr., and E.G. Frost. Ninth Thematic Conference on Geologic Remote Sensing, Pasadena, California. p.186-197.

Gastil, R.G., and J. Bushee 1961. Geology and geomorphology of eastern San Diego County. Field Trip no. 1, road log: Field Trip Guidebook, Geological Society of America, 57th Annual Meeting, Cordilleran Section, p. 8-22.

Gastil, R.G., R.P. Phillips, and E.C. Allison 1975. Reconnaissance geology of the State of Baja California. Geological Society of America, Memoir 140:1-170.

Gastil, R.G., G.J. Morgan, and D. Krummenacher 1981. The tectonic history of peninsular California and adjacent Mexico. In The Geotectonic Development of California, edited by W.G. Ernst. Prentice-Hall, Incorporated, New Jersey, Ruby Volume 1:284-305.

Gastil, R.G., and R.H. Miller 1984. Prebatholithic paleogeography of peninsular California and adjacent Mexico. In Geology of the Baja California Peninsula, edited by V.A. Frizzell, Jr. Society of Economic Paleontologists and Mineralogists, Pacific Section 39:9-16.

Gath, E.M., R.W. Ruff, R. McElwain, and M.L. Raub 1986. Geology of the Salton Trough field trip road log. In Geology of the Imperial Valley, California, edited by P.D. Guptil, E.M. Gath, and R.W. Ruff. South Coast Geological Society Annual Field Trip Guidebook 14:181-225.

Gibson, L.M. 1983. The configuration of the Vallecito-Fish Creek Basin, western Imperial Valley, California, as interpreted from gravity data. Master of Arts Thesis, Dartmouth College, New Hampshire. 87p.

Gibson, L.M., L.L. Malinconico, T. Downs, and N.M. Johnson 1984. Structural implications of gravity data from the Vallecito-Fish Creek Basin, western Imperial Valley, California. In The Imperial Basin, Tectonics, Sedimentation and Thermal Aspects, edited by C.A. Rigsby, Society of Economic Paleontologists and Mineralogists 40:15-29.

Gjerde, M.W. 1982. Petrology and geochemistry of the Alverson Formation, Imperial County, California. Master of Science Thesis, San Diego State University, California. 85p.

Grove, M. 1994. Contrasting Denudation Histories within the East-Central Peninsular Ranges Batholith (33°N). In Geological Investigations of an Active Margin, edited by S. F. McGill and T. M. Ross. Geological Society of America, Cordilleran Section Guidebook 27:235-240.

Hanna, G.D. 1926. Paleontology of Coyote Mountain, Imperial County, California. Proceedings of the California Academy of Sciences, 4th Series, 14:427-503.

Harbert, W. 1991. Late Neogene relative motions of the Pacific and North American plates. Tectonics 10:1-15.

Hart, M.W., and R.J. Dowlen 1974. Recent geologic and hydrologic studies, eastern San Diego County and adjacent areas. Field Trip Road Log. San Diego Association of Geologists, San Diego, California. p. 1-20.

Hawkins, J.W. 1970. Petrology and possible tectonic significance of late Cenozoic volcanic rocks, southern California and Baja California. Geological Society of America Bulletin 81:3323-3338.

Hoggatt, W.C. 1979. Geologic map of Sweeney Pass quadrangle, San Diego County, California. U.S. Geological Survey Open-File Report 79-754:1-34.

Hsu, K.J., and M.B. Cita 1973. The origin of the Mediterranean Evaporite. In Initial Reports of the Deep Sea Drilling Project, Leg XIII, 2, edited by W. B. F. Ryan and K. J. Hsu. U. S. Government Printing Office, Washington, D. C. 1203-1232.

Hsu, K.J., W.B.F. Ryan, and M.B. Cita 1973. Late Miocene desiccation of the Mediterranean. Nature 242:239-243.

Hudnut, K.W., and M.M. Clark 1989. New slip along parts of the 1968 Coyote Creek Fault rupture, California. Seismological Society of America Bulletin 79(2):451-465.

Hudnut, K.W., and L. Seeber 1986. Astroazimuth geodetic measurements of block rotation in the southern San Jacinto Fault Zone, California. EOS, Transactions of the American Geophysical Union 68:287.

Hudnut, K.W., L. Seeber, J. Pacheo, J. Armbruster, L. Sykes, G. Bond, and M. Kominz 1989. Cross faults and block rotation in southern California, earthquake triggering and strain distribution. Lamont-Doherty Geological Observatory Yearbook p. 44-48.

Hull, A.G. 1991. Pull-apart basin evolution along the northern Elsinore Fault Zone, southern California. Geological Society of America, Abstracts with Programs 23(5):A257.

Hutton, L.K., L.M. Jones, E. Hauksson, and D.D. Given 1991. Seismotectonics of southern California. In Neotectonics of North America, edited by D.B. Slemmons, E.R. Engdahl, M.D. Zoback and D.D. Blackwell, Geological Society of America, the Geology of North America Decade Map 1: 133-152.

Ingle, J.C. 1974. Paleobathymetric history of Neogene marine sediments, northern Gulf of California. In Geology of Peninsular California. American Association of Petroleum Geologists, Pacific Section, Guidebook for Field Trips p. 121-138.

Izett, G.A., and C.W. Naeser 1976. Age of the Bishop Tuff of eastern California as determined by the fission-track method. Geology 4(10):587-590.

Jackson, G., and G. Calderon 1988. Late Quaternary slip rates on the Clark Fault: inferences from offset fluvial landforms, SW ¼ Font's Point quadrangle, California. Field Studies in Geomorphology, University of Arizona, Tucson. Manuscript on File, Anza-Borrego Desert State Park, California 24p.

Johnson, N.M., C.B. Officer, N.D. Opdyke, G.D. Woodard, P.K. Zeitler, and E.H. Lindsay 1983. Rates of Cenozoic tectonism in the Vallecito-Fish Creek Basin, western Imperial County, California. Geology 11:664-667.

Keen, A.M., and H. Bentson 1944. Check list of California Tertiary marine mollusca. Geological Society of America, Special Paper 56:1-280.

Kerr, D.R. 1982. Early Neogene continental sedimentation, western Salton Trough, California. Master of Science Thesis, San Diego State University, California 138p.

Kerr, D.R., S. Pappajohn, and G.L. Peterson 1979. Neogene stratigraphic section at Split Mountain, eastern San Diego County, California. In Tectonics of the Juncture Between the San Andreas Fault System and the Salton Trough, Southeastern California, edited by J.C. Crowell and A.G. Sylvester, Geological Society of America Annual Meeting Guidebook p. 111-124.

Kerr, D.R., and S.M. Kidwell 1991. Late Cenozoic sedimentation and tectonics, western Salton Trough, California. In Geological Excursions in southern California and Mexico, edited by M. J. Walawender and B. B. Hanan. Geological Society of America Annual Meeting Guidebook, San

Diego, p. 379-416.

Kew, W.S.W. 1914. Tertiary echinoids of the Carrizo Creek region in the Colorado Desert. California University Publications, Bulletin of the Department of Geology 12(2):23-236.

Kidwell, S.M. 1987. Origin of macroinvertebrate shell concentrations in Pliocene shallow marine environments, Gulf of California. Geological Society of America Abstracts with Programs 19:726.

Kidwell, S.M. 1988. Taphonomic comparison of passive and active continental margins: Neogene shell beds of the Atlantic coastal plain and northern Gulf of California. Palaeogeography, Palaeoclimatology, Palaeoecology 63:201-223.

Kerr, D.R., and S.M. Kidwell 1991. Late Cenozoic sedimentation and tectonics, western Salton Trough, California. In Geological Excursions in Southern California and Mexico, edited by M.J. Walawender and B.B. Hanan, Geological Society of America Annual Meeting Guidebook, San Diego p. 379-416.

Lagenkamp, D., and J. Combs 1974. Microearthquake study of the Elsinore Fault Zone, southern California. Seismological Society of America Bulletin 64(1):187-203.

Lampe, C.M. 1988. Geology of the Granite Mountain area: implications of the extent and style of deformation along the southeast portion of the Elsinore Fault. Master of Science Thesis, San Diego State University, California 150p.

Lampe, C.M., M.J. Walawender, and T.K. Rockwell 1988. Contacts between La Posta-type plutonic rocks and stromatic migmatites offset across the Elsinore Fault, southern California. Geological Society of America Abstracts with Programs 20(3):174-175.

Lonsdale, P. 1989. Geology and tectonic history of the Gulf of California. In The Eastern Pacific Ocean and Hawaii, edited by E. L. Winterer, D. M. Hussong, and R. W. Decker. Geological Society of America. The Geology of North America N:499-521.

Lough, C.F. 1993. Structural Evolution of the Vallecito Mountains. In Colorado Desert and Salton Trough Geology, edited by J. Corones. San Diego Association of Geologists field trip guide. p. 91-109.

Lough, C.F., and A.L. Stinson 1991. Structural evolution of the Vallecito Mountains, southwest California. Geological Society of America Abstracts with Programs. p. A246.

Lowman, P.D., Jr. 1977. The Elsinore Fault of southern California: a re-evaluation based on orbital photography. Geological Society of America Abstracts with Programs 9:1076-1077.

Lucchitta, I. 1987. The mouth of the Grand Canyon and edge of the Colorado Plateau in the upper Lake Mead area, Arizona. In Rocky Mountain Section Centennial Field Guide, edited by S.S. Beus. Geological Society of America 2:365-370.

Mace, N.W. 1981. A paleomagnetic study of the Miocene Alverson Volcanics of the Coyote Mountains, western Salton Trough, California. Master of Science Thesis. San Diego State University, California. 142p.

Matti, J.C., and D.M. Morton 1993. Paleogeographic evolution of the San Andreas Fault in southern California: a reconstruction based on a new cross-fault correlation. In The San Andreas Fault System; Displacement, Palinspatic Reconstruction and Geologic Evolution, edited by R.E. Powell, R.J. Weldon, and J.C. Matti. Geological Society of America Memoir 178:107-160.

McDougall, K., C.L. Powell, II, J.C. Matti, and R.Z. Poore 1994. The Imperial Formation and the opening of the ancestral Gulf of California. Cordilleran Section. Geological Society of America Abstract with Programs 26(2):71.

Mendenhall, W.C. 1910. Geology of Carrizo Mountain, California. Journal of Geology 18:337-355.

Merriam, R.H., and O.L. Bandy 1965. Source of upper Cenozoic sediments in the Colorado delta region. Journal of Sedimentary Petrology 35:911-916.

Metzger, D.G. 1968. The Bouse Formation (Pliocene) of the Parker-Blythe-Cibola area, Arizona and California. U.S. Geological Survey Professional Paper 500-D:126-136.

Miller, D.E., and T. Kato 1990. Field evidence for late Cenozoic kinematic transition in the Coyote Mountains. Geological Society of America Abstracts with Programs 22(3):68.

Miller, R.H., and M.S. Dockum 1983. Ordovician conodonts from metamorphosed carbonates of the Salton Trough, California. Geology 11:410-412.

Miller, W.J. 1935. Geologic section across southern Peninsular Ranges of California. California Journal of Mines and Geology 31:115-142.

Minch, J.A., and P.L. Abbott 1973. Post batholithic geology of the Jacumba area, southeastern San Diego County, California. San Diego Society of Natural History Transactions 17(11):129-135.

Mitchell, E.D. 1961. A new walrus from the Imperial Pliocene of southern California: with notes on odobenid and otariid humeri. Los Angeles County Museum Contributions in Science 44:1-28.

Morton, D.M., and F.C. Miller 1987. K/Ar apparent ages of plutonic rocks from the northern part of the Peninsular Ranges Batholith, southern California. Geological Society of America, Abstracts with Programs 19:435.

Mueller, K. J., and T. K. Rockwell 1995. Late Quaternary activity of the Laguna Salada Fault in northern Baja California, Mexico. Geological Society of America Bulletin 107(1):8-18.

Muffler, L.P.J., and B.R. Doe 1968. Composition and mean age of detritus of the Colorado River delta in the Salton Trough, southeastern California. Journal of Sedimentary Petrology 38:384-399.

Nicholson, C., and L. Seeber 1989. Evidence for contemporary block rotation in strike-slip environments: examples from the San Andreas Fault system, southern California. In Paleomagnetic rotations and continental deformation, edited by C. Kissel and C.Laj, Kluwer Academic Publishers:247-280.

Norris, R. M., E. A. Keller, and G. L. Meyer 1979. Geomorphology of the Salton Basin, California: Selected Observations. In Geological Excursions in the Southern California Area, edited by P. L. Abbott. Department of Geological Sciences, San Diego State University, California. p.19-46.

----- 1993. Geomorphology of the Salton Basin, California: Selected Observations. In Colorado Desert and Salton Trough Geology, edited by J. Corones. San Diego Association of Geologists field trip guide, p. 37-67.

Onken, J.A., and S. L. Rathburn 1988. Fan fluvial system response to base level lowering. Geosciences 650, University of Arizona, Tucson. Manuscript on File, Anza-Borrego Desert State Park, California 22p.

Opdyke, N. D., E. H. Lindsay, N. M. Johnson, and T. Downs 1977. The paleomagnetism and magnetic polarity stratigraphy of the mammal-bearing section of Anza-Borrego State Park, California. Quaternary Research 7:316-329.

Peterson, G. L., R. G. Gastil, J. A. Minch, and C. E. Nordstrom 1968. Clast suites in the late Mesozoic-Cenozoic succession of the western Peninsular Ranges province, southwestern California and northwestern Baja California. Abstract. Geological Society of America Special Paper 115:177.

Peterson, G. L., and C. E. Nordstrom 1970. Sub-La Jolla unconformity in the vicinity of San Diego, California. American Association of Petroleum Geologists Bulletin 54:265-274.

Peterson, M. S. 1975. Geology of the Coachella Fanglomerate. In San Andreas Fault in Southern California, edited by J. C. Crowell. California Division of Mines and Geology, Special Report 118:119-126.

Pettinga, J.R. 1991. Structural styles and basin margin evolution adjacent to the San Jacinto Fault Zone, southern California. Geological Society of America Abstracts with Programs, San Diego p. A257.

Pinault, C.T. 1984. Structure, tectonic geomorphology, and neotectonics of the Elsinore Fault Zone between Banner Canyon and the Coyote Mountains, southern California. Master of Science Thesis, San Diego State University, California 231p.

Powell, C.L., II, 1986. Stratigraphy and bivalve molluscan paleontology of the Neogene Imperial Formation in Riverside County, California. Master of Science Thesis, California State University, San Jose. 275p.

----- 1987. Paleogeography of the Imperial Formation of southern California and its molluscan fauna: an overview. Western Society of Malacologists Annual Report 20:11-18.

Quinn, H.A., and T.M. Cronin 1984. Micropaleontology and depositional environments of the Imperial and Palm Spring Formations, Imperial Valley, California. In The Imperial Basin, Tectonics, Sedimentation and Thermal Aspects, edited by C. A. Rigsby, Society of Economic Paleontologists and Mineralogists 40:71-85.

Remeika, P. 1991a. Formational status of the Diablo redbeds; differentiation between Colorado River affinities and the Palm Spring Formation. Symposium on the Scientific Value of the Desert, Anza-Borrego Foundation, Borrego Springs, California. Abstracts. p. 12.

----- 1991b. A preliminary report on calcareous tufa deposits from the Palo Verde Wash area; evidence of a pre-Lake Cahuilla strandline in the Borrego Badlands, Anza-Borrego Desert State Park, California. Symposium on the Scientific Value of the Desert, Anza-Borrego Foundation, Borrego Springs, California. Abstracts p. 12.

----- 1992a. Preliminary report on the stratigraphy and vertebrate fauna of the middle Pleistocene Ocotillo Formation, Borrego Badlands, Anza-Borrego Desert State Park, California. Abstracts of Proceedings, 6th Annual Mojave Desert Quaternary Research Center Symposium. San Bernardino County Museum Association Quarterly 39(2):25-26.

----- 1992b. Paleontological program development at

Anza-Borrego Desert State Park, California. In Proceedings of The Third Conference on Fossil Resources in the National Park Service, edited by R. Benton and A. Elder. Abstract with Programs. Fossil Butte National Monument, Wyoming. Natural Resources Report NPS/NRFOBU/NRR 94:14:29.

----- 1993. Achieving paleogeology resource management strategies for General Plan development: a progress report on Anza-Borrego Desert State Park, California. Abstract with Programs. Paleontological Resource Management Symposium. Society of Vertebrate Paleontology 53rd Annual Meeting, Albuquerque, New Mexico 13(3):53A.

----- 1994. Lower Pliocene angiosperm hardwoods from the Vallecito-Fish Creek Basin, Anza-Borrego Desert State Park, California: deltaic stratigraphy, paleoclimate, paleoenvironment, and phytogeographic significance. Abstracts of Proceedings. 8th Annual Mojave Desert Quaternary Research Center Symposium. San Bernardino County Museum Association Quarterly 41(3):26-27.

Remeika, P. and L. Lindsay, 1992. Geology of Anza-Borrego: Edge of Creation. Sunbelt Publications, Kendall/Hunt Publishing Company, Dubuque, Iowa. 208p.

Remeika, P., and J.R. Pettinga 1991. Stratigraphic revision and depositional environments of the middle to late Pleistocene Ocotillo Conglomerate, Borrego Badlands, Anza-Borrego Desert State Park, California. Symposium on the Scientific Value of the Desert, Anza-Borrego Foundation, Borrego Springs, California. Abstracts p. 13.

Reynolds, R.E., and P. Remeika 1993. Ashes, Faults and Basins: the 1993 Mojave Desert Quaternary Research Center Field Trip. In Ashes, Faults and Basins, edited by R. E. Reynolds and J. Reynolds. San Bernardino County Museum Association Special Publication 93-1:3-33.

Robinson, J.W., and R.L. Threet 1974. Geology of the Split Mountain Area, Anza-Borrego Desert State Park, eastern San Diego County, California. In Recent Geologic and Hydrologic Studies, Eastern San Diego County and Adjacent Areas, edited by M. W. Hart and R. J. Dowlen. San Diego Association of Geologists Field Trip guidebook, California, p. 47-56.

Rockwell, T.K. 1989. Behavior of individual fault segments along the Elsinore-Laguna Salada Fault Zone, southern California and northern Baja California: implications for the characteristic earthquake model. Geological Society of America Abstracts with Programs 21(5):135.

Rockwell, T.K., and C.T. Pinault 1986. Holocene slip events on the southern Elsinore Fault, Coyote Mountains, Southern California. In Neotectonics and Faulting in Southern California, edited by P. Ehlig. Cordilleran Section. Geological Society of America Guidebook. p. 193-196.

Ruisaard, C.I. 1979. Stratigraphy of the Miocene Alverson Formation, Imperial County, California. Master of Science Thesis, San Diego State University, California. 125p.

Rymer, M.J. 1991. The Bishop Ash bed in the Mecca Hills. In Geological Excursions in Southern California and Mexico, edited by M.J. Walawender and B.B. Hanan. Geological Society of America Annual Meeting Guidebook p. 388-396.

Rymer, M.J., C.L. Powell, II, A.M. Sarna-Wojcicki, and J.A. Barron 1995. Late Miocene stratigraphic and paleogeographic seeting of Garnet Hill in the northwestern Salton Trough, California. AAPG, SEPM, SEG, AEG, SPWLA, DPA, DEG, EMD, AWG Pacific Section Convention, Schedule and Abstracts:43.

Rymer, M.J., A.M. Sarna-Wojcicki, C.L. Powell, II, and J.A. Barron 1994. Stratigraphic evidence for late Miocene opening of the Salton Trough in Southern California. Cordilleran Section. Geological Society of America Abstract with Programs 26(2):87.

Scheuing, D.F., L. Seeber, K.W. Hudnut, and N.L. Bogen 1988. Block rotation in the San Jacinto Fault Zone, southern California. EOS, Transactions of the American Geophysical Union 69:1456.

Schremp, L.A. 1981. Archaeogastropoda from the Pliocene Imperial Formation of California. Journal of Paleontology 55 (5):1123-1136.

Sharp, R.V. 1967. The San Jacinto Fault Zone in the Peninsular Ranges of Southern California. Geological Society of America Bulletin 78:705-730.

Sharp, R.V., and M.M. Clark 1972. Geologic evidence of previous faulting near the 1968 rupture on the Coyote Creek Fault. In The Borrego Mountain Earthquake of April 9, 1968. U.S. Geological Survey Professional Paper 787:131-140.

Sharpe, R.D., and G.G. Cork 1995. Geology and mining of the Miocene Fish Creek Gypsum in Imperial County, California. In 29th Forum on the Geology of Industrial Minerals: Proceedings; California Department of Conservation, Division of Mines and Geology, Special Publication 110:169-180.

Shaw, C.A. 1981. The middle Pleistocene El Golfo local fauna from northwestern Sonora, Mexico. Master of Science Thesis, Department of Biology, California State University, Long Beach 141p.

Smith, D.D. 1962. Geology of the northeast quarter of the Carrizo Mountain quadrangle, Imperial County, California. Master of Science Thesis, University of Southern California, Los Angeles. 89p.

Smith, J.T. 1991. Cenozoic marine mollusks and paleogeography of the Gulf of California. In The Gulf and Peninsular Province of the Californias, edited by J.P. Duaphin and B.R.T. Simoneit. American Association of Petroleum Geologists Memoir 47:637-666.

Stock, J.M., and K.V. Hodges 1990. Miocene to recent structural development of an extensional accomodation zone, northeastern Baja California, Mexico. Journal of Structural Geology 12:315-328.

Stock, J.M., A.B. Martin, V. Suarez, M.M. Miller 1991. Miocene to Holocene extensional tectonics and volcanic stratigraphy of NE Baja California, Mexico. In Geological Excursions in Southern California and Mexico, edited by M. J. Walawender and B. B. Hanan. Geological Society of America Guidebook, San Diego, California. p. 44-67.

Stout, B.W., and P. Remeika 1991. Status report on three major camelid tracksites in the lower Pliocene delta sequence, Vallecito-Fish Creek Basin, Anza-Borrego Desert State Park, California. Symposium on the Scientific Value of the Desert, Anza-Borrego Foundation, Borrego Springs, California p. 9.

Strand, C.L. 1980. Pre-1900 earthquakes of Baja California and San Diego County. Master of Science Thesis. San Diego State University, San Diego, California. 320p.

Stump, T.E. 1972. Stratigraphy and paleoecology of the Imperial Formation in the western Colorado Desert. Master of Science Thesis, San Diego State University, California. 132p.

Tarbet, L.A. 1951. Imperial Valley. American Association of Petroleum Geologists Bulletin 35:260-263.

Tarbet, L.A., and W.H. Holman 1944. Stratigraphy and micropaleontology of the west side of Imperial Valley. American Association of Petroleum Geologists Bulletin 28:1781-1782.

Theodore, T.G. 1967a. Structure and petrology of the gneisses and mylonites at Coyote Mountain, Borrego Springs, California. Ph.D. Dissertation, University of California, Los Angeles, 268 p.

----- 1967b. Structure and petrology of the gneisses and mylonites at Coyote Mountain, Borrego Springs, California. Geological Society of America Abstracts with Programs 28:954B.

----- 1968. High-grade mylonite zone in southern California. Geological Society of America Special Paper 101:220.

----- 1970. Petrogenesis of mylonites of high metamorphic grade in the Peninsular Ranges of southern California. Geological Society of America Bulletin 81:435-449.

Thomas, H.W., and L.G. Barnes 1993. Discoveries of fossil whales in the Imperial Formation, Riverside County, California: possible further evidence of the northern extent of the proto-Gulf of California. In Ashes, Faults and Basins, edited by R.E. Reynolds and J. Reynolds, San Bernardino County Museum Association Special Publication 93-1:34-36.

Todd, V.R., D.L. Kimbrough, and C.T. Herzig 1994. The Peninsular Ranges Batholith from Western Volcanic Arc to Eastern Mid-Crustal Intrusive and metamorphic Rocks, San Diego County, California. In Geological Investigations of an Active Margin, edited by S. F. McGill and T. M. Ross. Geological Society of America, Cordillera Section Guidebook 27:227-234.

Tucker, A.B., R.M. Feldmann, and C.L. Powell, II 1994. *Speocarcinus berglundi* n. sp. (Decapoda: Brachyura), a new crab from the Imperial Formation (late Miocene-late Pliocene) of southern California. Journal of Paleontology 68(4):800-807.

Umhoefer, P.J., R.J. Dorsey, and P. Renne 1994. Tectonics of the Pliocene Loreto Basin, Baja California Sur, Mexico, and evolution of the Gulf of California. Geology 22(7):649-652.

Vaughan, T.W. 1917. Significance of reef coral fauna at Carrizo Creek, Imperial County, California. Washington Academy of Science Journal 7:194.

Vaughan, F.E. 1922. Geology of the San Bernardino Mountains north of San Gorgonio Pass. University of California Department of Geological Sciences Bulletin 13:319-411.

Ver Planck, W.E. 1952. Gypsum in California: Fish Creek Mountains deposit, Imperial and San Diego Counties. California Division of Mines Bulletin 163:28-35.

Watkins, R. 1990a. Pliocene channel deposits of oyster shells in the Salton Trough region, California. Palaeogeography, Palaeoclimatology, Palaeoecology 79:249-262.

----- 1990b. Paleoecology of a Pliocene rocky shoreline, Salton Trough region, California. Palaois 5:167-

White, J.A., E.H. Lindsay, P. Remeika, B.W. Stout, T. Downs, and M. Cassiliano 1991. Society of Vertebrate Paleontology Field Trip Guide to the Anza-Borrego Desert. Society of Vertebrate Paleontology, Annual Meeting Guidebook. 23p.

Winker, C.D. 1987. Neogene stratigraphy of the Fish Creek-Vallecito section, southern California: implications for early history of the northern Gulf of California and Colorado delta. Doctoral Dissertation, University of Arizona, Tucson. 494p.

Winker, C.D., and S.M. Kidwell 1986. Paleocurrent evidence for lateral displacement of the Pliocene Colorado River delta by the San Andreas Fault system, southeastern California. Geology 14:788-791.

Woodard, G.D. 1963. The Cenozoic succession of the western Colorado Desert, San Diego and Imperial Counties, southern California. Doctoral Dissertation, University of California, Berkeley. 173p.

----- 1974. Redefinition of Cenozoic stratigraphic column in Split Mountain Gorge, Imperial Valley, California. American Association of Petroleum Geologists Bulletin 58:521-539.

Woodring, W.P. 1931. Distribution and age of the marine Tertiary deposits of the Colorado Desert. Carnegie Institution of Washington Publication 418:1-25.

Wright, L.H., and J.M. Bynum 1954. The Butterfield Overland Mail by Waterman L. Ormsby, only through passenger on the first westbound stage. The Huntington Library, San Marino.

# PRELIMINARY REPORT ON THE ECHINODERMATA FROM THE MIOCENE AND PLIOCENE OF THE COYOTE MOUNTAINS, SOUTHERN CALIFORNIA

Charles L. Powell, II

U.S. Geological Survey, Branch of Paleontology and Stratigraphy, MS-915, 345 Middlefield Road., Menlo Park, California 94025

## ABSTRACT

An abundant, subtropical to tropical, shallow water (intertidal to about 45 m) Echinodermata paleofauna of sixteen taxa is reported from the Miocene to Pliocene Imperial Formation in southern California. Five taxa are reported for the first time from the Imperial Formation. The paleofauna shows strong faunal ties to the Gulf of California, but is also similar to Caribbean faunas. Unusual is the co-occurrence of several taxa (some questionably identified) from the Imperial Formation (Gulf of California) and the Pliocene San Diego Formation (coastal southern California). These formations represent different ages, depositional environments, and were not connected during the Neogene.

Sediments attributed to the Imperial Formation have been informally divided into two units which differ in provenance, age, and molluscan macrofaunas: "northern" (Riverside County) and "southern" (Imperial and San Diego Counties, i.e., Coyote Mountains and Fish Creek Mountains). All taxa reported here were recovered from the "southern" exposures. While only five echinoderm taxa have been recovered from "northern" exposures, one not even to familial level. The large paleofauna results from excellent preservation and varied environments of deposition represented by the "southern" outcrops.

All taxa reported from the Imperial Formation were collected from the Coyote Mountains, except the genus *Brissus* which is known only from the Fish Creek Mountains. Several genera are reported in the literature from the Coyote Mountains, but the occurrence of many of these taxa has not been confirmed. Continued study will undoubtedly increase not only the number of, but the biostratigraphic and paleoecological utility of Echinodermata from the Imperial Formation.

## INTRODUCTION

The varied, well-preserved echinodermata paleofauna from the Imperial Formation lends new insights into the paleogeography and age of various echinoid groups. Sixteen taxa are reported here, five for the first time [*Astropecten* cf. *A. armatus* (Gray), *Toxopneustes* cf. *T. roseus* (Agassiz), *Lovenia* cf. *L. hemphilli* Israelsky, and the genera *Meoma* and *Brissus*]. They lived in a euhaline, shallow, tropical to subtropical sea with associated sandy and rocky sediments. Three species are endemic [i.e., *Tripneustes californicus* (Kew), *Echinoneus burgeri* Grant and Hertlein, and *Schizaster morlini* Grant and Hertlein] and show strong affinity to Gulf of California and Caribbean faunas. The regular echinoid *Toxopneustes* cf. *T. roseus* (Agassiz) is reported for the first time as a fossil, occurring at Travertine Point and from the southern Coyote Mountains. Also, the oldest fossil occurrences is recorded in the Imperial Formation from the Coyote Mountains of *Astropecten* cf. *A. armatus* (Gray), *Clypeaster bowseri* Weaver, *Encope tenuis* (Kew), and *Lovenia* cf. *L. hemphilli* Israelsky.

The occurrence of several Imperial Formation taxa in various parts of the proto-Gulf of California suggests a similar age and/or environment, but it is impossible to determine exact similarities due to paucity of data on proto-Gulf of California faunas. Although mollusk faunas are dissimilar, Echinodermata faunas from the Imperial Formation and Pliocene San Diego Formation (coastal southern California) are moderately similar with four taxa co-occurring in both formations [i.e., *Astropecten armatus* (Gray), *Eucidaris thouarsii* (Valenciennes), *Encope tenuis* (Kew), and *Lovenia hemphilli* Israelsky]. These formations are not correlatives because they represent different ages and environments, and were not connected during the Neogene.

The large number of specimens and taxa in the Imperial Formation is due to the excellent preservation of the varied paleoenvironments represented by the formation. Even today, after over a century of fossil collecting, there are places where sand dollar fragments

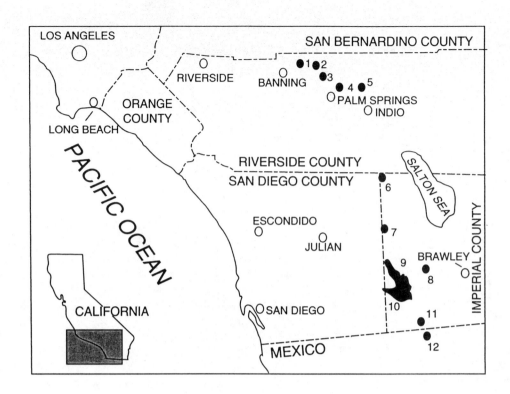

Figure 1. Index map showing distribution of the Imperial Formation in southern California and Baja California Norte. "Northern" outcrops include: 1-Cabazon; 2-Whitewater; 3-Garnet Hill; 4-Mt. Edom; 5-Willis Palms. "Southern" outcrops include: 6-Travertine Point; 7-Ocotillo Wells State Vehicular Recreation Area; 8-Superstition Mountains; 9-Split Mountain-Fish Creek Wash; 10-Coyote Mountains; 11-Yuha Buttes; 12-near Mexicali, Baja California Norte. Insert shows area of map.

litter the ground in such numbers that it is impossible to walk without stepping on them. In addition to the taxa listed below the presence of shallow, concave, sub-circular pits excavated in basement rocks exposed in the southern part of the Coyote Mountains are questionably attributed to regular echinoids (Watkins, 1990).

## PREVIOUS STUDIES

Although *Clypeaster bowseri* Weaver (1908) was the first echinoid described from the Imperial Formation, two years earlier Arnold (1906) lists *C. bowseri* from the Carrizo Creek beds (=Imperial Formation) and attributes it to Merriam. A comprehensive look at Imperial echinoids soon followed (Kew, 1914). Kew described four new species [i.e., *Encope tenuis* Kew, *Clypeaster deserti* Kew, *C. carrizoensis* Kew, and *Hipponoë californica* Kew (=*Tripneustes californicus* Kew)] and mentioned an undescribed *Cidaris* sp.[=*Eucidaris* sp. *E. thouarsii* (Valenciennes) of this report] and *Clypeaster bowseri* Weaver. Later he monographed Cretaceous and Cenozoic Echinoidea from the Pacific Coast of North America (Kew, 1920) and listed *Cidaris* indet. sp., Tripneustes (Hipponoë) *californicus* (Kew), *Clypeaster bowseri* Weaver, *C. carrizoensis* Kew, *C. deserti* Kew, and *Encope tenuis* Kew from the Imperial Formation (Carrizo Creek beds) in the Coyote Mountains. In their monograph of Mesozoic and Cenozoic echinoids of the United State, Clark and Twitchell (1915) mentioned only *C. bowseri* Weaver from the Imperial Formation. T.W. Vaughan (1917) described the coral fauna from the Imperial Formation and lists *Clypeaster bowseri* Weaver, *C. carrizoensis* Kew, *C. deserti* Kew, *Encope tenuis* Kew, *Hipponoë californica* Kew (=*Tripneustes californicus* (Kew) from the "Carrizo formation" (=Imperial Formation) in a faunal list. Hanna (1926)

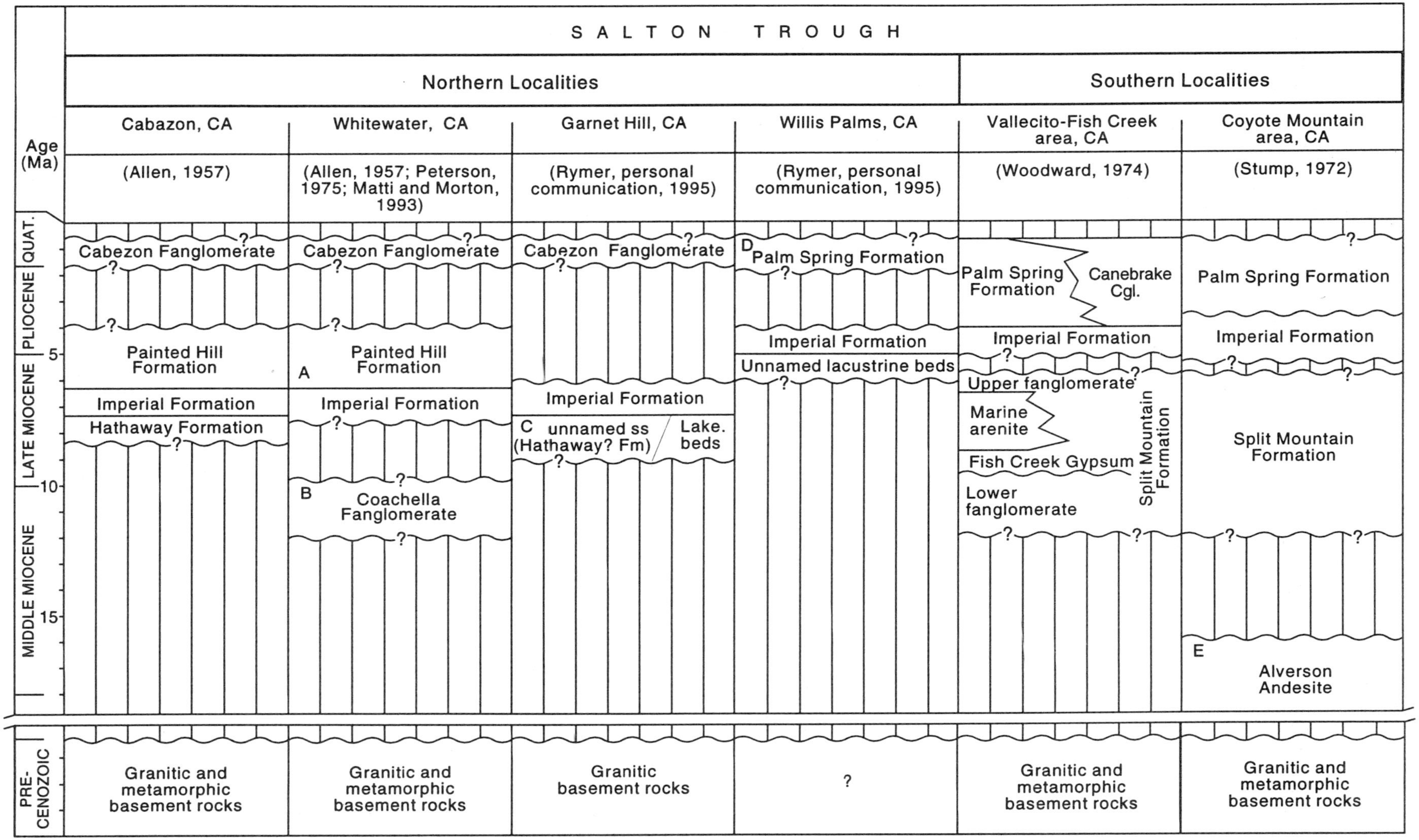

Figure 2. Correlation chart of the Imperial Formation and related lithologic units at selected sites in the Salton Trough. This chart shows the age difference between the "northern" and "southern" outcrops of the Imperial Formation. Numbers indicate age determinations. A - about 6 Ma (Matti, et al. 1985); B - about 10 Ma (Peterson, 1975); C - between 8 Ma and 7.6 Ma (Rymer, et al. 1994); D - 0.76 Ma (Rymer, 1994); E - between 24.8 Ma and about 15 Ma (Ruisaard, 1979).

monographed the Imperial Formation fauna from the southern Coyote Mountains and noted the occurrence of *Clypeaster bowseri* Weaver, *C. deserti* Kew, *Encope tenuis* Kew, *Hipponoë californica* Kew [=*Tripneustes californicus* (Kew)], and questionably *Metalia spatagus* (Linnaeus). Woodring (1931) discussed the age and distribution of marine Tertiary deposits (=Imperial Formation) in the Colorado Desert and noted the occurrence of two echinoids from "northern" outcrops (i.e., indeterminate *Clypeaster* sp. from Garnet Hill, and *Clypeaster* cf. *C. deserti* Kew from the Cabazon section). While reviewing west American Cenozoic Echinoidea Grant and Hertlein (1938) reported eight species, including one new species, from the Imperial Formation: *Eucidaris thouarsii* (Valenciennes), *Tripneustes californicus* (Kew), *Clypeaster bowseri* Weaver, *C. carrizoensis* Kew, *C. deserti* Kew, *Encope tenuis* Kew, *Echinoneus burgeri* Grant and Hertlein (described as new), and *Metalia spatagus* Linnaeus. In 1956, Grant and Hertlein described *Schizaster morlini* from a vague locality on the southern part of the Coyote Mountains.

Recent works that include Imperial Formation echinoids are in Stump (1972) and Albi (1992, 1994). Stump (1972) lists over 20 echinoid taxa from the Coyote Mountains: *Cidaris* sp. [=*Eucidaris* sp. cf. *E. thouarsii* (Valenciennes) of this report], *Centrostephanus* n. sp., *Arbacia incisa* (Agassiz), *Coelopleurus* n. sp., *Lytechinus* cf. *L. anamesus* Clark, *L.* sp. indet., *Tripneustes californicus* (Kew), *Strongylocentrotus purpuratus* (Stimpson), *Clypeaster bowseri* Weaver, *C. carrizoensis* Kew, *C. deserti* Kew, *C.* cf. *C. rotundus* (Agassiz), *Encope arcensis* Durham, *E. sverdrupi* Durham, *E. tenuis* Kew, *E.* n. sp., *Schizaster morlini* Grant and Hertlein, *Agassizia scroiculata* Valenciennes, *A.* n. sp., *Metalia spatagus* Linnaeus, and *Lovenia* cf. *L. hemphilli* Israelsky. Many of these taxa have not been found during the present study and their occurrence is questioned [i.e., *Centrostephanus* n. sp., *Arbacia incisa*, *Coelopleurus* n. sp., *Lytechinus* cf. *L. anamesus*, *L.* sp. indet., *Strongylocentrotus purpuratus*, *Clypeaster* cf. *C. rotundus*, *Encope arcensis*, *E. sverdrupi*, *E.* n. sp., and *Agassizia scrobiculata*]. Recently Albi (1992, 1994) recovered the genera *Arbacia* and *Agassizia* from the Coyote Mountains.

## GEOGRAPHIC DISTRIBUTION, AGE, AND STRATIGRAPHY

The Echinodermata lend no new evidence to the age of the Imperial Formation, although it does extend the age range of several taxa. Rocks attributed to the Imperial Formation in southern California (Figure 1) have a wide, patchy distribution from north of Palm Springs in Riverside County south to Baja California Norte (Woodring, 1931). Powell (1986, 1988) informally divided the Imperial Formation into two distinct rock units which differ in provenance of sediment, paleogeography in the proto-Gulf of California, age, and molluscan macrofauna. Echinodermata faunas from the "northern" and "southern" outcrops are not distinguishable, but this may change as a more extensive fauna from the "northern" outcrop is recovered.

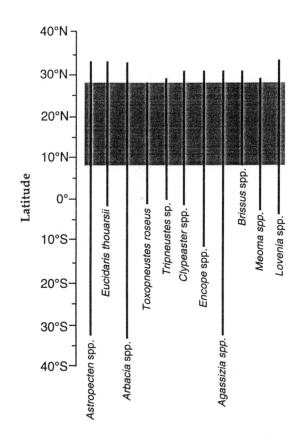

Figure 3. Latitudinal distribution of selected modern eastern Pacific Echinodermata taxa recovered from the Imperial Formation. The gray range bar shows the zone of overlap of distributions from 28°N to 8°N, or from 4° to 24° south of the latitude from which the fossils were collected.

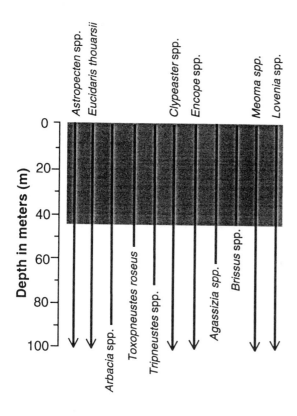

Figure 4. Depth distribution of selected modern eastern Pacific Echinodermata taxa recovered from the Imperial Formation. The gray range bar shows the zone of overlap of depth distributions from intertidal to about 45 m. The arrows indicate taxa which occur deeper than the maximum depth on the chart.

Current thinking about the age of the Imperial Formation is illustrated in Figure 2. The "southern" exposures (including Willis Palms in Riverside County) are considered to range from earliest Pliocene (Ingle, 1974; Stump, 1972) to late Pliocene [approximately 3.9 Ma (Johnson, et al. 1983)], while the "northern" outcrops are considered to be late Miocene [<10 Ma (Peterson, 1975); 8.3 to >6.04 Ma (McDougall, et al. 1994); <7.6-8 Ma (Rymer, et al. 1994); and >6 Ma (Matti, et al. 1985; Matti and Morton, 1993)].

A myriad of stratigraphic schemes have been proposed for the Imperial Formation (Kew, 1914;  Hanna, 1926; Woodring, 1931; Dibblee, 1954; Stump, 1972; Winker, 1987; Woodard, 1974). All have only local utility because the Imperial Formation, as now recognized, represents two distinct rock units and the same age rocks from different outcrops were deposited in small distinct basins with different depositional histories.

## PALEOGEOGRAPHY AND PALEOECOLOGY

Modern eastern Pacific representatives of Echinodermata from the Imperial Formation are subtropical to tropical in occurrence and are very similar to Caribbean taxa: the genus *Echinoneus* has not been previously recognized in the eastern Pacific, *Encope tenuis* Kew may prove to be the same as the Caribbean taxon *E. michelini* Agassiz, and the eastern Pacific *Meoma ventricosa grandis* Gray is nearly indistinguishable from the Caribbean species *M. ventricosa* Lamarck.

Possible faunal correlations between the Imperial Formation and other Neogene rock units in and around the Gulf of California is suggested based on the co-occurrence of several taxa. These are listed in the taxonomic section by species and may represent age or environmental similarities. More field studies are required before Gulf of California Echinodermata faunas are better understood.

Adding to this uncertainty is the occurrence of *Astropecten armatus* (Gray), *Eucidaris thouarsii* (Valenciennes), *Encope tenuis* (Kew) and *Lovenia hemphilli* Israelsky from the San Diego Formation and the Imperial Formation. These two formations represent different ages and environments are were not connected during the Neogene. Mollusks, which have been extensively studied from both formations, are not similar (i.e., Hanna, 1926; Hertlein and Grant, 1972; Powell, 1986, 1988) and suggest no connection between these formnations, making the co-occurrence of the above taxa an interesting, and as yet unresolved, problem.

Using latitudinal and depth range data for modern eastern Pacific echinoid taxa (Maluf, 1988) the Imperial Formation echinoid paleofauna shows overlap of latitudinal ranges between 28°N and 8°N (from the central Gulf of California south to the Gulf of Panama) (Figure 3), in water depths from the intertidal zone to about 45 m (figure 4). This data agrees with paleoecological interpretations utilizing mollusks from the same area (Powell, unpublished data, 1995).

# DISCUSSION OF IMPERIAL ECHINODERMS

## Class Asteroidea
### Family Astropectinidae

The starfish *Astropecten* is reported for the first time from the Imperial Formation in the central part of the Coyote Mountains. The genus occurs in the eastern Pacific from San Pedro, southern California, to Isla Lobos Afuera, Peru, from the intertidal zone to about 500 m except for the enigmatic species *A. benthophilus* Ludwig which lives to 1408 m (Maluf, 1988).

## Class Echinoidea
### Family Cidaridae

Numerous spines and three test fragments are questionably assigned to *Eucidaris thouarsii* (Valenciennes). All were collected from the Coyote Mountains. A specimen reported from the Miocene Whitewater section of the Imperial Formation in Riverside County could not be located during the present study (Grant and Hertlein, 1938), but spines questionably attributable to this species have been collected at Garnet Hill. Previous references to this species from the Imperial Formation (Kew, 1914, 1920; Stump, 1972) were attributed to *Cidaris* sp. indet. Other fossil occurrences of this species come from the Pliocene Marquer Formation (Durham, 1950), unspecified Pliocene rocks of Isla Carmen (Durham, 1950; Emerson and Hertlein, 1964), and Isla Esteban (Emerson and Hertlein, 1964). The modern occurrence of *Eucidaris thouarsii* (Valenciennes) is at Catalina Island, southern California, the Gulf of California south to Bahia Santa Elena, Panama, and the Galapagos Islands, from water depths of the intertidal zone to 150 m (Maluf, 1988).

### Family Arbaciidae

During recent field work several, small regular tests representing a new species of *Arbacia* sp. were recovered from the Coyote Mountains. The genus *Arbacia* is reported from the Coyote Mountains (Albi, 1992), and in the eastern Pacific from Newport Bay, southern California, to Isla Chincha, Peru, in water depths from intertidal to 90 M (Maluf, 1988).

### Family Toxopneustidae

Two taxa from the family Toxopneustidae occur in the Imperial Formation. *Toxopneustes* cf. *T. roseus* (Agassiz), is reported for the first time from the Imperial Formation, occurring at Travertine Point and the Coyote Mountains. Referred specimens are fragmentary and do not differ from modern specimens of this taxon. The modern occurrence is from Guaymas, Mexico, to Isla La Plata, Ecuador, in water depths from the intertidal zone to about 55 m (Maluf, 1988). The second taxon, *Tripneustes californicus* Kew, is restricted to the Imperial Formation. This genus is represented in the eastern Pacific Ocean today by *T. depressus* A. Agassiz and ranges from Midriff Island, Mexico, south to the Galapagos Islands, in water depths from the intertidal zone to about 73 m (Maluf, 1988).

### Family Echinoneidae

*Echinoneus burgeri* Grant and Hertlein is described only from the Imperial Formation in the northern part of the Coyote Mountains. Its occurrence here is the first record of the genus in the western Pacific Ocean (Grant and Hertlein, 1938). Although It is reported in the Atlantic Ocean from Europe to the West Indies and from the Indo-Pacific to Australia (Wagner and Durham, 1966).

### Family Clypeastridae

The family Clypeastidae is represented by three species described from the Imperial Formation. The first is *Clypeaster bowseri* Weaver, a large irregular, oval echinoid common in the Coyote Mountains. At Whitewater, similar large oval echnioids (observed in cross section), are questionably attributed to this species. *Clypeaster bowseri* occurs in unspecified Pliocene rocks west of San Felipe, Baja California Norte (Hertlein, 1968), from the mainland southeast of Isla San Marcos (Durham, 1950), from Punta Santa Antonita, lower California [California Academy of Sciences collections (CAS)], and from Isla Cerralvo, Baja California Sur (Emerson and Hertlein, 1964). The second taxon, *Clypeaster deserti* Kew, is easily distinguished from *C. bowseri* by its flatter shape and more pentate form. *C. deserti* occurs in the Coyote Mountains and is questionably reported from Whitewater. It is reported in the Gloria Formation from the Gulf of California (Durham, 1950), unspecified Pliocene rocks west of San Felipe, Baja California Norte (Hertlein, 1968), and from Cedros Island on the outer coast of Baja California (CAS collections). The third species is *Clypeaster carrizoensis* Kew, a small Clypeastid with a flattened oval outline similar to *C. subdepressus* (Gray) from the Caribbean. It is the rarest of the three Clypeastids from the Imperial Formation. It is reported only from the Coyote Mountains and questionably from Pliocene sediments west of San Felipe, Baja California Norte (Hertlein, 1968). Modern *Clypeaster* sp. in the eastern Pacific occur from Isla San Jorge, Mexico, south to Isla La Plata, Ecuador, and at the Galapagos Islands, from water depths of the intertidal zone to over 400 m (Maluf,

1988).

### Family Mellitidae

The most common echinoderm from the Imperial Formation is the sand dollar *Encope tenuis* (Kew). Most specimens are fragmented and their identification cannot be confirmed with certainty, but in over one hundred complete specimens examined no other taxa has been observed even though three other taxa have been reported by Stump (1972) from the Imperial Formation. *Encope tenuis* is also reported from the San Diego Formation (late Pliocene) of San Diego, California (Hertlein and Grant, 1960), and from unspecified Pliocene rocks west of San Felipe, Baja California Norte (Hertlein, 1968). The genus *Encope* is represented in the eastern Pacific from San Diego, California, to Bahia Santa Elena, Ecuador, in water depths of the intertidal zone to 134 m (Maluf, 1988).

### Family Schizasteridae

Two taxa of the family Schizasteridae are represented in the Imperial Formation. Two specimens of indeterminate *Agassizia* sp. have been found in collections of the Los Angeles County Museum of Natural History (LACM) from the Coyote Mountains. Albi (1994) separated this species from the modern *A. scrobiculata* Valenciennes by a number of anatomical features. This genus is monotypic in the modern eastern Pacific, represented by *A. scrobiculata* Valenciennes which occurs from Baja California Norte south to Peru, in water depths from intertidal to about 62 m (Maluf, 1988). *Schizaster morlini* Grant and Hertlein is known from a few specimens collected during the 1950's at an obscure locality cited as "...from the southern slope, approximately midway east to west about two thirds of the way to the top of the Coyote Mountains...". These specimens are compressed and misshapened making identification and comparison with related taxa difficult. The genus *Schizaster* is cosmopolitan in tropical seas.

### Family Brissidae

Three Brissid taxa occur in the Imperial Formation. The first is the genus *Brissus* represented by two poorly preserved specimens from Barrett Canyon, in the Fish Creek Mountains. The genus *Brissus*, represented by *B. obesus* Verrill, occurs from Puerto Penasco, Mexico, to Islas Secas, Panama, in water depths from the intertidal zone to about 45 m (Maluf, 1988). The second taxon is a poorly preserved specimen referred to *Metalia spatagus* (Linnaeus) by Hanna (1926) from the Coyote Mountains. Further examination is warranted but at the present time this specimen is attributed to *Metalia*. The genus *Metalia* occurs from Bahia San Felipe, Baja California Norte, south to Panama Bay, Panama, in water depths from the intertidal zone to about 20 m (Maluf, 1988). The third taxon, *Meoma* sp. is represented by two test fragments from the Coyote Mountains. The fragments show the deeply impressed, curved petal characteristic of *Meoma* sp. Similar fossil tests also occur in the Marquer Formation (Durham, 1950), from unspecified Pliocene rocks west of San Felipe, Baja California Norte (Durham, 1950), and at Isla San Jose, Baja California Sur (CAS collections). The genus *Meoma* ranges in the eastern Pacific from Isla Midriff, Mexico, south to Ecuador, in water depths of the intertidal zone to about 200 m (Maluf, 1988).

### Family Loveniidae

Well preserved specimens, questionably referred to *Lovenia hemphilli* Israelsky from the Coyote Mountains, are incomplete but very well preserved including the spines on most specimens, and appear identical to those from the San Diego Formation. They are questionably referred to the above taxa because they lack the oral surface and are commonly covered with spines. The eastern Pacific Holocene analog *L. cordiformis* A. Agassiz occurs from Santa Barbara, southern California, to northern Peru, in water depths of the intertidal zone to about 200 m (Maluf, 1988). *Lovenia hemphilli* Israelsky is also reported from the San Diego Formation (Hertlein and Grant, 1960), the Carmen Formation (Durham, 1950) and unspecified Pliocene rocks from Isla Monserrate, in the Gulf of California (CAS).

## ACKNOWLEDGMENTS

I thank Richard Mooi (CAS) for assistance with identification, Kris McDougall U.S. Geological Survey (USGS), Mary McGann (USGS), Richard Mooi (CAS), and William Sliter (USGS) for review of the manuscript. I also thank Paul Remeika of Anza-Borrego Desert State Park (ABDSP) for assistance in obtaining temporary collecting permits, guidance to ABDSP and Bureau of Land Management cooperative agreement localities, assistance in the field, and for many informative discussions.

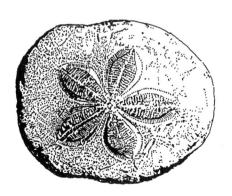

**LITERATURE CITED**

Albi, Y. 1992. A newly found echinoid, *Arbacia,* from the Imperial Formation, California. Bulletin of the Southern California Paleontological Society 24(7/8):74-77.

----- 1994. An addition of *Agassizia* sp. to the echinoids found in the Imperial Formation, southern California. Bulletin of the Southern California Paleontological Society 27(11/12):95-104.

Arnold, R., 1906. The Tertiary and Quaternary Pectens of California. U.S. Geological Survey Professional Paper 47:1-264.

Allen, C.R. 1957. San Andreas Fault Zone in San Gorgonio Pass, southern California. Bulletin of the Geological Society of America 68(3):315-350.

Bell-Countryman, P. 1984. Environments of deposition of the Pliocene Imperial Formation, Coyote Mountains, southwest Salton Trough. In The Imperial Basin - Tectonics, Sedimentation and Thermal Aspects, edited by C.A. Rigsby. Pacific Section, Society of Economic Paleontologists and Mineralogists:45-70.

Bramkamp, R.A. 1935. Stratigraphy and molluscan fauna of the Imperial Formation of San Gorgonio Pass, California. Ph.D. Dissertation, University of California, Berkeley, 371 p.

Christensen, A.D. 1957. Part of the geology of the Coyote Mountain area, Imperial County, California. M.A. Thesis, University of California, Los Angeles, 158 p.

Clark, W.B., and M.W. Twitchell 1915. The Mesozoic and Cenozoic Echinodermata of the United States. U.S. Geological Survey Monographs, LIV:1-341.

Dibblee, T.W., Jr. 1954. Geology of the Imperial Valley region, California. In Geology of Southern California, edited by R.H. Jahns. California Division of Mines and Geology Bulletin 170(2):21-28.

Downs, T., and J.A. White 1968. A vertebrate faunal succession in superimposed sediments from late Pliocene to middle Pleistocene in California. Report of the XXIII Seesion, International Geological Congress 10:41-47.

Durham, J.W. 1950. 1940 E.W. Scripss cruise to the Gulf of California. Part II, Megascopic paleontology and marine stratigraphy. Geological Society of America Memoir 43:1-216.

Emerson, W.K., and L.G. Hertlein 1964. Invertebrate megafossil of the Belvedere Expedition to the Gulf of California. Transactions of the San Diego Society of Natural History 13(17):333-368.

Grant, U.S., IV, and L.G. Hertlein 1938. The west American Cenozoic Echinoidea. University of California at Los Angeles Publications in Mathematical and Physical Sciences 2:1-225.

----- 1956. *Schizaster morlini*, a new species of Echinoid from the Pliocene of Imperial County, California. Bulletin of the Southern California Academy of Sciences 55(2):107-109.

Hanna, G.D. 1926. Paleontology of Coyote Mountains, Imperial County, California. Proceedings of the California Academy of Sciences, Fourth Series, 14(18):427-503.

Hertlein, L.G. 1968. Three late Cenozoic molluscan faunules from Baja California, with a note on diatomite from west of San Felipe. Proceedings of the California Academy of Sciences, Fourth Series, 30(19):401-405.

Hertlein, L.G., and U.S. Grant, IV, 1960. The geology and paleontology of the marine Pliocene of San Diego, California. Paleontology (Coelenterata, Bryozoa, Brachiopoda, Echinodermata). Memoirs of the San Diego Society of Natural Histroy, 2(2a):73-133.

----- 1972. The geology and paleontology of the marine Pliocene of San Diego, California. Paleontology (Pelecypods). Memoirs of the San Diego Society of Natural History, 2(2b):135-411.

Ingle, J.C. 1974. Paleobathymetric history of Neogene marine sediments, northern Gulf of California. Geology of Peninsular California. A guidebook for the 49th Annual Meeting of the Pacific Sections, AAPG-SEG-SEPM:121-138.

Johnson, N.M., C.B. Officer, N.D. Opdyke, G.D. Woodard, P.K. Zeitler, and E.H. Lindsay 1983. Rates of late Cenozoic tectonism in the Vallecito-Fish Creek Basin, western Imperial Valley, California. Geology 11(11):664-667.

Kew, W.S.W. 1914. Tertiary echinoids of the Carrizo Creek region in the Colorado Desert. University of California, Publications of the Department of Geology 8(5):39-60.

----- 1920. Cretaceous and Cenozoic Echinoidea of the Pacific Coast of North America. University of California Publications, Bulletin of the Department of Geology 12(2):23-236.

Maluf, L.Y. 1988. Composition and distribution of the central eastern Pacific Echinoderms. Natural History Museum of Los Angeles County, Technical Reports 2:1-242.

Matti, J.C., and D.M. Morton 1993. Paleogeographic evolution of the San Andreas Fault in southern California: a reconstruction based on a new cross-fault correlation. In The San Andreas Fault System: Displacement, Palinspastic Reconstruction, and Geologic Evolution, edited by R.E. Powell, R.J. Weldon, II, and J.C. Matti. Geological Society of America Memoir 178:107-159.

Matti, J.C., D.M. Morton, and B.F. Cox 1985. Distribution and geologic relations of fault systems in the vicinity of the Central Transverse Ranges, southern California. U.S. Geological Survey Open-File Report 85-365:1-23.

McDougall, K., C.L. Powell, II, J.C. Matti, and R.Z. Poore 1994. The Imperial Formation and the opening of the ancestral Gulf of California. Abstracts with Programs, 90th Annual Cordilleran Section, Geological Society of America 26(2):71.

Mount, J.D. 1974. Molluscan evidence for the age of the Imperial Formation, southern California. Abstracts of Program, Annual Meeting, Southern California Academy of Sciences:29.

Orcutt, C.R. 1890. The Colorado Desert. California Journal of Mines and Geology 10:899-919.

Peterson, M.S. 1975. Geology of the Coachella Fanglomerate. In San Andreas Fault in Southern California - A Guide to San Andreas Fault from Mexico to Carrizo Plain, edited by J.C. Crowell. California Division of Mines and Geology Special Report 118:119-126.

Powell, C.L., II. 1986. Stratigraphy and bivalve molluscan paleontology of the Neogene Imperial Formation in Riverside County, California. M.S. Thesis, San Jose State University, 325 p.

----- 1988. The Miocene and Pliocene Imperial Formation of southern California and its molluscan fauna: an overview. The Western Society of Malacologists, Annual Report 20:11-18.

Quinn, H.A., and T.M. Cronin 1984. Micropaleontology and depositional environments of the Imperial and Palm Spring Formations, Imperial Valley, California. In The Imperial Basin - Tectonics, Sedimentation and Thermal Aspects, edited by C.A. Rigsby. Pacific Section, Society of Economic Paleontologists and Mineralogists:71-85.

Ruisaard, C.I. 1979. Stratigraphy of the Miocene Alverson Formation, Imperial County, California. Master of Science Thesis, San Diego State University, 124 p.

Rymer, M.J. 1994. Quaternary fault-normal thrusting in the northwestern Mecca Hills, southern California. In Geological Investigations of an Active Margin, edited by S.F. McGill and T.M. Ross. Geological Society of America Cordilleran Section Meeting and Guidebook, San Bernardino County Museum Association:325-329.

Rymer, M.J., A.M. Sarna-Wojcicki, C.L. Powell, II, and J.A. Barron 1994. Stratigraphic evidence for late Miocene opening of the Salton Trough in southern California. Abstracts with Programs, 90th Annual Cordilleran Section, Geological Society of America 26(2):87.

Smith, D.D. 1962. Geology of the northeast quarter of the Carrizo Mountain quadrangle, Imperial County, California. M.A. Thesis, University of Southern California, Los Angeles, 89 p.

Stump, T.E. 1972. Stratigraphy and paleontology of the Imperial Formation in the western Colorado Desert. Master of Science Thesis, San Diego State University, 132 p.

Vaughan, F.E. 1922. Geology of San Bernardino Mountains north of San Gorgonio Pass. University of California Publications, Bulletin of the Department of Geological Sciences 13(9):319-411.

Vaughan, T.W. 1900. The Eocene and Oligocene coral fauna of the United States. U.S. Geological Survey Monograph 39:1-263.

----- 1917. The reef-coral fauna of Carrizo Creek, Imperial County, California and its significance. Shorter contributions to general geology, U.S. Geological Survey Professional Paper 98T:355-376.

Wagner, C.D., and J.W. Durham 1966. Holectypoids. In Treatise on Invertebrate Paleontology, edited by R.C. Moore. Echinodermata 3, 2:U1-U695.

Watkins, R. 1990. Paleoecology of a Pliocene rocky shoreline, Salton Trough region, California. Palaios 5:167-175.

Weaver, C.W. 1908. New echinoids from the Tertiary of California. University of California Publications, Bulletin of the Department of Geology 5(17):271-274.

Winker, C.D. 1987. Neogene stratigraphy of the Fish Creek-Vallecito section, southern California: implications for early history of the northern Gulf of California and Colorado Desert. Ph.D Dissertation, University of Arizona, Phoenix, 494 p.

Woodring, W.P. 1931. Distribution and age of the marine Tertiary deposits of the Colorado Desert. Carnegie Institute of Washington Publication 418:1-21.

Woodard, G.D. 1974. Redefinition of Cenozoic stratigraphic column in Split Mountain Gorge, Imperial Valley, California. American Association of Petroleum Geologists Bulletin 58:521-539.

# CRETACEOUS PALYNOFLORA AND NEOGENE ANGIOSPERM WOODS FROM ANZA-BORREGO DESERT STATE PARK, CALIFORNIA: IMPLICATIONS FOR PLIOCENE CLIMATE OF THE COLORADO PLATEAU AND AGE OF THE GRAND CANYON

**Paul Remeika**
California Park Service, Colorado Desert District, Anza-Borrego Desert State Park, California 200 Palm Canyon Drive, Borrego Springs, California 92004

**R. Farley Fleming**
United States Geological Survey, Branch of Paleontology and Stratigraphy, MS-919, P. O. Box 25046, Denver, Colorado 80225

## ABSTRACT

The Vallecito-Fish Creek Basin of Anza-Borrego Desert State Park includes a continuous time-stratigraphic sequence of Pliocene prodelta, delta-front and delta-plain deposits from the ancestral Colorado River. Exposed along Fish Creek Wash, these extralimital fine-grained sediments yield silicified angiosperm fossil wood (dicotyledons and monocotyledons) of the Carrizo local flora and reworked Cretaceous pollen. This collaborative research study involved detailed stratigraphic, sedimentologic, and paleobotanical investigations in conjunction with detailed palynological investigations. We document: (1) the presence of Cretaceous pollen in Pliocene sediments and use the presence of wood to support a wetter paleoenvironment than today, and (2) use the stratigraphic distribution of Cretaceous pollen to suggest erosional events on the Colorado Plateau.

Based on tracheid cell structure, seven families are recognized in the paleoflora, representing eleven taxa. Eight are new for the area. Families include the Lauraceae, Salicaceae, Oleaceae, Hippocastanaceae, Arecaeae, Juglandaceae, and the Cupressaceae. Climatic data inferred from this riparian association and on tree-ring growth analyses suggest that the paleoclimate of the northern Gulf of California was more temperate than now with winter rainfall dominant.

Assemblages of Cretaceous pollen, also discovered in the deltaic deposits, contain reworked *Proteacidites* spp. and *Aquilapollenites* spp., which have known stratigraphic distributions from the Upper Cretaceous Mancos Shale of the western interior of North America. In the southern Colorado Plateau, erosion of Cretaceous rocks that preserve *Proteacidites* spp. but lack *Aquilapollenites* spp. began by the early Pliocene. Pollen of *Aquilapollenites* spp. from the northern Colorado Plateau first appear in the Vallecito-Fish Creek Basin at about 3.9 Ma. This indicates erosion and transportation of Cretaceous sediments from the northern half of the plateau did not commence until the mid Pliocene. This evidence suggests rapid and extensive erosion of the Colorado Plateau occurred during the Pliocene, rather than earlier in the Tertiary. Thus, the combined paleobotanical and palynological data are in accord and indicate that the Pliocene climate of the southwestern United States was much wetter than today, with increased precipitation and cooler temperatures.

## INTRODUCTION

The primary goal of PRISM (Pliocene Research, Interpretation, and Synoptic Mapping) is to produce a paleoclimatic map for the world at the time of a warm interval identified in the mid-Gauss part of the Pliocene (Dowsett and Poore, 1991). As part of this effort, the Pliocene Continental Climates Project within PRISM is attempting to reconstruct paleoclimatic conditions in western North America during the Pliocene. This report contains stratigraphical, paleobotanical and palynological data from the Vallecito-Fish Creek Basin (VFCB) of Anza-Borrego Desert State Park (ABDSP). This Pliocene section has excellent rock exposures in a high-quality temporal framework (Opdyke, et al. 1977; Johnson, et al. 1983).

During the lower Pliocene (Hemphillian-

Blancan LMA), and extending into the middle Pleistocene (Irvingtonian LMA), the ancestral Colorado River deposited an enormous amount of extraregional fine-grained clastic sediments from the Colorado Plateau directly into the diffuse transtensional northern Gulf of California structural depression (Winker and Kidwell, 1986; Lucchitta, 1987). These sediments are informally referred to as the Colorado River group (CRG) (White, et al. 1991; Reynolds and Remeika, 1993). They represent a clearly recognizable sedimentary succession of deltaic clay, silt and calcarenite lithofacies that is today geographically restricted in outcrop to two subordinate alluvial paleobasins along the western margin of the aggrading Salton Trough rift basin in ABDSP (Figure 1). These areas include the VFCB in the southern end of the state park, and the Borrego-San Felipe Basin (BSFB) in the north end.

The indurated deltaic sediments of both paleobasins are particularly noteworthy for the presence of palynofloras (Fleming, 1994), leaf impressions (*Populus trichocarpa* and *Salix gooddingii*) and petrified wood (Remeika, et al. 1988; Remeika, 1994). Informally named the Carrizo local flora (CLF) (Remeika, 1994), the floral assemblage is represented by ten well-preserved silicified and calcified dicots and monocots (Table 1) that were transported into a wide deltaic-marine catchment apron that developed and matured from the mouth of the perennial Colorado River during the Plio-Pleistocene. The small to medium diameter woody debris was uprooted and transported bayward by suspension-load/bed-load floodwaters, became water-logged, sank and was differentially buried, petrified and preserved, oriented parallel to flow direction, in the subaqueous deltaic sediment-water interface. Over the course of

Figure 1. Map of part of southwestern North America showing the outline of the Colorado Plateau (shaded area), and the course of the Colorado River and its major tributaries (solid, heavy black lines). Anza-Borrego Desert State Park is located along the western Salton Trough of southern California. Approximate extent of the Salton Trough is shown by dotted line. Location of the Vallecito-Fish Creek Basin is marked by large black dot.

millennia, silica and calcium carbonate minerals replaced the original woody material, retaining the internal structure. The anatomical structure of the majority of the woods indicate a species rich, mixed dicotyledonous angiosperm flora typical of the Madro-Tertiary Geoflora (Axelrod, 1958), with a complex standard tree regime that has affinities with extant *Umbellularia california* and *Persea podadenia* (Lauraceae), *Juglans californica* (Juglandaceae), *Populus* sp. and *Salix* sp. (Salicaceae), *Fraxinus oregona* (Oleaceae), and *Aesculus californica* (Hippocastanaceae). All but *Persea podadenia* (Sierra Madrean Woodland Element) are hydrophtyic mixed-evergreen understory hardwoods of the California Woodland Element (e.g. central California coastline [Big Sur]).

Historically, relatively little is known of the woods. Paleobotanical evidence is discussed matter-of-factly by early investigators with less attention than to vertebrate and invertebrate faunas. Schaffer (1857) first noted the occurrence of petrified wood in the Colorado Desert. Woodring (1931) postulated fragments of silicified wood to be generally attributed to desert ironwood *Olneya tesota* Gray, a hitherto unrecognized assumption at the time, later shared by Dibblee (1954) and Woodard (1963) (pers. comm., 1981). Pinault (1984) added oak, *Quercus* spp., without any supportive evidence. Despite being a neglected resource of study, Remeika, et al. (1988) identified and described three dicotyledon genera in the paleoflora, providing the first accurate accessment of the local deltaic hardwood assemblage and paleoclimate of proto-ABDSP during the Pliocene. The most common dicots include ancestral forms of California laurel, *Umbellularia salicifolia* and California walnut, *Juglans pseudomorpha*, both showing affinities to the modern species *Umbellularia californica* and *Juglans californica*, respectively. In addition,

## TABLE 1. CARRIZO LOCAL FLORA TAXONOMIC LIST

| FOSSIL SPECIES | ALLIED LIVING SPECIES | MADRO-TERTIARY GEOFLORA[a] | | |
|---|---|---|---|---|
| | | 1 | 2 | 3 |
| **Arecaceae** | | | | |
| Gen. et. sp. indet. | monocot (palm) | | *✓ | |
| *Sabal* cf. *S. miocenica* Axelrod | *S. urseana* (palm) | | *✓ | |
| **Hippocastanaceae** | | | | |
| *Aesculus* sp. | *A. californica* (Calif. buckeye) | *✓ | | |
| **Juglandaceae** | | | | |
| *Juglans pseudomorpha* Condit | *J. californica* (Calif. walnut) | * | | |
| **Lauraceae** | | | | |
| *Persea coalingensis* Axelrod | *P. podadenia* (avocado) | | *✓ | |
| *Umbellularia salicifolia* Axelrod | *U. californica* (bay-laurel) | * | | |
| **Oleaceae** | | | | |
| *Fraxinus caudata* Dorf | *F. oregona* (Oregon ash) | *✓ | | |
| **Salicaceae** | | | | |
| *Populus* sp. indet. | *Populus* sp. (cottonwood) | * | * | |
| cf. *P. alexanderi* Dorf | *P. trichocarpa* (black cottonwood) | *✓ | *✓ | |
| *Salix* sp. indet. | *Salix* sp. (willow) | * | | |
| *S. gooddingii* Ball | *S. gooddingii* (Dudley willow) | *✓ | | |
| **Cupressaceae** | | | | |
| Gen. et. sp. Indet. | softwood (cedar or juniper) ? | | | *✓ |

Distribution of fossil taxa comparable to allied living species and occurrence (*) in the Madro-Tertiary Geoflora. [a]1 = California Woodland Element; 2 = Sierra Madrean Woodland Element; 3 = Conifer Woodland Element. ✓ = new to Anza-Borrego.

parallochthonous macrodetritus of palm fronds with attached petiole fragments resemble the modern monocotyledon genus *Sabal* (J. Cornett, pers. comm., 1990), mark the first occurrence of fossil herbaceous material from ABDSP (Remeika, 1991). Additional monocot debris, including stem, seed and disaggreagate vascular bundles, may also be referrable to *Sabal*. They range throughout the delta-plain sediments.

The presence of Cretaceous pollen in these same sediments of the VFCB indicates that these fossils have been reworked (i.e., they were originally deposited in Cretaceous sediments that later eroded from the Colorado Plateau then transported, via the Colorado River, and redeposited in Pliocene deltaic sediments of the northern Gulf of California). Two kinds of Cretaceous pollen are discussed in this study. One is *Proteacidites* spp., and the other is *Aquilapollenites* spp. (including *Mancicorpus* spp., which is closely related to *Aquilapollenites* spp.). Together *Aquilapollenites* spp. and *Mancicorpus* spp. are referred to as triprojectate pollen. These fossils are common in Cretaceous rocks of the western interior of North America but have restricted stratigraphic and paleobiogeographic ranges.

In this study, the stratigraphic distributions of these fossils in Pliocene sediments exposed in Fish Creek Wash are used to date erosional events on the Colorado Plateau and thus constrain the age of the Grand Canyon. The erosional history that emerges from reworked pollen evidence has implications for Pliocene climatic reconstruction of southwestern North America.

## STRATIGRAPHIC OCCURRENCE

The Diablo formation (informal name of Remeika, 1991; Remeika and Lindsay, 1992), part of the CRG, represents a prominent non-marine, high-energy delta-floodplain arenitic deposit. Woodard (1963) originally designated the basal 300 m-thick Diablo member of the Palm Spring Formation (Woodring, 1931) as fine reddish-brown, massive crossbedded arenaceous and argillaceous sandstones with less gypsiferous claystones that grade from the Burrobend formation (new informal name, this volume) of the Imperial group (new informal name, this volume). In the study area, sediments are composed of light-colored, fine to very fine-grained, friable, massive quartzitic arenite paleochannel sands, locally polleniferous, wood-bearing and concretionary, with subordinate brown ripple-laminated overbank claystones. Sediment velocities were high enough to scour channel bottoms. Well-preserved wood is a common occurrence in the Diablo delta-plain deposits, occurring exclusively in concretionary arentitic point bar and channel sands as isolated, disarticulated logs, branches and trunk sections. Specimens also occur as float eroded from parent material. The presence of reworked wood, aligned on strike in the sediments, absence of upright trunks, a herbaceous leaf-litter component not preserved in the paleobotanical assemblage and absence from low-energy overbank splay deposits, indicate they were transported by overall outgrowth (progradation) of an extensive delta system and not buried in place.

Petrographic investigations by Merriam and Bandy (1965), Muffler and Doe (1968), and recent palynofacies research of the internal architecture by Fleming (1994) and Fleming and Remeika (1994) conclude that the flood-generated Diablo sandstones debouched into ABDSP via the Colorado River. This is based on the presence of reworked Cretaceous palynomorphs from the Colorado Plateau, some of which have restricted biostratigraphic ranges and paleobiogeographic distributions in the western interior of North America. Triprojectate pollen *Aquilapollenites* spp. and *Mancicorpus* spp. range from Campanian to Maastrichtian, and *Proteacidites* spp. ranges from Coniacian to Maastrichtian (Fleming, 1993a,b). They became extinct at the Cretaceous-Tertiary (K/T) boundary. Their presence, along with the Eocene pollen *Pistillipollenites* spp. reworked from the Green River Formation of Utah-Wyoming indicates rapid floodwater discharge for the Diablo formation.

In the VFCB, Tarbet (1951) recognized three informal members for the Imperial group. Stump (1972) recognized two members. Winker (1987) revised the stratigraphy depositionally, subdividing it into 10 informal units. Along Fish Creek Wash, only the polleniferous, wood-bearing portion of the upper Imperial group, represented by the Deguynos and Camels Head members (informal names of Woodard, 1963; Winker, 1987) of the Burrobend formation, is exposed. This genetic, brackish-marine regime is composed of high-constructive, upward-fining deltaic stratigraphy that prograded into the drowned eastern portion of proto-ABDSP, under tectonically induced subsidence, by the ancestral Colorado River that infilled protected relatively flat-profiled intertidal embayments of the marginal northern Gulf of California. Progradation resulted from a high rate of fine-grained sedimentation into an area having a low regional bathymetric gradient of 10-15 m water depth (Meldahl, 1993).

The Deguynos member is a clay-dominated, intertidal-bedded progradational rhythmite sequence deposited in a very shallow subtidal to lowest intertidal paleoenvironment (Winker, 1987). The high suspended sediment concentrations are characterized by a vertical organization of thin-bedded yellow-green claystone-siltstone couplets that reflect daily delta-front depositional cycles generated by the dynamics of mixed, semidiurnal tidal currents and spring tides. These typically equate to ranges similar to conditions that exist today along the shallow-water northern Gulf of California (Fürsich and Flessa, 1987).

This rock unit is very fossiliferous with rare woody debris (strictly *Juglans pseudomorpha*) and an impoverished endolithic molluscan infauna that is restricted in numbers and kinds of taxa. Species are represented overwelmingly by the monospecific oyster *Dendostrea? vespertina*, clam *Anomia subcostata*, and pectinid *Argopecten deserti* macroinvertebrates that display low species diversity, less equitability, but high population densities (coquinas) scattered throughout the section (Hanna, 1926). Taxonomic, paleoecologic and taphonomic criteria (Kidwell, 1988) suggest biostromal deposition by storm related events into the shallow-water embayment. In addition, intertidal zones are virtually overrun with well-bedded turritellid and echinoid marker genera, arranged one above the other separated by clayey stratum that preserve ephemeral mammalian ichnofossils along the basin-margin (Stout and Remeika, 1991). Infaunal evidence indicates a Panamic Province affinity (Powell, 1987).

Up-section, the delta-front grades into a thin, distal paleofacies of the Camels Head member. Poorly-preserved fossil driftwood has been recovered from the Deguynos-Camels Head tidal flats along Fish Creek Wash. The Camels Head member grades upward into the major wood-bearing Diablo formation (White, et al. 1991; Remeika and Lindsay, 1992). Studied in outcrop, both genetically-related rock units reveal a complex lateral and vertical juxtaposition of basinal lithofacies, grading from subaqueous delta-front deposits into tidal flat to subaerial delta-plain basin and basin-margin depositional relationships. This, in turn, grades into coeval, locally-derived syndeposited Canebrake Conglomerate (Dibblee, 1954) along the basin-margins (undifferentiated proximal alluvial fan conglomerates and sandstones), and is abruptly overlain by the Palm Spring Formation (Woodring, 1931; Woodard, 1963) in the basin-center (floodplain silts and sands). This is consistent with a westward stratigraphic thickening in the developing VFCB asymmetric half-graben during the Plio-Pleistocene, beginning approximately 2.5 MYA (see roadlog, this volume).

**TEMPORAL FRAMEWORK**

A review of ABDSP's paleontologic (see bibliography, Volume II) and magnetostratigraphic literature (Opdyke, et al. 1977; Johnson, et al. 1983) establishes the base of the Diablo formation gradational contact with the Burrobend formation at approximately 3.8 Ma, with continued deposition spanning 1.2 Ma in the VFCB.

Downs and White (1968) published the first comprehensive list of fossil vertebrates from the VFCB and established a biostratigraphic zonation for the upper part of the Burrobend formation and the Diablo formation. They divided the stratigraphic section into three local faunas (LF). From oldest to youngest these are the Layer Cake LF, the Arroyo Seco LF, and the Vallecito Creek LF. Downs and White (1968) estimated the Land Mammal Age (LMA) of the Layer Cake LF to be early Blancan, the Arroyo Seco LF to be late Blancan, and the Vallecito Creek LF to be Irvingtonian. Based on the chronology of Kurtén and Anderson (1980) these LMA's indicate the western part of the VFCB is Pliocene to middle Pleistocene in age.

Using the work of Downs and White (1968) for biostratigraphic and geographic control, Opdyke et al. (1977) published the first paleomagnetic stratigraphy for Plio-Pleistocene rocks in the VFCB. They collected samples from 150 sites through approximately 3800 m

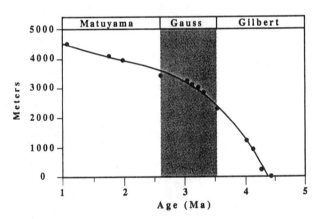

Figure 2. Age model for the Vallecito-Fish Creek Basin section. Age (Ma) is plotted versus sediment accumulation. Previous paleomagnetic studies established a well-controlled magneto-stratigraphy for the section. Radiometric dates on ashes calibrated the magnetostratigraphy. The black dots mark the stratigraphic positions in the section of the boundaries between the paleomagnetic Magnetochrons. The shaded area represents the Gauss Normal Polarity Magnetochron. The Matuyama and Gilbert Reversed Polairty Magnetochrons are represented without shading. Ages for palynology samples collected in this study were interpolated using this model.

of a stratigraphically continuous interval in the Burrobend, Diablo and Palm Spring formations. Using the pattern of declination values for the statistically valid sites, Opdyke et al. (1977) derived a magnetic stratigraphy for the VFCB section. By calibrating the magnetic stratigraphy with the vertebrate zonation of Downs and White (1968), they identified the Gilbert (Reversed Polarity), Gauss (Normal Polarity), and Matuyama (Reversed Polarity) Magnetochrons. Within the Gilbert they detected a single event that they interpreted to be the Cochiti (Normal Polarity) Magnetosubchron. Within the Gauss they identified the Mammoth and Kaena (Reversed Polairty) Magnetosubchrons. They detected one event within the Matuyama and the base of another event at the top of the sequence. They suggested that the lower event could be either the Reunion or Olduvai (Normal Polarity) Magnetosubchrons and the upper event could be the base of either the Olduvai or the Jaramillo (Normal Polarity) Magnetosubchrons.

Johnson, et al. (1983) remeasured the database of Opdyke, et al. (1977) after thermal demagnetization of the original samples. In addition, they obtained a fission-track date of 2.3 Ma $\pm$ 0.4 Ma on an air-fall tuff that occurs in the Palm Spring Formation. With these new data, they refined the magnetic stratigraphy of the VFCB, identifying the lower event in the Matuyama as the Olduvai and the event at the top of the section as the base of the Jaramillo. Repenning (1992) disagreed with the conclusion of Johnson, et al. (1983) that the event at the top is the base of the Jaramillo; he maintains that the upper event is the base of the Brunhes. Repenning based his conclusion on the presence of *Pitymys meadensis* below the youngest normal polarity zone in the VFCB. This species is not know before .85 Ma (Repenning, 1992). For the purposes of this study, modifying the age model according to Repenning would make little difference. Most of the samples collected are below the tuff (dated at 2.3 Ma) and the magnetostratigraphy proposed by Johnson, et al. (1983) is valid below the stratigraphic level of the tuff.

In constructing the age model for this study, we first developed a stratigraphic framework for the 86 samples collected during two field excursions in 1992 and for the 25 samples provided by T. Cronin. As discussed above, the sample positions were intergrated with the stratigraphic section of Opdyke, et al. (1977) and Johnson, et al. (1983). Using the Burrobend-Diablo formational contact as a datum, all of the sample positions for this study were intergrated within a single strtatigraphic section.

The age model for the VFCB was derived from the paleomagnetic stratigraphy as developed in Johnson et al. (1983) and the currently revised geomagnetic polarity time scale of Cande and Kent (1992). Johnson, et al. (1983) also published a figure showing sediment accumulation versus age. We revised this curve with the new age values from Cande and Kent (1992) and fit a third-order polynomial to the resulting plot (Figure 2). The polynomial has an $R2=0.995$ and the curve closely matches that published by Johnson, et al. (1983). We used this equation to calculate the sample ages, which are listed in Figure 3.

## METHODOLOGY: COLLECTING THE MATERIAL: (SILICIFIED WOOD)

A reasonable amount of fossil wood specimens were systematically field collected by hand from a wide sampling transect across the deltaic stratigraphic interval exposed from along Fish Creek Wash westward (up-section) to Arroyo Seco del Diablo. Reconnaissance samples were collected from sections 7, 8,9,16,17,18, 19,20, and 21 of T14S, R8E, and sections 12 and 13 of T14S, R7E of the VFCB, during 1990-1993. Outcrops within the overall Fish Creek Badlands have a moderately low relief and consist of dendritically-eroded gullies and mudhills interspersed with ephemeral arroyo drainages. Generally, the sediments alternate between thick, resistant cream-colored quartzose channel sandstones (forming strike ridges) and thick, differentially weathered soft, reddish-brown overbank claybeds.

There seems to be no wood bias in deposition or mode of preservation of the remains. The amount of wood material depends on the concentrations and preservation quality within the sediments. In a few instances, wood samples (fragments and chips) littered the ground as float. Streaks of muted red limonite/hematite, secondarily concentrated staining the surface, indicate where trunk-molds filled with crypto-crystalline iron-bearing minerals, subjected to oxidation, had differentially weathered out. Only those hand-picked individual specimens that occurred in situ, and were large enough by field analyses to reveal the desired visible structural detail on rough transverse sections to be identifiable macroscopically, were collected from the matrix. Recovered samples were primarily tan-gray to brown in coloration. Each sample was appropriately wrapped and bagged in order to avoid damage.

The state of preservation was variable. In many instances, the wood underwent considerable post-burial decomposition before mineralization. Often, the woody grain on the surface that appeared to be bark turned out to be the weathering pattern of the mineralized wood. Also, many recovered samples reveal a certain amount of post-deposition breakage. Size fractionation samples measure from between 1-8 cm in diameter and 5-10 cm long to large logs 0.5 m in diameter and up to 26 m cm in length. To augment this database, additional collections were made from similar wood-bearing deltaic exposures (Diablo formation) in the Borrego Badlands, Volcanic Hills, and in the Yuha Badlands.

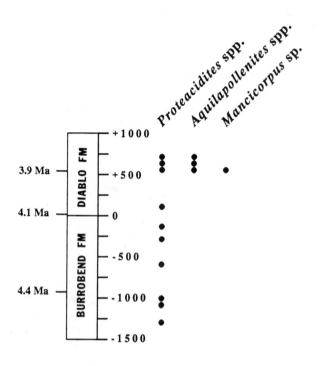

Figure 3. Stratigraphic distribution of reworked Cretaceous pollen in the Vallecito-Fish Creek Basin section. Sample position is given in meters above (positive) or below (negative) the contact between the Burrobend and Diablo formations.

## PREPARATION

In order to be suitable for examination by conventional hand-lens, transmitted-light microscopy or scanning electron microscopy (SEM), each sample was secured in a vice, sectioned at right angles to the grain by a Covington 18-inch diameter vertically-mounted metal core diamond bladed rock saw. The cut produces a transverse surface that could be polished to show, in detail, the fossil internal tree ring growth structural anatomy of the wood. This treatment requires no prior laboratory preparation of the material except removal of the adhering matrix by rinsing with tap water.

Samples revealing adequate preservation for further study were wet sanded on a Covington 16-inch diameter horizontal tapered iron plate abrasive grinding wheel rotating at 200 rpm single-speed. First sandings used a coarse 80 and 220 silicon carbide grit slurry to remove all scractches and to obtain a smooth transverse optical surface. Second sandings using 400 and 600 grit sizes improved the surface texture so that the material was acceptable for preliminary identification and study with a 10-power hand lens or high-powered microscope. In some cases a final polishing (Kenrick and Edwards, 1988) was obtained using a wet cerium oxide paste on the disc. Thin-sections were prepared by grinding on the disc using carborundum powder.

Follow-up investigations may also include using a modified cellulose acetate peel chemical technique (Joy, et al. 1956) whereby the sections are etched in hydrofluoric acid to produce an acetate film impression of the thin- and thick-walled tracheid structures. For this study, 182 samples were collected, 164 were sectioned, and of those only 140 have adequate preservation for determining identification and further study. They consist of 125 dicot specimens from 6 species, 15 monocot specimens, and 1 softwood specimen. These samples are deposited within the ABDSP paleobotanical collections.

## DESCRIPTION AND IDENTIFICATION

All hand-polished transverse sections of the fossil wood samples were preliminary examined with a 10-power hand lens. Hand lens magnifications determined that the majority of samples comprise hardwoods with only a small percentage of monocot representatives. The hardwoods closely resemble wood from extant broad-leaved riparian forest trees of the western United States.

Following the example of Remeika, et al. (1988), only the best preserved samples were sub-divided into the three well-defined groups: semi-ring porous, diffuse porous and ring-porous, based on the macroscpoic growth ring pore change observed across the growth increment on the polished transverse sections.

The classification of angiosperm hardwoods to family or genus mentioned herein follows keys for wood identifcation in Wood Structure and Identification (Core, et al. 1979), Textbook of Wood Technology (Panshin and de Zeeuw, 1980), the Computer-aided OPCN Wood Identification database (Wheeler, et al. 1986; LaPasha and Wheeler, 1987), and the International Association of

Wood Anatomists List of Features (IAWA Committee, 1989).

Affinities to fossil woods were determined by comparison with extant material collected by permission in 1989 along the Big Sur River through Pfeiffer Big Sur State Park and Andrew Molera State Park along the central California coastline south of the Monterey Peninsula.

Further examination was made of selected representative samples using transmission light microscopy on thin-sections with a Zeiss polarizing microscope. A few samples turned out to be poorly preserved although some cell structure was evident, or were so obliterated by crystallization and compression distortion that there was not recognizable wood structure.

The general pattern of cell elements was sufficient enough to allow the placement of each of the hardwood groups into existing families and genera of riparian forest trees extant in the state of California. Characteristic photomicrographs of samples from some of the groups were made. The microstructure of the thin-sections in polarized light was examined and photographed. X-ray crystallographic analysis confirmed some of the results of the petrographic analysis (Remeika, et al. 1988).

**SYSTEMATIC DESCRIPTIONS:**

### CLASS ANGIOSPERMAE
### SUBCLASS DICOTYLEDONES
### ORDER RANALES
### Family LAURACEAE
### Genus UMBELLULARIA Nuttall
### Umbellularia salicifolia (Lesquereux) Axelrod

*Laurus salcifolia* Lesquereux, U.S. Geol. Surv. Terr., Vol. 8: 251, pl. 58, figs. 4,5, 1883.
*Salix?* sp. Knowlton, U.S. Geol. Surv. Ann. Rept. 21 (2): 213-214, pl. 30, fig. 14, 1900.
*Umbellularia dayana* (Knowlton) Berry, U.S. Geol. Surv. Prof. Paper 186: 174, pl. 52, fig. 13, 1937.
*Umbellularia oregonensis* Dorf, Carnegie Inst. Wash. Pub. No. 412: 94, pl. 10, fig. 2, 1930; E. Oliver, Carnegie Inst. Wash. Pub. No. 455, I: 21, 1934; B.W. Brooks, Ann. Carnegie Mus., Vol. 24, pl. 19, fig. 6, 1935; C. Condit, Carnegie Inst. Wash. Pub. No. 476, V: 261, 1938.
*Umbellularia salicifolia* (Lesquereux) Axelrod, Carnegie Inst. Wash. Pub. 516: 102-103, pl. 8, 1939; Remeika, et al., Review of Palaeobotany and Palynology 56:190-191, pl. I, figs. 1-4, 1988.

### Description
**Wood:** Diffuse-porous.
**Growth rings:** Distinct, delineated by a dark band of dense summerwood (fibrous zone).
**Pore arrangement:** Small, barely visible to the naked eye, distant, evely distributed throughout the growth ring, solitary and in multiples of 2 to several, encircled by a sheath of vasicentric parenchyma.
Xylem rays: Fine, not distinct to the naked eye but plainly visible under 10-power magnification. Uniformly spaced.
**Repository:** ABDSP paleobotanical collection.
**Horizon and age:** Diablo formation, Colorado River group. Pliocene.
**Present range:** The extant species *Umbellularia californica* (Hook and Arn.) Nuttall, with which this fossil shows a very close resemblance, is restricted to moist slopes and ravines of the Laguna Mountains in Oriflamme Canyon, and in the Santa Ana Mountains in Trabuco Canyon of southern California, where ground water is available throughout the year. It is common along the seaward and northward slopes of the Coast Ranges of southwestern Oregon and California, residing as the principal understory species with *Lithocarpus densiflorus*, *Acer macrophyllum* and *Alnus rubra* of the redwood forest botanical community. It ranges from a dry climate in southern California to a rainy climate along the central-north coast. It also occurs on lower slopes of the Sierra Nevada from Shasta County southward, in the Upper Sonoran and Transition Zones (Fowels, 1965; Grinnell, 1908; Howell, 1970; Klyver, 1931; Rockwell and Stocking, 1969; and Griffin and Critchfield, 1972).

### Genus PERSEA
### Persea coalingensis (Dorf) Axelrod

*Persea coalingensis* (Dorf) Axelrod, Carnegie Inst. Wash. Pub. 553, V:132, 1944; Condit, Carnegie Inst. Wash. Pub. 553, III:79, pl. 16, fig. 6, 1944; Axelrod, Carnegie Inst. Wash. Pub. 553, VII:201, pl. 38, figs. 1, 3, 1944; Axelrod, Carnegie Inst. Wash. Pub. 590, II:61, pl. 3, figs. 5,6, 1950; Axelrod, Carnegie Inst. Wash. Pub. 590, III:105, pl. 2, fig. 7, 1950; Axelrod, Carnegie Inst. Wash. Pub. 590, IV:148, pl. 3, fig. 9, 1950.

### Description
**Wood:** Diffuse-porous.
**Growth rings:** Indistinct ring margins.

**Pore arrangement:** Pore structure is smaller than in *Umbellularia*. Solitary. Pores are scattered throughout the growth ring.
**Xylem rays:** Uniform, less distinct, and very hard to see with the naked eye.
**Repository:** ABDSP paleobotanical collection.
**Horizon and age:** Diablo formation, Colorado River group. Pliocene.
**Present range:** The fossil avocado *Persea coalingensis* is one of the most commonest Pliocene species in California. The living equivalent, *Persea podadenia* Black, lives in large numbers along stream banks and lake borders of the Sierra Madre Occidental in Sonora and Durango, Mexico.

## ORDER SALICALES
## Family SALICACEAE
## Genus POPULUS Linné
## Populus alexanderi Dorf

*Populus trichocarpa* Torrey and Gray, Hooker Icon: 878, pl. 9, 1852; Axelrod, Univ. Calif. Publ. Geol. Sci., Vol. 60: 65, pl. 8, figs. 1-3, 1966.
*Populus alexanderi* Dorf, Carnegie Inst. Wash. Pub. 412: 75-77, pl. 6, figs. 10, 11; pl. 7, figs. 2, 3, 1930; Chaney, Carnegie Inst. Wash. Pub. 476, IV: 215, pl. 6, figs. I, 5; pl. 7, fig. 2, 1938; Axelrod, Carnegie Inst. Wash. Pub. 516: 92, pl. 6, fig. 6, 1939; Condit, Carnegie Inst. Wash. Pub. 553, II: 40, 1944; Axelrod, Carnegie Inst. Wash. Pub. 553, V: 129, 1944; Axelrod, Carnegie Inst. Wash. Pub. 553, VI: 160, 1944; Axelrod, Carnegie Inst. Wash. Pub. 553, VII: 193, 1944; Axelrod, Carnegie Inst. Wash. Pub. 553, X: 281, pl. 48, fig. 4, 1944; Brown, Journ. Wash. Acad. Sci., Vol. 39: 226, fig. 19, 1949.

### Description
**Wood:** Semi-ring to diffuse-porous.
**Growth rings:** Distinct, but inconspicuous, narrow to wide.
**Pore arrangement:** Small, uniformly and closely spaced in radial multiples of 2 to several. Numerous, decreasing gradually in size through the summerwood.
**Xylem rays:** Very fine, homocellular, not visible to the naked eye. Normally spaced.
**Repository:** ABDSP paleobotanical collection.
**Horizon and age:** Diablo formation, Colorado River group. Pliocene.
**Present range:** *Populus trichocarpa* of the western United States, which closely resembles the fossil species, is a widely distributed tree along riparian woodlands in California, ranging north into Alaska and the northern Rocky Mountains. Along the Pacific coast ranges of Washington and Oregon it forms extensive bottomland forests along stream courses. In southern California and Baja California, it is most abundant along coastal stream banks at low to moderate elevations (Fowells, 1965; Griffin and Critchfield, 1992). This species is also represented by one leaf impression recovered from the Borrego Formation.

## Genus SALIX Linné
## Salix sp. Dorf

*Salix* sp. Dorf, Carnegie Inst. Wash. Pub. 412, 78-80, pl. 8, fig. 3, 1930.

### Description
**Wood:** Semi-ring (semi-diffuse) porous.
**Growth rings:** Inconspicuous, narrow to wide.
**Pore arrangement:** Pores small, closely spaced, numerous, solitary and in multiples of 2 to several. Pores decrease gradually in size through the springwood.
**Xylem rays:** Narrow, heterocellular, not visible to the unaided eye. Normally spaced.
**Repository:** ABDSP paleobotanical collection.
**Horizon and age:** Diablo formation, Colorado River group. Pliocene.
**Present range:** The fossil form may be equivalent to *Salix hesperia* Knowlton or *S. wildcatensis* Axelrod. The latter species occurs locally in the Pliocene of Chula Vista (Axelrod and Deméré, 1984). The living willow descendants *Salix lasiolepis* Bentham and *S. Lasiandra* Bentham range from Alaska and Canada to southern California on the Pacific Coast, and eastward to the southern Rocky Mountains and high plateaus of the southwestern United States. They are common members of riparian vegetation in oak woodlands.

### Salix gooddingii Ball

*Salix truckeana* Chaney, Carnegie Inst. Wash. Pub. 553, XI,:316, pl. 52, figs. 2-6, 1944.
*Salix gooddingii* Ball, Bull. Agric. and Mech. Coll. Texas, 4th ser. V. 2, No. 5, p. 376, 1931.

### Description
Leaves long lanceolate, 5-10 cm long and .4-1.1 cm wide; base rounded cuneate; petiole stout; midrib rather heavy at base of leaf; long, alternate secondaries leaving midrib at angles of from 50° to 80°, curving

upward to the outer third of the blade, then extending nearly parallel to the midrib, each vein running along outside the one above; tertiaries an irregular, angular mesh, becoming percurrent in the outer part of the blade, forming many connections between the long secondaries; teeth small, serrate, evenly and rather widely spaced, inconspicuous in the basal part of the leaf; texture thin to firm (Chaney, 1944).
**Repository:** ABDSP paleobotanical collection.
**Horizon and age:** Borrego Formation, Colorado River group. Plio-Pleistocene.
**Present range:** *Salix gooddingii* Ball has a wide distribution from the Central Valley of California southeastward through the desert region into southern Arizona, New Mexico, and northern Mexico. Associated along streams in association with desert, desert grassland, and oak woodland vegetation.

## ORDER GENTIANALES
## Family OLEACEAE
## Genus FRAXINUS (Tourn.) Linné
## Fraxinus velutina Torrey

*Fraxinus caudata* Dorf, Carnegie Inst. Wash. Pub. 412, I,: 106-107, pl. 13, fig. 8, 1930; Chaney, Carnegie Inst. Wash. Pub. 553, pp. 350-351, 1944.
*Fraxinus velutina* Torrey, in Emory, Notes Military Rec.,: 149, 1848; Axelrod, Univ. Calif. Pub. Geol. Sci. 60: 74, pl. 14, figs. 6,7, 1966.

### Descritpion
**Wood:** Ring-porous.
**Growth rings:** Narrow.
**Pore arrangement:** Springwood pores quite large and visible to the unaided eye, 2-4 pores wide. Transition to summerwood is abrupt. Summerwood zone usually narrow due to growth conditions. Pores widely scattered, solitary, barely visible to the unaided eye, in radial multiples of 2-4.
**Xylem rays:** Not visible to the eye, normally spaced.
**Repository:** ABDSP paleobotanical collections.
**Horizon and age:** Diablo formation, Colorado River group. Pliocene.
**Present range:** *Fraxinus velutina* Torrey is a typical riparian tree that is rare in the fossil record. Today it inhabits stream-border areas o finterior southern California, ranging from oak woodland to the lower margins of the yellow (Ponderosa) pine forest (Axelrod, 1967).

## ORDER JUGLANDALES
## Family JUGLANDACEAE
## Genus JUGLANS
## Juglans pseudomorpha Condit

*Juglans nevadensis* Berry, Journ. Wash. Acad. Sci., Vol. 18: 158, fig. 1, 1928.
*Juglans Beaumontii* Axelrod, Carnegie Inst. Wash. Pub. 476, III: 171, pl. 4, fig. 12, 1937.
*Juglans pseudomorpha* Condit, Carnegie Inst. Wash. Pub. 553, II: 42, pl. 4, figs. 4, 5, 1944; Remeika, et al., Review of Palaeobotany and Palynology 56: 194-195, pl. III, figs. 1-4, 1988.

### Description
**Wood:** Semi-ring porous.
**Growth rings:** Distinct, sharply defined by a dramatic difference in size between pores of the outer summerwood and those in the springwood of adjacent ring.
**Pore arrangement:** Scattered. Springwood pores are readily visible to the naked eye, decreasing gradually in size toward the outer margin of the ring. Solitary, and in radial multiples of 2 to several. Tyloses common.
**Xylem rays:** Fine, indistinct to the naked eye.
**Repository:** ABDSP paleobotanical collection.
**Horizon and age:** Burrobend formation (Deguynos member-Camels Head member) - Diablo formation, Colorado River group. Pliocene.
**Present range:** *Juglans californica* Watson, the living equivalent, is common to the cooler, maritime savanna-oak woodlands of the western Santa Ana Mountains and Puente Hills, where it shows a preference for moist soils along stream courses. It ranges northwestward to the coastal Santa Ynez Mountains of Santa Barbara County in southern California, and extends into the Coast Ranges of central California (Axelrod, 1937).

## ORDER HIPPOCASTANALES
## Family HIPPOCASTANACEAE
## Genus AESCULUS
## Aeculus sp.

### Description
**Wood:** Diffuse-porous.
**Growth rings:** Not visible or barely visible.
**Pore arrangement:** Pores numerous, extremely small and constant in size. Not visible to the unaided eye. Solitary, evenly distributed, and in multiples.

**Xylem rays:** Extremely narrow (uniseriate) and very fine, usually storied forming fine ripple marks on the tangential surface.
**Repository:** ABDSP paleobotanical collections.
**Horizon and age:** Diablo formation, Colorado River group. Pliocene.
**Present range:** Most closely related to *Aesculus californica* (Spach) Nutt., which is the only species of buckeye endemic to the state.. Buckeyes appear along both stream beds and open dry slopes to wooded canyons of the Central Valley and Coast Ranges (Monterey County).

## SUBCLASS MONOCOTYLEDONES
## Family ARACACEAE: CORYPHOIDEAE
## Genus SABAL Adanson ex Guersent
## Sabal sp. cf. S. Miocenica Axelrod

*Sabal miocenica* Axelrod, Carnegie Inst. Wash. Pub. 516:88, pl. 6, fig. 8, 1939; Axelrod, Carnegie Inst. Wash. Pub. 590:197-198, pl. 3, 1950.

### Description

Many fragmentary palm frond specimens are too incomplete to permit accurate description and assignment as a fossil species. The material is, therefore, designated by the non-committal term, monocot.

The permineralized monocotyledonous vegetative remains include sabaloid unarmed petiole and costapalmate leaf architectural impressions (J. Cornett, pers. comm., 1991) that show affinities to the modern species *Sabal uresana* Trelease. These include isolated petiole fragments (which have smooth rather than toothed margins), palm rays which are attached to tapering hastulas, and stem material that are morphologically and anatomically similar to one another, indicating they represent one type. Each stem has numerous collateral vascular bundles scattered throughout the aerenchymatous ground tissue. In tranverse cross-section, the bundles display a range of variation in size and construction and lack diagnostic anatomical characters that can be used to identify them to a specific species. Identification is based on leaf impressions associated with additional monocot remains.

The fossil material does not resemble the extant palm *Washingtonia filifera* Wendland of the Colorado Desert. It does, however, compare favorably to *Sabal miocenica* Axelrod of the Neotropical-Tertiary Geoflora from the Miocene Tehachapi Flora (Axelrod, 1939) and from the middle Pliocene Piru Gorge Flora (Axelrod, 1950) of the western Mojave Desert.

**Present range:** The New World genus *Sabal* is common throughout subtropical forests of the Caribbean Basin, from Mexico to Panama, where rainfall is at least 80 inches annually and where the climate is uniformly warm throughout the year. It is declining in abundance (Gentry, 1942). It also thrives in anthropogenic habitats from Bermuda to Sonora, Mexico, and from Texas to Trinidad (Zona, 1990). The broad-leaved evergreen extant species *Sabal uresana* Trelease is restricted to northwestern Mexico, from the Sonoran Desert coastal plain into the xerophyllous woodlands (foothills) of the Sierra Madre Occidental in Sonora and Sinoloa (Axelrod, 1950) and Chihuahua, Mexico. The nearest occurrence of *S. uresana* is in Carbo, Sonora, Mexico. Occurs in thorn forest and oak forest along watercourses and valleys from sea level to 1500 feet.

### (POLLEN)

86 samples of pollen from 33 localities were collected along Fish Creek Wash, with the sample positions tied to the stratigraphic framework of Downs and White (1968), Opdyke, et al. (1977). In addition to samples collected specifically for palynological analysis, T. Cronin supplied an additional 25 samples originally collected for foraminifera and ostracodes from the VFCB (Quinn and Cronin, 1984). Samples were processed using standard palynological processing procedures (Doher, 1980). Initially samples with promising lithologies from the interval of interest (i.e., Gauss Magnetochron) were processed. These samples produced sparse palynological assemblages or were barren--some were reprocessed using large sample sizes and alternative processing techniques but results were not significantly improved. Out of 86 samples collected, the 25 most promising samples were processed with generally poor results.

### PLIOCENE PALYNOLOGY

Palynological assemblages from the Burrobend and Diablo formations are sparse, with most samples barren of palynomorphs. Marginally adequate assemblages were recovered from the lower part of the section (i.e., the upper part of the Burrobend formation [Deguynos and Camels Head members] and the lower part of the Diablo formation). Preliminary counts were attempted but the samples were too sparse and several slides with numerous transects were necessary to generate a minimally

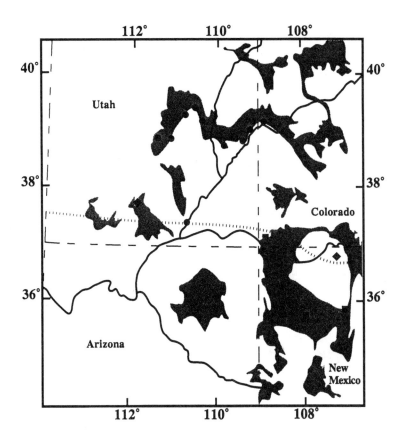

Figure 4. Map of Colorado Plateau with Cretaceous exposures shaded. Black circles are localities containing *Proteacidites* spp. and triprojectate pollen. Black diamonds are localities containing *Proteacidites* spp. but only rare triprojectates. Black squares are localities containing *Proteacidites* spp. and no triprojectates. Dotted line marks inferred southernmost extent of triprojectate pollen during the Cretaceous. The stratigraphic distribution of reworked pollen reflects this paleobiogeographic distribution. Dot near the confluence of the Green and Colorado Rivers marks the position where erosion of the Colorado River reached rocks containing all three genera at about 3.9 Ma.

adequate number of specimens. Because the samples proved to be so sparse, we did not regard the counts as statistically reliable and the raw and relative abundance data are not presented.

All samples from the upper part of the section are barren, including eighteen samples from the Gauss Magnetochron. Because this interval was the focus of this study, we did not process additional samples from the VFCB section. These results indicate that the lithologies through the upper part of this section are unsuitable for paleoclimatic analysis using Pliocene palynomorphs.

Some samples produced assemblages containing three palynomorph categories that are potentially useful for making paleoclimatic interpretations. The first category is algal. Two samples (D7697-83TC38 and D7697-83TC39) from the lowermost Diablo formation produced assemblages that are overwhelmingly dominated by coenobia of *Pediastrum* spp.; the assemblages also contain coenobia of *Scenedesmus* spp. In modern environments *Pediastrum* spp. and *Scenedesmus* spp. occur as associated phytoplankton in freshwater lakes and ponds. Fossil coenobia of these genera are used as paleoenvironmental indicators (Fleming, 1989) and their presence in sediments of the Diablo formation indicates that conditions were wet enough for ponds or lakes to develop. This suggests that conditions in the VFCB at about 4 Ma was wetter than today.

The second category is represented by *Picea*

spp. Pollen of *Picea* spp. in Burrobend formation samples (D7697-83TC19 and D7697-83TC20) have age estimates of 4.42 Ma. Adam (1973) reported *Picea* spp. from Pleistocene sediments near Lake Tahoe and discussed its climatic significance. He concluded that *Picea* spp. is partly dependent on adequate summer rainfall. Its presence in the VFCB section provides additional evidence for wetter conditions that today in ABDSP.

The third category is reworked Cretaceous palynomorphs. The significance of these fossils in terms of paleoclimate will be discussed in the following section.

## REWORKED CRETACEOUS POLLEN

Palynological assemblages from the upper part of the Imperial group (Burrobend formation), and the lower part of the Colorado River group (Diablo formation) contain reworked Cretaceous pollen, spores, and dinoflagellates. A suit of reworked palynomorphs occurs in eleven samples collected. Taxa recognized are: *Proteacidites* spp., *Aquilapollenites* spp., *Mancicorpus* sp., *Tricolpites interangulus*, *Pistillipollenites* sp. cf. *P. mcgregorii*, *Corollina* sp., *Appendicisporites* sp., *Cicastricosisporites* sp., *Camarazonosporites* sp., *Dinogymnium* sp., and *Palaeohystrichophora infusorioides*. In some samples, reworked fossils compose a signifcant part of the assemblage.

The occurence of reworked palynomorphs in the samples are listed in Appendix 1. The suite includes the distinctive Cretaceous genera *Proteacidites*, *Aquilapollenites*, and *Mancicorpus*. *Proteacidites* spp. occurs in most of the Burrobend and Diablo samples that yield palynomorphs; triprojectate pollen (*Aquilapollenites* spp. and *Mancicorpus* sp.) first appears in the upper part of the section, in the lower part of the Diablo formation. The first appearance of triprojectate pollen is in a sample with abundant reworked Cretaceous fossils. Two lines of evidence suggest that this first appearance is not related to the abundance of reworked palynomorphs. First, several samples from the lower part of the section also contain abundant reworked Cretaceous palynomorphs but lack triprojectate pollen (e.g., D7697-83TC30). Second, samples above the first appearance horizon that contain meager amounts of Cretaceous pollen also contain triprojectate pollen. This distribution suggests that *Proteacidites* spp. was reworked into the VFCB before *Aquilapollenites* spp. and *Mancicorpus* sp.

Merriam and Bandy (1965) demonstrated that reworked Cretaceous foraminifera in the Burrobend and Diablo formations originated from the upper Cretaceous Mancos Shale. The Mancos Shale and its stratigraphic equivalents are exposed on the Colorado Plateau and are known to contain abundant Cretaceous palynomorphs (Thompson, 1972; Franczyk, et al. 1990; Cushman and Nichols, 1992). The available sedimentological and paleontological evidence indicates that reworked Cretaceous palynomorphs in the VFCB section were also derived from the Mancos Shale and its equivalents on the Colorado Plateau.

The stratigraphic and paleobiogeographic ranges of fossil pollen in the Cretaceous of the western interior of North America provide insight into the stratigraphic distribution of reworked Cretaceous pollen in the Pliocene of ABDSP. In the western interior, the stratigraphic range of *Proteacidites* spp. is from the Coniacian to the Maastrichtian and the range of triprojectate pollen is from Campanian to the Maastrichtian (Nichols, et al. 1982). In addition to being stratigraphically restricted, triprojectate pollen was also restricted biogeographically. *Aquilapollenites* spp. and *Mancicorpus* sp. have been documented from southwestern Colorado and southeastern Utah (Lohrengel, 1969; May, 1972; Fouch, et al. 1983; Franczyk, et al. 1990; Cushman and Nichols, 1992). They become less abundant farther south, and are extremely rare or absent in northern Arizona and northern New Mexico (Anderson, 1960; Thompson, 1972; Tschudy, 1973; Jameossanaie, 1987). Their southernmost extent forms a line that approximately parallels the Arizona-Utah and Colorado-New Mexico borders (Figure 4). This paleobiogeographic boundary roughly divides the Colorado Plateau into a northern half (southeastern Utah and southwestern Colorado) and a southern half (northeastern Arizona and northwestern New Mexico). The Mancos Shale and its equivalents in the northern part of the Colorado Plateau contain triprojectate pollen (*Aquilapollenites* spp. and *Mancicorpus* sp.); the Mancos and its equivalents in the southern half lack triprojectate pollen. In contrast, *Proteacidites* spp. is a common constituent of upper Cretaceous sedimentary rocks throughout the western interior, including Arizona and New Mexico (Anderson, 1960; Orlansky, 1971; Tschudy, 1973; Jameossanaie, 1987; Nichols, et al. 1982).

The relative stratigraphic distribution of reworked *Proteacidites* spp. and triprojectate pollen in Pliocene rocks of the VFCB section reflects the pattern of erosion on the Colorado Plateau. If the pattern of erosion was uniform across the plateau and exposed younger rocks before older rocks, then reworked triprojectate pollen would be present in the lowest Pliocene assemblages containing reworked Cretaceous palynomorphs. However, their actual distribution suggests that erosion on the plateau during the Pliocene proceeded from south to north rather than uniformly from younger to older rocks. Initial erosion of Cretaceous rocks containing *Proteacidites* spp. from the southern Colorado Plateau preceded erosion of Cretaceous rocks containing both *Proteacidites* spp. and triprojectate pollen from the northern Colorado Plateau.

**EROSIONAL HISTORY**

The stratigraphic distribution of Cretaceous pollen in the Burrobend and Diablo formations can be used to aid in reconstructing the erosional history of the Colorado Plateau. Erosion of rocks containing *Proteacidites* spp. began by 4.5 Ma. This is recorded in the lower part of the Pliocene sequence where *Proteacidites* spp. first appears without triprojectate pollen in a sample from the Burrobend formation with an estimated age of 4.49 Ma. At this time, erosion apparently exposed the Mancos Shale or its stratigraphic equivalents in the southern part of the Colorado Plateau, allowing transport and deposition of specimens of *Proteacidites* spp. The presence of *Proteacidites* spp. and absence of triprojectate pollen in the upper part of the Burrobend and lowermost part of the Diablo formations provide evidence that rocks containing triprojectate pollen were not contributing sediment to the Colorado River from the northern part of the plateau at this time.

Erosion did not reach rocks on the Colorado Plateau that contained *Proteacidites* spp., *Aquilapollenites* spp., and *Mancicopus* sp. until about 3.9 Ma. This is indicated by the first appearance of triprojectate pollen in a sample from the lower part of the Diablo formation (Figure 3). This record provides timing constraints on the sequence of the erosional history of the Colorado River. This indicates that at 3.9 Ma approximately 2000+ m of Jurassic, Cretaceous, and Tertiary rocks were still present in the northern part of the Colorado Plateau. The Colorado River eroded triprojectate-bearing Mancos Shale rocks and transported the detritus from this erosion to the northern Gulf of California where it was deposited in the ABDSP region. Based on radiometric ages of lava flows across its course, the Colorado River probably reached a position close to its present level by approximately 1.2 Ma (Hamblin, 1989; Elston and Young, 1991). These interpretations suggest that much of the Grand Canyon was cut relatively recently, during the Pliocene.

**PALEOCLIMATIC IMPLICATIONS**

The data on paleobotanical woods and reworked pollen have important implications for Pliocene climate in the southwestern part of the United States. The erosional history of the Colorado Plateau indicates the Colorado River eroded and exported a large volume of rock from the Plateau between 3.9 and 1.2 Ma. This requires discharge of volumes for the Colorado River significantly higher than today. Higher discharge volumes implies precipitation levels on the Colorado Plateau much greater than today, thus suggesting a much wetter climate.

The stratigraphic distribution of reworked fossils in Pliocene sediments of the VFCB suggests that erosion of Cretaceous rocks containing *Proteacidites* spp., but lacking *Aquilapollenties* spp. and *Mancicorpus* sp., began at least by 4.5 Ma. This is reflected in the lower part of the sequence where *Proteacidites* spp. first appears without triprojectate pollen. At this time, erosion on the Colorado Plateau had exposed the Mancos Shale in Arizona and Mexico. In the northern part of the plateau, Mancos Shale containing triprojectate pollen is not exposed at this time. Erosion into rocks containing triprojectate pollen began by 3.9 Ma, as reflected in the first appearance of these forms in the Diablo formation. This suggest the Mancos Shale was not exposed in southern Utah and southern Colorado until about 3.9 Ma. The timing of this erosional event indicates that a considerable amount of erosion on the Colorado Plateau occurred after 3.9 Ma. This is consistent with Lucchitta's (1972, 1979, 1987) conclusions that cutting of the Grand Canyon occurred during the Pliocene.

The data corroborates results reported from significant studies on the Pliocene climatic evolution of the southwestern United States. Thompson (1991) summarized paleoclimatic proxy data from the western interior of the United States showing levels of effective moisture were higher than modern conditions. Smith, et al. (1993) concluded, on the basis of stable-isotopic compositions of paleosol carbonates from Arizona, that a Pliocene wet period occurred in western North America. Winograd, et al. (1985) suggests uplift of the Sierra Nevada and Transverse Ranges during the Pleistocene blocked inland-bound Pacific storm systems that provided moisture to the southwestern United States during the Pliocene. Remeika, et al. (1988) documented the presence of temperate tree taxa in the VFCB section during the Pliocene that are indicative of a more temperate climate with maritime influences from the Gulf of California and the Pacific Ocean, with dominance of winter precipitation.

## CONCLUSIONS

The CLF recovered from the Diablo formation indicates that the Pliocene climate in the ABDSP region was much wetter and cooler than today. The stratigraphic distribution of reworked Cretaceous *Proteacidites* spp. and *Aquilapollenites* spp. in Pliocene rocks from the VFCB reflects the paleobiogeographic distribution of these fossils in upper Cretaceous rocks of the Colorado Plateau. The timing and appearances of reworked Cretaceous pollen in the VFCB indicates that most of the erosion of the Colorado Plateau and incision of the Grand Canyon occurred during the Pliocene. The erosion and transport of large amounts of material from the Colorado Plateau requires higher discharge volumes from the Colorado River than its rate of discharge (in historical times), and also suggests the Pliocene climate of the southwestern United States was much wetter than today.

## ACKNOWLEDGMENTS

This work was supported, in part, by the PRISM Project of the U.S. Geological Survey Global Change Program and the James Brainerd Memorial Grant. We thank Dr. Wilfred A. Côté, Jr., Scott Zona, Irwin Fischbein, Steven Fischbein, Bob Thompson, Fred Peterson, Doug Nichols, Ivo Lucchitta, Karen Franczyk, Bob Cushman, and Tom Ager for informative discussions regarding reworking of sediments and related implications. We also express appreciation to Michael J. Arct at Loma Linda University for assistance in lapidairy work.

Appendix 1. Reworked palynomorphs collected from deltaic sediments of the Vallecito-Fish Creek Basin section. Presence of a species is indicated by an "X".

| APPENDIX 1 (PART 1) | | | | |
|---|---|---|---|---|
| Sample Number | D7697-83TC16 | D7697-83TC19 | D7697-83TC20 | D7697-83TC23 |
| Position (meters) | -1261 | -1015 | -999 | -903 |
| Age | 4.49 | 4.42 | 4.42 | 4.39 |
| *Proteacidites* | X | X | X | |
| *Aquilapollenites* | | | | |
| *Mancicorpus* | | | | |
| *Tricolpites interangulus* | | | | |
| *Pistillipollenites* | | | | |
| *Corollina* | | X | | |
| *Appendicisporites* | | X | | |
| *Cicatricosisporites* | | X | | X |
| *Camarazonosporites* | | | | |
| *Dinogymnium* | | X | | |
| *Palaeohystrichophora infusorioides* | | | | |

| APPENDIX 1 (PART 2) | | | | |
|---|---|---|---|---|
| Sample Number | D7697-83TC26 | D7697-83TC30 | D7697-83TC33 | D7697-83TC38 |
| Position (meters) | -679 | -279 | -167 | 95 |
| Age | 4.33 | 4.22 | 4.19 | 4.11 |
| *Proteacidites* | X | X | X | X |
| *Aquilapollenites* | | | | |
| *Mancicorpus* | | | | |
| *Tricolpites interangulus* | | | | |
| *Pistillipollenites* | | | | |
| *Corollina* | | X | | |
| *Appendicisporites* | | X | | |
| *Cicatricosisporites* | | | X | |
| *Camarazonosporites* | | | | |
| *Dinogymnium* | | | | |
| *Palaeohystrichophora infusorioides* | | | | |

| APPENDIX 1 (PART 3) | | | |
|---|---|---|---|
| Sample Number | D7868-FF11 | D7868-FF12 | D7868-FF14 |
| Position (meters) | 633 | 683 | 713 |
| Age | 3.92 | 3.90 | 3.89 |
| Proteacidites | X | X | X |
| Aquilapollenites | X | X | X |
| Mancicorpus | X | | |
| Tricolpites interangulus | | | X |
| Pistillipollenites | X | | |
| Corollina | X | | |
| Appendicisporites | | | |
| Cicatricosisporites | X | | |
| Camarazonosporites | | X | |
| Dinogymnium | | X | |
| Palaeohystrichophora infusorioides | X | | |

## REFERENCES CITED

Axelrod, D.I. 1937. A Pliocene flora from the Mount Eden beds, southern California. Carnegie Institute of Washington Publications 476: 125-183.

----- 1939. A Miocene flora from the western border of the Mojave Desert. Carnegie Institute of Washington Publications 516: 1-129.

----- 1950. The Anaverde flora of southern California. Carnegie Institute of Washinton Publications 590: 119-158.

----- 1958. Evolution of the Madro-Tertiary Geoflora. Bot. Rev. 24: 433-509.

----- 1967. The Pleistocene Soboba flora of southern California. University of California Publications in Geological Sciences 60: 1-108.

Axelrod, D.I., and T.A. Deméré 1984. A Pliocene flora from Chula Vista, San Diego County, California. Transactions of the San Diego Society of Natural History 20(15):277-300.

Chaney, R.W. 1944. The Dalles flora. Carnegie Institute of Washington Publications 553: 316.

Core, H.A., W.A. Côté, and A.C. Day 1979. Wood structure and identification. Second Edition. Syracuse University Press, Syracuse, New York, 182 p.

Dibblee, T.W., Jr. 1954. Geology of the Imperial Valley region, California. In Geology of Southern California, edited by R.H. Jahns. Calif. Div. Mines and Geology Bull. 170 (2,2):21-81.

Downs, T., and J.A. White 1968. A vertebrate faunal succession in superposed sediments from late Pliocene to middle Pleistocene in California. In Tertiary/Quaternary Boundary, International Geological Congress 23, Prague 10:41-47.

Fleming, R.F. 1993a. Palynological data from the Imperial and Palm Spring Formations, Anza-Borrego Desert State Park, California. U.S. Geological Survey Open-File Report 93-678:1-29.

----- 1993b. Cretaceous pollen and Pliocene climate. The American Association of Stratigraphic Palynologists Annual Meeting, Program and Abstracts, 26:24.

Fleming, R.F., and P. Remeika 1994. Pliocene climate of the Colorado Plateau and age of the Grand Canyon: evidence from Anza-Borrego Desert State Park, California. 4th National Park Service Conference on Fossil Resources, Colorado Springs, Colorado (unpaginated).

Fürsich, F.T., and K.W. Flessa 1987. Taphonomy of tidal flat molluscs in the northern Gulf of California: paleoenvironmental analysis despite the perils of preservation. Palaios 2: 543-559.

Gentry, H.S. 1942. Rio Mayo plants. Carnegie Institute of Washington Publication 527:1-328.

Griffin, J.R., and W.B. Critchfield 1972. The distribution of forest trees in California. USDA Gorest Service Research Paper PSW-82:1-118.

Grinnel, J. 1908. The biota of the San Bernardino Mountains. Univ. Calif. Pub. Zool. 5:1-170.

Hanna, G.D. 1926. Paleontology of Coyote Mountains, Imperial County, California. Calif. Acad. of Science Proceedings, 4(14):51-186.

Howell, J.T. 1970. Marin Flora. 2nd Ed. Univ. Calif. Press, Berkeley and Los Angeles, 1-366.

IAWA Committee, 1989. IAWA list of microscopic features for hardwood identification. IAWA Bulletin 10:219-332.

Johnson, N.M., C.B. Officer, N.D. Opdyke, G.D. Woodard, P.K. Zeitler, and E.H. Lindsay 1983. Rates of late Cenozoic tectonism in the Vallecito-Fish Creek Basin, western Imperial County, California. Geology 11:664-667.

Joy, K.W., A.J. Willis, and W.S. Lacey 1956. A rapid cellulose peel technique in paleobotany. Ann. Bot. 20:635-637.

Kendrick, P., and D. Edwards 1988. The anatomy of lower Devonian *Gosslingia breconensis* Heard based on pyritized axes, with some comments on the permineralized process. Bot. J. Linn. Soc. 97:95-123.

Kidwell, S.M. 1988. Taphonomic comparison of passive and active continental margins: Neogene shell beds of the Atlantic Coastal Plain and northern Gulf of California. Palaeogeography, Palaeoclimatology, Palaeoecology, 63:201-223.

Klyver, F.D. 1931. Major plant communities in a transect of the Sierra Nevada Mountains of California. Ecology 12:1-17.

LaPasha, C.A., and E.A. Wheeler 1987. A microcomputer based system for computer-aided wood identification. IAWA Bulletin 8(4):347-354.

Lucchitta, I. 1987. The mouth of the Grand Canyon and edge of the Colorado Plateau in the upper Lake mead area, Arizona. In Rocky Mountain Section Centennial Field Guide, edited by S.S. Beus. Geological Society of America 2:365-370.

Meldahl, K.H. 1993. Geographic gradients in the formation of shell concentrations: Plio-Pleistocene marine deposits, Gulf of California. Palaeogeography, Palaeoclimatology, Palaeoecology 101(1,2):1-25.

Merriam, R.H., and O.L. Bandy 1965. Source of upper Cenozoic sediments in the Colorado delta region. Journal of Sedimentary Petrology 35:911-916.

Muffler, L.P.J., and B.R. Doe 1968. Composition and mean age of detritus of the Colorado River delta in the Salton Trough, southeastern California. Journal of Sedimentary Petrology 38:384-399.

Opdyke, N.D., E.H. Lindsay, N.M. Johnson, and T. Downs 1977. The paleomagnetism and magnetic polarity stratigraphy of the mammal-bearing section of Anza-Borrego State Park, California. Quaternary Research 7:316-329.

Panshin, A.J., and C. deZeeuw 1980. Textbook of wood technology. Fourth Edition. McGraw-Hill Publishing Company, New York, 722 p.

Pinault, C.T. 1984. Structure, tectonic geomorphology and neotectonics of the Elsinore Fault Zone between Banner Canyon and the Coyote Mountains, southern California. Master of Science Thesis, Department of Geology, San Diego State University, California, 231 p.

Remeika, P. 1991a. Formational status of the Diablo redbeds; differentiating between Colorado River affinities and the Palm Spring Formation. Abstracts with Programs. Symposium on the Scientific Value of the Desert, Anza-Borrego Foundation, Borrego Springs, p. 12.

----- 1991b. Additional contributions to the Neogene paleobotany of the Vallecito-Fish Creek Basin and vicinity, Anza-Borrego Desert State Park, California. Abstracts with Programs. Symposium on the Scientific Value of the Desert. Anza-Borrego Foundation, Borrego Springs, p. 13.

----- 1992. Preliminary report on the stratigraphy and vertebrate fauna of the middle Pleistocene Ocotillo formation, Borrego Badlands, Anza-Borrego Desert State Park, California. Mojave Desert Quaternary Research Symposium, Abstracts of Proceedings, San Bernardino County Museum Association Quarterly 39(2):25-26.

----- 1994. Lower Pliocene angiosperm hardwoods from the Vallecito-Fish Creek Basin, Anza-Borrego Desert State Park, California: deltaic stratigraphy, paleoclimate, paleoenvironment, and phytogeographic significance. Abstracts of Proceedings. 1994 Mojave Desert Quaternary Research Symposium. San Bernardino County Museum Association Quarterly 41(3):26-27.

Remeika, P., I.W. Fischbein, and S.A. Fischbein 1988. Lower Pliocene petrified wood from the Palm Spring Formation, Anza-Borrego Desert State Park, California. Review of Palaeobotany and Palynology 56:183-198.

Remeika, P., and G.T. Jefferson 1993. The Borrego Local Fauna: revised basin-margin stratigraphy and paleontology of the western Borrego Badlands, Anza-Borrego Desert State Park, California. In Ashes, Faults and Basins, edited by R.E. Reynolds and J. Reynolds. San Bernardino County Museum Association Special Publication 93(1):90-93.

Remeika, P., and L. Lindsay 1992. Geology of Anza-Borrego: Edge of Creation. Dubuque, Kendall/Hunt Publishing Company. 208p.

Reynolds, R.E., and P.Remeika 1993. Ashes, Faults and Basins: the 1993 Mojave Desert Quaternary Research Center field trip. In Ashes, Faults and Basins, edited by R.E. Reynolds and J. Reynolds. San Bernardino County Museum Association Special Publication 93-1:3-33.

Rockwell, J.A., and S.K. Stocking 1969. Checklist of the flora, Sequoia and Kings Canyon National Parks. Sequoia Nat. Hist. Assoc., Three Rivers, California, 1-97.

Schaeffer, G.C. 1857. Description of the structure of fossil wood from the Colorado Desert. In 1855 Report of the Explorations and Survey for a Railroad Route from the Mississippi to the Pacific, U.S. Congress, 2nd Session, Senate Executive Document 91, 5(2):338-339.

Stout, B.W., and P. Remeika 1991. Status report on three major camelid tracksites in the lower Pliocene delta sequence, Vallecito-Fish Creek Basin, Anza-Borrego Desert State Park, California. Abstracts with Programs. Symposium on the Scientific Value of the Desert. Anza-Borrego Foundation, Borrego Springs, p. 9.

Stump, T.E. 1972. Stratigraphy and paleontology of the Imperial Formation in the western Colorado Desert. Master of Science Thesis, Department of Geology, San Diego State University, 128p.

Tarbet, L.A. 1951. Imperial Valley. Am. Assoc. Pet. Geol. Bull. 35:260-263.

Tarbet, L.A., and W.H. Holman 1944. Stratigraphy and micropaleontology of the west side of Imperial Valley. Am. Assoc. Pet. Geol. Bull. 28:1781-1782.

Wheeler, E.A., R.G. Pearson, C.A. LaPasha, T. Zack, and W. Hatley 1986. Computer-aided wood identification. The North Carolina Agricultural Research Service Bulletin 474:1-160.

White, J.A., E.H. Lindsay, P. Remeika, B.W. Stout, T. Downs, and M. Cassiliano 1991. Society of Vertebrate Paleontology field trip guide to the Anza-Borrego Desert. Society of Vertebrate Paleontology fifty-first Annual Meeting and Field Trip Guide, San Diego, 23p.

Winker, C.D., and S.M. Kidwell 1986. Paleocurrent evidence for lateral displacement of the Pliocene Colorado River delta by the San andreas Fault system, southeastern California. Geology 14:788-791.

Woodard, G.D. 1963. The Cenozoic succession of the western Colorado Desert, San Diego and Imperial Counties, Southern California. Doctoral Dissertation, University of California, Berkeley, 173p.

Woodring, W.P. 1931. Distribution and age of the Tertiary deposits of the Colorado Desert. Carnegie Institute of Washington Publication, 418:1-25.

Zona, S. 1990. A monograph of *Sabal* (Arecaceae: Coryphoideae). ALISO 12(4):583-666.

# FOSSIL VERTEBRATE FAUNAL LIST FOR THE VALLECITO-FISH CREEK AND BORREGO-SAN FELIPE BASINS, ANZA-BORREGO DESERT STATE PARK AND VICINITY, CALIFORNIA

**Paul Remeika, George T. Jefferson, and Lyndon K. Murray**
California Park Service, Colorado Desert District, Anza-Borrego Desert State Park, 200 Palm Canyon Drive, Borrego Springs, California 92004

## INTRODUCTION

Within the Vallecito-Fish Creek Basin (VFCB) of southern Anza-Borrego Desert State Park (ABDSP), vertebrate fossils have been recovered from four major depositional sequences: **(1)** The late Miocene/early Pliocene, marine and deltaic deposits in the Imperial group (Imperial Formation of other authors) yield marine fish and marine and terrestrial mammals. Hemphillian age vertebrates occur in both the lower, Latrania formation (new informal name, this guidebook) (= Latrania Sands or Lycium member of previous authors) and the upper, Burrowbend formation (= Coyote Mountain Clays or Deguynos member of previous authors). **(2)** the early Pliocene ancestral Colorado River delta-plain deposits, of the Diablo formation (informal name of Remeika, 1991) with interbedded alluvial tongues of the basin-margin Olla member (informal name of Winker, 1987) of the Canebrake Conglomerate (Dibblee, 1954), yield terrestrial mammals; **(3)** the late Pliocene through mid-Pleistocene, basin-margin alluvial fan and floodplain deposits of the Palm Spring Formation (Woodring, 1931) have produced freshwater fish, amphibians, reptiles, birds, and land mammals, which are also found in; **(4)** the mid through late Pleistocene, alluvial fan, floodplain and paralymnitic deposits of the Ocotillo formation (informal name of Remeika, 1992) (= Ocotillo Conglomerate or upper Palm Spring Formation of some authors). In the VFCB, the Palm Spring Formation conformably overlies the Diablo formation which, in turn, conformably overlies the uppermost unit (Burrobend formation) of the Imperial group.

Although the age of the Imperial group is not firmly established, previous age estimates range from 7.6 to 3.8 Ma with most falling between 5.5 and 4.5 Ma. Paleomagnetic correlations suggest an age of 4.3 Ma, within the middle of the Gilbert (Reversed) Magnetochron. A discussion of the age, stratigraphy and vertebrate paleontology of the Imperial group is presented by Deméré (1993) and Remeika (field trip road log, this guidebook).

Within the Burrobend, Diablo and Palm Spring formations, the Loop Wash, Layer Cake, Arroyo Seco and Vallecito Creek paleofaunas occur in a superposed, conformable biochronologic sequence. The Loop Wash assemblage is late Hemphillian in age and comprises the oldest terrestrial vertebrate material recovered from ABDSP (Burrobend formation/lower half of Diablo formation). This assemblage underlies the early Blancan Layer Cake Local Fauna (LF), and is presumed to fall within the middle/ upper Gilbert Reversed-Polarity Magnetochron, about 4.0 Ma. Strata (Diablo formation) which yield the Layer Cake LF fall within the Gilbert (Reversed) Magnetochron, and range from the base of the Conchiti Magnetosubchron to the Gilbert/Gauss Magnetochron boundary, or between about 3.9 and 3.4 Ma. The Arroyo Seco LF (Diablo formation) is mid-Blancan in age and directly follows the Layer Cake LF at about 3.4 Ma. The transition between the Arroyo Seco LF and the following Vallecito Creek LF occurs before the end of the Gauss Normal-Polarity Magnetochron at $2.3 \pm 0.4$ Ma. The date is based on volcanic ash fission track analyses (Johnson, et al.1983). Strata (Palm Spring Formation) which produce the late Blancan through mid-Irvingtonian age Vallecito Creek LF fall within the Matuyama Reversed-Polarity Magentochron, and range from its base to the Jaramillo Magnetosubchron, or from about 2.3 to 0.9 Ma. For a discussion of the VFCB faunal sequence and chronology, see Downs and White (1968), Kurtén and Anderson (1980), Lindsay, et al. (1987), and Lundelius, et al. (1987).

The mid-Irvingtonian and possibly early Rancholabrean age Borrego LF, from the Borrego and Coyote Badlands of the Borrego-San Felipe Basin (northern ABDSP), ranges in age from about 1.25 to as young or younger than 0.37 Ma. It overlaps chronologically with the range of the Vallecito Creek LF. The terrestrial and lacustrine Ocotillo formation, which yields the Borrego LF, falls within the Matuyama (Reversed) and Brunhes (Normal) Magnetochrons. The stratigraphic section also includes the Bishop Tuff dated at about 0.74 Ma (Rymer 1991). Remeika and Jefferson (1993) provide a further treatment of the Borrego LF. A single ichnotaxon has been recovered from late Pleistocene, presumably Rancholabrean age terrestrial deposits that overlie the Ocotillo formation.

In 1981, cloven-hoofed (llama-sized) tetrapod footprints were discovered at Camel Ridge in the Vallecito Badlands (Stout and Remeika, 1991). Since then, additional ichnites have been located, restricted to the stratigraphically continuous (Hemphillian-Blancan) Canebrake Conglomerate I (Olla member) and Diablo formation basin-margin interbeds in the VFCB. Ephemeral ichnites include various-sized camelids, plus canid, felid, equid and avian footprints and trackways. In the BSFB, Irvingtonian age ichnites (camelids and mammoth?) have recently been found in the Ocotillo formation.

## ACKNOWLEDGMENTS

This detailed vertebrate faunal list has been compiled from existing literature, from collections records of the Natural History Museum of Los Angeles County and the Colorado Desert District, ABDSP, and archived personal communications including E. Anderson, T. Cavender, T. Downs, H. Howard, B. Kurtén, G. Miller, J. White, and R. Zakrzewski. We are grateful to C. Bell and D. Long of U.C. Berkeley for providing assistance in literature searches.

Explanation: * = type specimens from ABDSP; † = extinct taxon; the following local fauna and assemblage abbreviations are listed in ascending stratigraphic order, CM = Coyote Mountains marine assemblage, LW = Loop Wash assemblage, LC = Layer Cake LF, AS = Arroyo Seco LF, VC = Vallecito Creek LF, BL = Borrego LF, and BQ = Borrego Badlands late Quaternary assemblage. Taxonomic names have been revised to conform with current usage.

## SYSTEMATIC LIST

### Class Chrondricthyes

Order Galeomorpha
    Family Cetorhinidae
        *Cetorhinus* sp. (basking shark) CM
    Family Carcharinidae
        *Carcharodon carcharias* (Linnaeus, 1758) (white shark) CM
        *Hemipristus serra* Agassiz, 1843 (shark) CM
Order Batoidea
    Family Myliobatidae
        Gen. et sp. indet. (eagle ray) LC

### Class Osteichthyes

    Order indet. (bony fish) AS, LC
Order Salmoniformes
    Family Salmonidae
        ? *Salmo* sp. (salmon) VC
Order Cypriniformes
    Family Catostomidae
        Gen. et sp. indet. (minnow, stickleback, or sucker) AS, VC, BL
        *Xyrauchen* sp. cf. *X. texanus* (Abbott, 1861) (humpback sucker) VC, BL
Order Culpeiformes
    Family Culpeidae
        Gen. et sp. indet. (herring) AS

Order Syngnathiformes
    Family Syngnathidae
        *Hipposyngnathus imporcitor* Fritzsche, 1980 (sea horse) CM

## Class Amphibia

Order Anura
    Family indet. (frog or toad) AS
    Family Bufonidae
        *Bufo* sp. (toad) VC
    Family Ranidae
        *Rana* sp.

## Class Reptilia

Order Testudines
    Family Kinosternidae
        *Kinosternon* sp. (mud turtle) VC
    Family Testudinidae
        Family indet. (tortoise) BL
        *Geochelone* sp. (giant tortoise) AS, VC
        *Gopherus agassizii* (Cooper, 1863) (desert tortoise) LC, VC
    Family Chelonidae
        Gen. et sp. indet. (sea turtle) CM
    Family Emydidae
        *Clemmys* sp. (pond turtle) AS
        *Clemmys marmorata* (Baird and Girard, 1852) (western pond turtle) VC, BL
        *Trachemys scripta* (Scheopff, 1792) (common slider) VC
Order Squamata
  Suborder Lacertilia
    Family Iguanidae
        *Callisaurus* sp. (zebra-tailed lizard) VC
        *Crotaphytus* sp. (collared lizard) AS
        *Dipsosaurus dorsalis* Hallowell, 1854 (desert iguana) AS
        *Gambelia corona* Norell, 1989 * † (crowned leopard lizard) AS, VC
        *Iguana iguana* (Linnaeus, 1758) (green iguana) AS, VC
        *Phrynosoma* sp. (horned lizard) AS
        *Phrynosoma anzaense* Norell, 1989 * † (Anza horned lizard) AS, VC
        *Pumilia novaceki* Norell, 1989 * † (Novacek's small iguana) LC, AS
        *Sceloporus* sp. A (spiny lizard) AS, VC
        *Sceloporus* sp. B (spiny lizard) AS, VC
        *Sceloporus magister* Hallowell, 1854 (desert spiny lizard) VC
        *Uta stansburiana* Baird and Girard, 1852 (side-blotched lizard) VC
    Family Teiidae
        *Ameiva* sp. or *Cnemidophorus* sp. (ground lizard or whiptail) LC, AS, VC
        *Cnemidophorus tigris* Baird and Girard, 1852 (western whiptail) VC
    Family Scincidae
        *Eumeces* sp. (skink) AS, VC
    Family Xantusiidae
        *Xantusia downsi* Norell, 1989 * † (Downs' night lizard) AS, VC
        *X. vigilis* Baird, 1859 (desert night lizard)
    Family Anguidae
        *Gerrhonotus multicarinatus* (Blainville, 1853) (southern alligator lizard) AS, VC
  Suborder Serpentes

Family Colubridae
- *Hypsiglena* sp. (night snake) VC
- *Lampropeltis getulus* (Linnaeus, 1766) (common king snake) VC
- *Masticophis flagellum* (Shaw, 1802) (coachwhip snake) VC
- *Thamnophis* sp. (garter snake) VC

Family Crotalidae
- *Crotalus* sp. (rattlesnake) VC

## Class Aves

Order Gaviformes
    Family Gaviidae
- *Gavia* sp. (loon) VC

Order Podicipediformes
    Family Podicipedidae
- *Podiceps* sp. (grebe) VC
- *Podiceps nigricollis* Brehm, 1831 (eared grebe) VC

Order Procellariiformes
    Family Procellariidae
- *Puffinus* sp. (shearwater) ?LW

Order Pelecaniformes
    Family Pelecanidae
- *Pelicanus* sp. (pelican) ?LW

Order Ciconiiformes
    Family Teratornithidae
- *Teratornis* sp. † (teratorn) VC
- *Teratornis incredibilis* Howard, 1963 † (incredible teratorn) LC, VC

    Family Vulturidae (Cathartidae)
- *Cathartes aura* Linnaeus, 1758 (Turkey vulture) BL
- *Gymnogyps* sp. (condor) VC

Order Pheonicopteriformes
    Family Pheonicopteridae
- *Phoenicopterus* sp. (flamingo) BL

Order Anseriformes
    Family indet. (duck) BL
    Family Anatidae
- *Aix* sp. ? *A. sponsa* (Linnaeus, 1758) (wood duck) VC
- *Anas* sp. ? *A. acuta* Linnaeus, 1758 (pintail duck) VC
- *A. clypeata* Linnaeus, 1758 (shoveller duck) VC
- *Anser* sp. (large goose) VC
- *Branta* sp. cf. *B. canadensis* (Linnaeus, 1758) (Canada goose) VC
- *Brantadorna downsi* Howard, 1963 * † (Downs' gadwall duck) VC
- *Bucephala fossilis* Howard, 1963 * † (fossil goldeneye) VC
- *Chen rossii* (Cassin, 1861) (Ross' goose) or *Branta bernicula* (Linnaeus, 1758) (brant)
- *Cygnus* sp. ? *C. paloregonus* Cope, 1878 † (Oregon swan) VC
- ? *Lophodytes cucullatus* (Linnaeus, 1758) (hooded merganser) VC
- *Melanitta* sp. ? *M. persipicillata* (Linnaeus, 1758) (surf scoter) VC
- *Mergus* sp. (merganser) VC
- *Oxyura bessomi* Howard, 1963 * † (Bessom's stiff-tailed duck) VC
- *O. jamaicensis* (Gmelin, 1789) (ruddy duck) VC

Order Accipitriformes
    Family Accipitridae
- Gen. et sp. indet. (hawk) VC

    *Accipiter cooperii* (Bonaparte, 1828) (Cooper's hawk) BL
    *A. striatus* (Vieillot, 1807) (sharp-shinned hawk) BL
    *Aquila* sp. ? *A. chrysaëtos* (Linnaeus, 1758) (golden eagle) VC, BL
    *Buteo* sp. cf. *B. jamaicensis* (Gmelin, 1788) (red-tailed hawk) BL
    *Buteo* sp. ? *B. lineatus* (Gmelin, 1788) (red-shouldered hawk) VC
    *Neophrontops vallecitoensis* Howard, 1963 * † (Vallecito neophron) VC
  Family Falconidae
    Gen. et sp. indet. (kestrel, merlin or falcon) VC
Order Galliformes
  Family Phasianidae
    Gen. et sp. indet. (pheasant, partridge, turkey, quail or grouse) VC
    *Meleagris anza* (Howard, 1963) * † (Anza turkey) VC
   Subfamily Odontophorinae
    *Callipepla* sp. ? *C. californica* (Shaw, 1798) (California quail) VC
    *Callipepla gambelii* (Gambel, 1843) (Gambel's quail) VC
Order Gruiformes
  Family Gruidae
    *Grus canadensis* (Linnaeus, 1758) (sandhill crane) VC
  Family Rallidae
    *Fulica americana* (Gmelin, 1789) (American coot) VC
    *F. hesterna* Howard, 1963 * † (yesterday's coot) VC
    *Gallinula* sp. (gallinule) VC
    *Rallus* sp. (rail) VC
    *Rallus limicola* Vieillot, 1819 (Virginia rail) VC
Order Charadriiformes
  Family Charadriidae
    *Charadrius vociferus* Linnaeus, 1758 (killdeer) VC
Order Strigiformes
  Family Strigidae
    Gen. et sp. indet. (owl) VC
    *Asio* sp. (eared owl) AS, VC
Order Piciformes
  Family Picidae
    Gen. et sp. indet. (woodpecker or wryneck) VC
Order Passeriformes
  Family Corvidae
    *Corvus* sp. (crow) VC
  Family Fringillidae
   Subfamily Carduelinae
    Gen. et sp. indet. (finch) VC
  Family Emberizidae
   Subfamily Embrizinae
    Gen. et sp. indet. (sparrow) VC

## Class Mammalia

Order Insectivora
  Family Soricidae
    Gen. et sp. indet. (shrew) LC
    *Notiosorex jacksoni* Hibbard, 1950 † (Jackson's desert shrew) AS, VC
    *Sorex* sp. (red-toothed shrew) AS, VC
  Family Talpidae
    *Scapanus malatinus* Hutchison, 1987 (mole) VC

Order Chiroptera
    Family Vespertilionidae
        Gen. et sp. indet. (bat) VC
        *Anzanycteris anzensis* White, 1967 * † (Anza bat) AS
        *Myotis* sp. (bat) VC

Order Xenarthra
    Family Megalonychidae
        *Megalonyx* sp. (ground sloth) AS
        *Megalonyx jeffersoni* (Desmarest, 1822) † (Jefferson's ground sloth) VC, BL
        *M. wheatleyi* Cope, 1871 † (Wheatley's ground sloth) VC
    Family Megatheriidae
        *Nothrotheriops* sp. cf. *N. shastensis* (Sinclair, 1905) † (Shasta ground sloth) VC, BL
    Family Mylodontidae
        *Paramylodon* sp. ? *P. harlani* † (giant ground sloth) VC, BL

Order Lagomorpha
    Family Leporidae
    Subfamily Archaeolaginae
        Gen. et sp. indet. (rabbit) LC, AS
        *Hypolagus edensis* Frick, 1921 † (Eden rabbit) AS
        *H. regalis* Hibbard, 1939 † (royal rabbit) LC, AS
        *H. vetus* (Kellogg, 1910) † (ancient rabbit) LC, AS
    Subfamily Leporinae
        *Lepus* sp. cf. *L. californicus* Gray, 1837 (black-tailed jackrabbit) VC, BL
        *Lepus callotis* Wagler, 1830 (white-sided jackrabbit) VC
        *Nekrolagus* sp. ? *N. progressus* (Hibbard, 1939) † (progressive rabbit) AS
        *Pewelagus dawsonae* White, 1984 * † (Dawson's rabbit) AS, LC, VC
        *Sylvilagus* sp. (cottontail rabbit) VC
        *Sylvilagus audubonii* (Baird, 1858) (desert cottontail) BL
        *S. hibbardi* White, 1984 * † (Hibbard's cottontail rabbit) AS, VC

Order Rodentia
    Family Sciuridae
        Gen. et sp. indet. (squirrel) LC, AS
        *Ammospermophilus leucurus* (Merriam, 1889) (white-tailed antelope squirrel) BL
        *Eutamias* sp. (chipmunk) AS
        *Spermophilus* sp. (ground squirrel) VC
    Family Geomyidae
        *Geomys* sp. (pocket gopher) BL
        *Geomys anzensis* Becker and White, 1981 * † (Anza pocket gopher) AS, VC
        *G. garbanii* White and Downs, 1961 † (Garbani's pocket gopher) VC
        *Thomomys* sp. (pocket gopher) BL
        *Thomomys bottae* (Eydoux and Gervais, 1836) (Botta's pocket gopher) VC
    Family Heteromyidae
        *Dipodomys* sp. (A) * † (kangaroo rat) LC, AS, VC
        *Dipodomys* sp. (B) * † (kangaroo rat) AS
        *Dipodomys* sp. (kangaroo rat) BL
        *Dipodomys* sp. cf. *D. minor* Gidley, 1922 † (small kangaroo rat) AS
        *D. compactus* True, 1889 (gulf coast kangaroo rat) AS, VC
        *D. hibbardi* Zakrzewski, 1981 † (Hibbard's kangaroo rat) AS, VC
        *Microdipodops* sp. † (kangaroo mouse) VC
        *Perognathus* sp. (pocket mouse) AS, VC
        *Perognathus hispidus* Baird, 1858 (hispid pocket mouse) LC, AS
    Family Castoridae

*Castor* sp. (beaver) VC
Family Cricetidae
    Gen. et sp. indet. † (microtine) ?AS
    *Baiomys* sp. (pygmy mouse) VC
    *Calomys (Bensonomys)* sp. † (mouse) VC
    *Microtus* sp. (meadow vole) VC
    *Microtus californicus* (Peale, 1848) (California meadow vole) VC, BL
    *Nelsonia* sp. † (pygmy woodrat) AS
    *Neotoma* sp. (woodrat) AS, VC
    *Neotoma (Hodomys)* sp. (A) † (woodrat) AS, VC
    *Neotoma (Hodomys)* sp. (B) † (woodrat) LC, AS, VC
    *Onychomys* sp. (grasshopper mouse) AS, VC
    *Peromyscus* sp. (white-footed mouse) AS, VC
    *Peromyscus maniculatus* (Wagner, 1845) (deer mouse) BL
    *Pliopotamys minor* (Wilson, 1933) † (pygmy muskrat) VC
    *Pitymys meadensis* Hibbard, 1944 † (Mead's vole) VC
    *Reithrodontomys* sp. (harvest mouse) AS, VC
    *Sigmodon curtisi* Gidley, 1922 † (Curtis cotton rat) VC
    *S. hispidus* Say and Ord, 1825 (hispid cotton rat) BL
    *S. lindsayi* Martin and Prince, 1989 * † (Lindsay's cotton rat) LC, AS
    *S. medius* Gidley, 1922 and/or *S. minor* Gidley, 1922 † (intermediate and/or small cotton rat) AS
    *Synaptomys anzaensis* Zakrezewski, 1972 * † (Anza bog lemming) VC
Family Erethizontidae
    *Coendou stirtoni* White, 1968 * † (Stirton's coendou) VC
Family Hydrochoeridae
    *Hydrochoerus* sp. (capybara) VC
Order Carnivora
  Family Canidae
    *Borophagus* sp. † (bone-eating dog) VC, BL
    *Canis dirus* Leidy, 1858 † (dire wolf) VC, BL
    *C. edwardii* Gazin, 1942 or *C. priscolatrans* Cope, 1899 † (Edward's dog or wolf-coyote) VC
    *Canis latrans* Say, 1823 (coyote) AS, ?VC, BL
    *C. lupus* Linnaeus, 1758 (wolf) VC
    *Urocyon* sp. (gray fox) AS, VC
    *Vulpes* sp. (fox) AS, VC
  Family Ursidae
    *Arctodus* sp. ? *A. simus* (Cope, 1879) † (short-faced bear) VC, BL
    *Tremarctos* sp. cf. *T. floridanus* (Gidley, 1928) † (Florida cave bear) AS, VC
    *Ursus* sp. (bear) BL
    *Ursus americanus* Pallas, 1780 (black bear) VC
  Family Procyonidae
    *Bassaricus* sp. (ring-tailed cat) VC
    *Bassaricus* sp. cf. *B. casei* Hibbard, 1952 † (Case's ringtail) LC
    *Nasua* sp. (coatimundi) VC
    *Procyon lotor* (Linnaeus, 1858) VC
  Family Mustelidae
    *Gulo* sp. (wolverine) VC
    *Mustela* sp. cf. *M. frenata* Lichtenstein, 1831 (long-tailed weasel) AS, VC
    *Martes* sp. (martin or fisher) VC
    *Satherium* sp. ? *S. piscinarium* (Leidy, 1873) † (Blancan otter) VC

        *Spilogale* sp. cf. *S. putorius* (Linnaeus, 1758) (spotted skunk) VC
        *Taxidea* sp. ? *T. taxus* (Schreber, 1778) (badger) VC, BL
        *Trigonictis sp.* † (grison) VC
    Family Felidae
        *Acinonyx* sp. (cheetah) VC
        *Felis concolor* Linnaeus, 1758 (mountain lion) BL
        *F. rexroadensis* Stephens, 1959 † (Rexroad cat) VC
        *Lynx rufus* (Schreber, 1777) (bob cat) VC, BL
        *Homotherium* sp. † (scimitar cat) ?AS
        *Smilodon fatalis* † (Leidy, 1868) (sabertooth) ?VC, BL
        *S. gracilis* Cope, 1880 † (gracile sabertooth) AS, VC, ?BL
    Family Odobenidae
        *Valenictus imperialensis* Mitchell, 1961 * † (Imperial walrus) CM
Order Cetacea
    Family Balaenopteridae
        Gen. et sp. indet. (baleen whale) CM
    Family Cetotheriidae
        Gen. et sp. indet. (whale-bone whale) ?VC
Order Sirenia
    Family Dugongidae
        Gen. et sp. indet. (dugong) CM
Order Proboscidea
    Family Mammutidae
        *Mammut* sp. ? *M. americanum* (Kerr, 1791) † (mastodon) VC
    Family Gomphotheriidae
        *Cuvieronius* sp. or *Stegomastodon* sp. † (Cuvier's gomphothere or stegomastodont) VC
    Family Elephantidae
        *Mammuthus columbi* (Falconer, 1857) † (Columbian mammoth) BL
        *M. imperator* (Leidy, 1858) † (imperial mammoth) BL
Order Perissodactyla
    Family Equidae
        cf. *Dinohippus* sp. † (Pliocene horse) LC, AS
        *Equus* (*Asinus* or *Hemionus*) sp. (ass or half-ass) ?AS, VC
        *Equus* (Equus) sp. A † (medium-sized horse) VC
        *Equus* (*Equus*) sp. B † (medium-sized horse) VC
        *Equus* (? *Hemionus*) sp. (half-ass) ?AS, VC, BL
        *Equus bautistensis* Frick, 1921 † (Bautista horse) BL
        *E. enormis* Downs and Miller 1994 † (giant zebra) VC, BL
        *E.* (*Dolichohippus*) sp. cf. *E.* (*D.*) *simplicidens* Cope, 1892 † (American zebra) AS, VC, ?BL
        *Equus* sp. ? *E. pacificus* Leidy, 1869 † (Pacific horse) VC
        *Hippidion* sp. † (South American equid) VC
    Family Tapiridae
        *Tapirus* sp. (tapir) VC
        *Tapirus merriami* Frick, 1921 † (Merriam's tapir) VC
Order Artiodactyla
    Family Tayassuidae
        *Mylohyus* sp. † (long-nosed peccary) VC
        *Platygonus vetus* Leidy, 1882 † (Leidy's peccary) AS, VC
    Family Camelidae
        Gen. et sp. indet. † (camel or llama) CM
        *Blancocamelus* sp. ? *B. meadei* Dalquest, 1975 † (Meade's camel) AS, VC

    *Camelops* sp. † (camel) VC
    *Camelops* sp. cf. *C. hesternus* (Leidy, 1873) † (yesterday's camel) BL
    *Camelops* sp. ? *C. huerfanensis* (Cragin, 1892) † (Huerfano camel) BL
    *Hemiauchenia* sp. † (llama) LC, BL
    *Hemiauchenia* sp. ? *H. blancoensis* (Meade, 1945) † (Blanco llama) AS, VC
    *Hemiauchenia* sp. cf. *H. macrocephala* (Cope, 1893) † (large-headed llama) AS, VC
    *Megatylopus* sp. † (large camel) LW, LC
    *Titanotylopus* sp. † (giant camel) VC, BL
  Family Cervidae
    Gen. et sp. indet. (deer) LC, AS
    *Cervus elaphus* Linnaeus, 1758 (elk) VC, BL
    *Odocoileus* sp. (deer) VC, BL
    *Odocoileus* sp. cf. *O. virginianus* (Zimmermann, 1780) (white-tailed deer) VC
    *Navahoceros* sp. † (mountain deer) VC
  Family Antilocapridae
    *Antilocapra* sp. (pronghorn) VC
    *Capromeryx* sp. † (diminutive pronghorn) AS, VC, BL
    *Tetrameryx* sp. † (four-horned pronghorn) AS, VC, BL
  Family Bovidae
    ? *Euceratherium* sp. † (shrub-oxen) VC, BL


## ICHNITE LIST

### Division Vertebratichnia
### Subdivision Avipedia

Order Charadriiformepeda
  Family Avipedea
    *Avipeda* sp. cf. *Calidris* sp. (sanderling track) AS
    *Avipeda* sp. cf. *Tringa* sp. (sandpiper track) AS

### Subdivision Mammalipedia

Order Lagomorphipeda
  Family Archaeolagipedidae
    *Archaeolagipeda* sp. cf. *Hypolagus* sp. † (rabbit track) AS
Order Carnivoripeda
  Family Carvivoripedae
    Subfamily Canipedinae
      *Canipeda* sp. cf. *Borophagus* sp. † (bone-eating dog track) LC
      *Canipeda* sp. cf. *Canis latrans* (coyote track) BQ
    Subfamily Felipedinae
      *Felipeda* sp. cf. *Felis* sp. (cat track) AS
      *Felipeda* sp. ? *Smilodon* sp. † (sabertooth track) AS
Order Proboscidipedida
  Family Proboscipedidae
    *Proboscipeda* sp. cf. *Mammuthus* sp. † (mammoth track) BL
Order Perissodactipedida
  Family Hippipedidae
    *Hippipeda* sp. cf. *Dinohippus* sp. † (Pliocene horse track) LC
Order Artiodactipedida
  Family Pecoripedidae
    *Pecoripeda* sp. cf. *Blancocamelus* sp. † (camel track) LC

*Pecoripeda* sp. cf. *Camelops* sp. † (camel track) BL
*Pecoripeda* sp. cf. *Hemiauchenia* sp. † (llama track) LC, AS
*Pecoripeda* sp. cf. *Megatylopus* sp. † (large camel track) LW, LC
*Pecoripeda* sp. cf. *Titanotylopus* sp. † (giant camel track) BL

## REFERENCES CITED

Anderson, E. 1984. Review of the small carnivores of North America during the last 3.5 million years. In Contributions in Quaternary Vertebrate Paleontology: A Volume in Memorial to John E. Guilday, edited by H.H. Genoways and M.R. Dawson, Carnegie Museum of Natural History Special Publication 8:257-266.

Becker, J.J., and J.A. White 1981. Late Cenozoic geomyids (Mammalia: Rodentia) from the Anza-Borrego Desert, southern California. Journal of Vertebrate Paleontology 1:211-218.

Bell, C.J. 1993. Fossil lizards from the Elsinore Fault Zone, Riverside County, California, with comments on the Neogene fossil history of the legless lizard, *Anniella*. Abstracts of Proceedings Desert Symposium, San Bernardino County Museum Association Quarterly 40(2):20-21.

Cunningham, G.D. 1984. The Plio-Pleistocene Dipodomyinae and geology of the Palm Spring Formation, Anza-Borrego Desert, Califronia. Master of Science, Department of Geology, Idaho State University, Pocatello 193 p.

Demére, T.A. 1993. Fossil mammals from the Imperial Formation (upper Miocene/lower Pliocene, Coyote Mountains, Imperial County, California. In Ashes, Faults and Basins, edited by R.E. Reynolds and J. Reynolds, San Bernardino County Museum Association Special Publication MDQRC 93(1):82-85.

Downs, T. 1957. Late Cenozoic vertebrates from the Imperial Valley region, California. Bulletin of the Geological Society of America Abstract 68(12-2):1822-1823.

----- 1965. Pleistocene vertebrates of the Colorado Desert, California. International Association for Quaternary Research, VII International Geological Congress General Session, Abstracts with Program p. 107.

Downs, T., and G.J. Miller 1994. Late Cenozoic equids from the Anza-Borrego Desert of California. Natural History Museum of Los Angeles County Contributions in Science 440:1-90.

Downs, T., and J.A. White 1966. The vertebrate faunal sequence of the Vallecito-Fish Creek area, western Colorado Desert. Society of Vertebrate Paleontology Field Trip Guide 5 p.

----- 1968. A vertebrate faunal succession in superposed sediments from late Pliocene to middle Pleistocene in California. 23rd International Geological Congress 10:41-47.

Howard, H.H. 1963. Fossil birds from Anza Borrego Desert. Los Angeles County Museum Contributions in Science 73:1-33.

Hutchison, J.H. 1987. Moles of the *Scapanus latimanus* group (Talpidae, Insectivora) from the Pliocene and Pleistocene of California. Natural History Museum of Los Angeles County Contributions in Science 386:1-15.

Jefferson, G.T. 1989. Late Cenozoic tapirs (Mammalia: Perissodactyla) of western North America. Natural History Museum of Los Angeles County Contributions in Science 406:1-21.

Jefferson, G.T., and P. Remeika 1994. The mid-Pleistocene stratigraphic co-occurrence of *Mammuthus columbi* and *M. imperator* in the Ocotillo Formation, Borrego Badlands, Anza-Borrego Desert State Park, California. Current Research in the Pleistocene 11:89-92.

Johnson, N., N.D. Opdyke, and E. Lindsay 1975. Magnetic polarity stratigraphy of Pliocene-Pleistocene terrestrial deposits and vertebrate fauna, San Pedro Valley, Arizona. Geological Society of America Bulletin 86:5-11.

Johnson, N.M., N.D. Opdyke, G.D. Woodard, P.K. Zeitler, and E.H. Lindsay 1983. Rates of Cenozoic tectonism in the Vallecito-Fish Creek

Basin. Geology 11:664-667.

Kurtén, B., and E. Anderson. 1980. Pleistocene mammals of North America. Columbia University Press, New York 442 p.

Lindsay, E.H., N.M. Johnson, and N.D. Opdyke 1975. Preliminary correlation of North American land mammal ages and geomagnetic chronology. In Studies on Cenozoic Paleontology and Stratigraphy, Claude W. Hibbard Memorial Volume 3, University of Michigan Papers in Paleontology 12:111-119.

Lindsay, E.H., N.M. Johnson, N.D. Opdyke, and R.F. Butler 1987. Mammalian chronology and the magnetic time scale. In Cenozoic Mammals of North America, edited by M.O. Woodburne, University of California Press, Berkeley p. 269-284.

Lundelius, E.L., Jr., T. Downs, E.H. Lindsay, H.A. Semken, R.J. Zakrzewski, C.S. Churcher, C.R. Harrington, G.E. Schultz, and S.D. Webb 1987. The North American Quaternary sequence. In Cenozoic Mammals of North America, Geochronology and Biostratigraphy, edited by M.O. Woodburne, University of California Press, Berkeley p. 211-235.

Martin, R.A. 1993. Late Pliocene and Pleistocene cotton rats in the southwestern United States. In Ashes, Faults and Basins, edited by R.E. Reynolds and J. Reynolds, San Bernardino County Museum Association Special Publication MDQRC 93(1):88-89.

Martin, R.A., and R.H. Prince 1989. A new species of early Pleistocene cotton rat from the Anza-Borrego Desert of southern California. Southern California Academy of Sciences Bulletin 88(2):80-78.

Miller, G.J. 1985. A look into the past of the Anza-Borrego Desert. Environment Southwest 510:12-17.

Miller, G.J., P. Remeika, J.D. Parks, B. Stout, and V. Waters 1991. A preliminary report on half-a-million year old cut marks on mammoth bones from the Anza-Borrego Desert Irvingtonian. Imperial Valley College Museum Society Occasional Paper 8:1-47.

Miller, W.E. 1980. The late Pliocene Las Tunas local fauna from southernmost Baja California, Mexico. Journal of Paleontology 54:762-805.

Norell, M.A. 1989. Late Cenozoic lizards of the Anza-Borrego Desert, California. Natural History Museum of Los Angeles County Contributions in Science 414:1-31.

Opdyke, N.D., E.H. Lindsay, N.M. Johnson, and T. Downs 1977. The paleomagnetism and magnetic polarity stratigraphy of the mammal-bearing section of Anza-Borrego State Park, California. Quaternary Research 7(3):316-329.

Remeika, P. 1992. Preliminary report on the stratigraphy and vertebrate fauna of middle Pleistocene Ocotillo Formation, Borrego Badlands, Anza-Borrego Desert State Park, California. Abstracts of Proceedings 6th Annual Mojave Desert Symposium, San Bernardino County Museum Association Quarterly 39(1):25-26.

Remeika, P., and G.T. Jefferson 1993. The Borrego local fauna: revised basin-margin stratigraphy and paleontology of the western Borrego Badlands, Anza-Borrego Desert State Park, California. In Ashes, Faults and Basins, edited by R.E. Reynolds and J. Reynolds, San Bernardino County Museum Association Special Publication MDQRC 93(1):90-93.

Repenning, C.A. 1992. Allophaiomys and the age of the Olyor Suite, Krestovka sections, Yakutia. U.S. Geological Survey Bulletin 2037:1-98.

Rymer, M.J. 1991. The Bishop ash bed in the Mecca Hills. In Geological excursions in southern California and Mexico, edited by M.J. Walawender and B.B. Hanan, Guidebook, 1991 Annual Meeting, Geological Society of America, San Diego State University, California p.388-396.

Savage, D.E., and D.E. Russell 1983. Mammalian Paleofaunas of the World. Addison-Wesley Publishing Company, Reading, Massachusetts 432 p.

Stout, B.W., and P. Remeika 1991. Status report on three major camel track sites in the lower Pliocene delta sequence, Vallecito-Fish Creek Basin, Anza-Borrego Desert State Park, California. Symposium on the Scientific Value of the Desert, Anza-Borrego Foundation, Borrego Springs, California p. 9.

Werdelin, L. 1985. Small Pleistocene felines of North America. Journal of Vertebrate Paleontology 5(3):194-210.

White, J.A. 1964. Kangaroo rats (Family Heteromyidae) of the Vallecito Creek Pleistocene of California. Geological Society of AmericaSpecial Paper 82:288-289.

----- 1965. Late Cenozoic vertebrates of the Anza-Borrego Desert area, southern California. American Association for the Advancement of Science, Abstracts with Program, Section E Berkeley, California (unpaginated).

----- 1967. Late Cenozoic bats (Subfamily

Nyctophylinae) from the Anza-Borrego Desert of California. University of Kansas Museum of Natural History Miscellaneous Publications 51:275-282.

----- 1968. A new porcupine from the middle Pleistocene of the Anza-Borrego Desert of California. Los Angeles County Museum of Natural History Contributions in Science 136:1-15.

----- 1969. Late Cenozoic bats (Subfamily Nyctophylinae) from the Anza-Borrego Desert of California. University of Kansas Museum of Natural History Miscellaneous Publications 51:275-282.

----- 1970. Late Cenozoic porcupines (Mammalia, Erethizontidae) of North America. American Museum Novitates 241:1-15.

----- 1984. Late Cenozoic Leporidae (Mammalia, Lagomorpha) from the Anza-Borrego Desert, southern California. Special Publication of the Carnegie Museum of Natural History 9:41-57.

----- 1987. The Archaeolaginae (Mammalia, Lagomorpha) of North America, excluding *Archaeolagus* and *Panolax*. Journal of Vertebrate Paleontology 7(4):425-450.

----- 1991. North American Leporinae (Mammalia: Lagomorpha) from late Miocene (Clarendonian) to latest Pliocene (Blancan). Journal of Vertebrate Paleontology 11(1):67-89.

White, J.A., and T. Downs 1961. A new *Geomys* from the Vallecito Creek Pleistocene of California. Los Angeles County Museum Contributions in Science 42:1-34.

----- 1965. Vertebrate microfossils from the Canebrake Formation of the Imperial Valley region, California. Society of Economic Mineralogists and Paleontologists, Pacific Section, Abstracts with Program, Los Angeles, California (unpaginated).

White, J.A., E.H. Lindsay, P. Remeika, E.W. Stout, T. Downs, and M. Cassiliano 1991. SVP field trip to Anza-Borrego Desert. Unpublished manuscript on file Anza-Borrego Desert State Park, Borrego Springs, California 23 p.

Zakrzewski, R.J. 1972. Fossil microtines from late Cenozoic deposits in the Anza-Borrego Desert, California, with a description of a new subgenus of *Synaptomys*. Los Angeles County Museum of Natural History Contributions in Science 221:1-12.

# AN ADDITIONAL AVIAN SPECIMEN REFERABLE TO *TERATORNIS INCREDIBILIS* FROM THE EARLY IRVINGTONIAN, VALLECITO-FISH CREEK BASIN, ANZA-BORREGO DESERT STATE PARK, CALIFORNIA

### George T. Jefferson
California Park Service, Colorado Desert District, 200 Palm Canyon Drive, Borrego Springs, California 92004

## INTRODUCTION

The remains of very large late Pliocene and Pleistocene teratornithid birds, presently referred to *Teratornis incredibilis* (Howard 1952, 1963, 1972a) are very rare in the North American fossil record. These animals are estimated to have had a wingspread of about 5.25 m (Howard 1952). Their paleozoogeographic distribution is limited to the southwestern US. Described materials include a late Pliocene (Blancan) specimen, and an early Pleistocene (Irvingtonian) specimen both from the Vallecito-Fish Creek Basin (VFCB), Anza-Borrego Desert State Park (ABDSP), California, and two late Pleistocene (Rancholabrean) specimens recovered from cave sites in central eastern Nevada and central western Utah (Table 1).

A fifth teratorn specimen, ABDSP(IVCM) 519/5660 (IVCM denotes the Imperial Valley College Museum portion of the ABDSP collection), was recognized recently during routine curation of the ABDSP collection. The material, which consists of the proximal articular end and proximal portion of the diaphysis of a left ulna, was recovered from the Vallecito member of the Palm Spring Formation (Woodard, 1963) in the June Wash area of the VFCB. Assemblages from this portion of the stratigraphic section comprise the Vallecito Creek Local Fauna (Downs and White, 1968; White, et al, 1991) of Irvingtonian age (Lindsay, et al, 1987). Strata here fall between the Olduvai and the Jaramillo Magnetosubchrons within the upper Matuyama Reversed-Polarity Magnetochron, and range in age from about 1.8 to 0.9 Ma.

The specimen is tentatively referred to *Teratornis* sp. cf. *T. incredibilis*. Its large size precludes an assignment to any other avian taxon presently described from the North American record of this period. No other ulnae of *T. incredibilis* are known.

## DISCUSSION

ABDSP (IVCM) 519/5660 is crushed and compressed in an anconal-palmar direction. The morphology of the articular surfaces and the bicipital crest region of the specimen are comparable to that exhibited by the late Pleistocene *Teratornis merriami* sample from Rancho La Brea. However, poor preservation in this region of the VFCB specimen does not permit detailed anatomical comparisons.

The maximum width of the proximal articular surfaces in ABDSP(IVCM) 519/5660 measures approximately 58 mm, about 57% larger than the average (36.9 mm) of five measured specimens of *Teratornis merriami* from Rancho La Brea (Table 2). With respect to the maximum width, including the bicipital ridge (or attachment) and the articular surfaces in the Rancho La Brea specimens, the dimension of ABDSP(IVCM) 519/5660 (63 mm) exceeds their average (39.0 mm) by about 62%.

The distance between individual secondary papillae on the proximal portion of the ulnar shaft in *T. merriami* tends to increase distally. These dimensions are not changed by crushing which, in ABDSP(IVCM) 519/5660, is directed perpendicular to the long axis of the specimen. The average distance between individual anterior secondary papillae in the Rancho La Brea specimens (Table 2) is 15.5 mm and ranges from 13.0 to 17.9 mm. With respect to this dimension, ABDSP(IVCM) 519/5660 is 52% larger than the sample average of *T. merriami*.

The type specimen of *Teratornis incredibilis*, LACM(CIT) 251/5067 (Los Angeles County Museum California Institute of Technology collection), is typified as being 43% larger than *T. merriami* (Howard 1952). Individual measurements of the hypotype, LACM 1318/3803 (Howard 1963), range from 29 to 55% larger than in *T. merriami*, and LACM 6747/26697 is 40% larger (Howard 1972b). Even discarding the proximal width measurements of ABDSP(IVCM) 519/5660, which may reflect some distortion (Table 2), the lengths between secondary papillae fall at the upper end of the

**TABLE 1.** Described specimens of *Teratornis incredibilis*. Explanation: LACM = Los Angeles County Museum; LACM(CIT) = Los Angeles County Museum, California Institute of Technology.

### Blancan

LACM 6747/26697, Vallecito-Fish Creek Basin
    *Teratornis* sp. cf. *T. incredibilis*, hypotype (tentative) Howard (1972a, 1972b)
    anterior rostrum (mandible)
    measurements and figures, Howard (1972b)

### Irvingtonian

LACM 1318/3803, June Wash, Vallecito-Fish Creek Basin
    *Teratornis incredibilis*, hypotype, Howard (1963, 1972a)
    distal articular end and diaphysis of right radius
    measurements and figures, Howard (1963)
ABDSP(IVCM) 519/5660, June Wash, Vallecito-Fish Creek Basin
    *Teratornis* sp. cf. *T. incredibilis*, this paper
    proximal articular end and proximal diaphysis of left ulna
    measurements, this paper

### Rancholabrean

LACM(CIT) 251/5067, Smith Creek Cave, Nevada
    *Teratornis incredibilis*, holotype, Howard (1952, 1972a)
    cuneiform (ulnare)
    measurements and figures, Howard (1952)
Utah Locality Number 42Md620V, Crystal Ball Cave, Utah
    cf. *Teratornis incredibilis*, Heaton (1984)
    vertebra
    "very large," measurements not provided by Heaton (1984)

**TABLE 2.** Comparative measurements in mm of ABDSP(IVCM) 519/5660 and *Teratornis merriami* from Rancho La Brea. Explanation: LACMHC = Natural History Museum of Los Angeles County Hancock Collection; n = number of measurements between individual papillae per specimen; 1. = average and range of the individual distances between the proximal secondary papillae; 2. = maximum width proximal articular surfaces; 3. maximum width including bicipital ridge and articulations; * = approximate, reconstructed measurement.

| specimen number | 1. | 2. | 3. |
|---|---|---|---|
| ABDSP(IVCM) 5660 | 23.5, 21.2-24.8 (n = 3) | 58* | 63* |
| LACMHC B439 | 14.1, 12.0-15.5 (n = 4) | 36.6 | 37.0 |
| LACMHC B572 | 13.0, 13.0-13.0 (n = 5) | 36.3 | 39.6 |
| LACMHC B1026 | 15.0, 14.3-16.5 (n = 5) | 35.7 | 39.3 |
| LACMHC B1192 | 13.4, 11.8-14.4 (n = 5) | 36.6 | 37.0 |
| LACMHC G6911 | 16.3, 14.6-17.9 (n = 4) | 39.2 | 42.2 |

known size range for *T. incredibilis*.

Although the preservation of ABDSP(IVCM)519/5660 and LACM 1318/3830 is essentially identical, and both were recovered from the same portion of the stratigraphic section in June Wash, western VFCB, aerial photograph field records indicate that the two specimens were collected about 600 m from one another and do not come from the same horizon. Additional LACM material from VFCB, LACM 1111/104811, tentatively identified as *Teratornis* is best referred to *Gymnogyps*. Therefore, the minimum number of individual specimens presently referred to *T. incredibilis* includes 1 Blancan, 2 Irvingtonian and 2 Rancholabrean age specimens.

## ACKNOWLEDGMENTS

K. Campbell and C. Shaw of the George C. Page Museum, Natural History Museum of Los Angeles County kindly provided access to the Rancho La Brea collections. The original, incorrect species identification of ABDSP 519/5660 was brought to my attention by P. Remeika of ABDSP.

## REFERENCES CITED

Downs, T., and J.A. White 1968. A vertebrate faunal succession in superposed sediments from the late Pliocene to middle Pleistocene in California. 23rd International Geological Congress 10:41-47.

Heaton, T.H. 1984. Preliminary report on the Quaternary vertebrate fossils from Crystal Ball Cave, Millard County, Utah. Current Research in the Pleistocene 1:65-67.

Howard, H. 1952. The prehistoric avifauna of Smith Creek Cave, Nevada, with a description of a new gigantic raptor. Southern California Academy of Sciences Bulletin 51:50-54.

----- 1963. Fossil birds from the Anza-Borrego Desert. Los Angeles County Museum Contributions in Science 73:1-33.

----- 1972a. Type specimens of avian fossils in the collections of the Natural History Museum of Los Angeles County. Natural History Museum of Los Angeles County Contributions in Science 228:1-27.

----- 1972b. The incredible teratorn again. Condor 74(3):341-344.

Lindsay, E.H., N.M. Johnson, N.D. Opdyke, and R.F. Butler 1987. Mammalian chronology and the magnetic polarity time scale. In Cenozoic mammals of North America, edited by M.O. Woodburne, University of California press, Berkeley p. 269-284.

Opdyke, N.H., E.H. Lindsay, N.M. Johnson, and T. Downs 1977. The paleomagnetism and magnetic polarity stratigraphy of the mammal-bearing section of Anza-Borrego State Park, California. Quaternary Research 7:316-329.

White, J.A., E.H. Lindsay, P. Remeika, E.W. Stout, T. Downs, and M. Cassiliano 1991. SVP field trip to the Anza-Borrego Desert. Field Trip Log Society of Vertebrate Paleontology meetings in San Diego, California 23 p.

Woodard, G.D. 1963. The Cenozoic stratigraphy of the western Colorado Desert, San Diego and Imperial counties, southern California. Doctoral Dissertation, University of California, Berkeley 223 p.

# THE BORREGO LOCAL FAUNA: REVISED BASIN-MARGIN STRATIGRAPHY AND PALEONTOLOGY OF THE WESTERN BORREGO BADLANDS, ANZA-BORREGO DESERT STATE PARK, CALIFORNIA

**Paul Remeika** and **George T. Jefferson**
California Park Service, Colorado Desert District, Anza-Borrego Desert State Park, 200 Palm Canyon Drive, Borrego Springs, California 92004

## INTRODUCTION

The Borrego Local Fauna (BLF) (Remeika, 1992) is a diverse assemblage of primarily large terrestrial vertebrates recovered from the fluvial-floodplain syndepositional sedimentary sequence throughout the mid-Pleistocene Ocotillo formation (informal name of Remeika, 1992). Vertebrate-bearing strata (Figure 1) exposed throughout the westernmost half of the Borrego Badlands (BB) along the northwestern basin-margin of the Borrego-San Felipe Basin (BSFB) (Reynolds and Remeika, 1993) measures 377 m in thickness. Based on recent geologic mapping, the stratoype for the BLF has been located in the Arroyo Otro-Mammoth Cove sedimentary section, above and below the Ocotillo Rim and north of the left-lateral Inspiration Point Fault (Remeika, 1992). The BLF presently includes over 5 k specimens that represent 49 vertebrate taxa (Table 1). Within the region, the BLF most closely compares taxonomically with a similar age fauna recovered from the Coyote Badlands of the BSFB, the El Golfo local fauna of the northern Gulf of California (Shaw, 1981), and the Pauba Formation, Riverside County (Reynolds and Reynolds, 1990; Reynolds, et al. 1991), in the Murrieta area 90 km northwest of the BB.

Under the direction of C. Frick of the American Museum of Natural History (AMNH), G. Hazen first discovered vertebrate fossils in the BB in 1935. Field studies initiated in the 1970's by T. Downs and H. Garbani of the Natural History Museum of Los Angeles County (LACM), and continued during the late 1970's and 1980's by G. Miller of the Imperial Valley College Museum, and in the 1990's by P. Remeika and G. Jefferson, Anza-Borrego Desert State Park (ABDSP), have led to periodic recovery of significant vertebrate remains. Assemblages representing the BLF are presently housed in the AMNH, LACM, and in the Vertebrate Paleontology Collection of ABDSP.

## GEOLOGIC SETTING

Fossiliferous basinal sediments of the Ocotillo formation exposed in the BB lie along the westernmost edge of the BSFB within the active seismogenic San Jacinto Fault Zone. These deposits record Neogene basin-margin deposition in response to a fault-induced subsiding structural trough situated between two parallel mountain ranges: the Santa Rosa Mountains to the northeast and Coyote Mountain to the northwest. Quaternary deformation is dominated by dip-slip faulting (Coyote Creek Fault) with subordinate NW-SE right-lateral, strike-slip motion on the Clark Fault, and by NE-SW left-lateral cross-faults that display significant crustal shortening and rotation orthogonal to the main faults. Block rotation behavior has been discussed by Scheuing, et al. (1988) and Nicholson and Seeber (1989). Trends of *en echelon* fold axes imply progressive deformation of the sedimentary package by a complex basement-cover *decollement* (Pettinga, 1991).

## STRATIGRAPHY AND AGE

The Ocotillo formation is composed primarily of locally-derived fluvial and lacustrine deposits (part of the informal Anza-Borrego group of Reynolds and Remeika, 1993). Five informal genetically interrelated sedimentary depositional subenvironments that are typical of topographically low basin centers have been stratigraphically delineated (Remeika and Pettinga, 1991; Remeika, 1992). In ascending order they include: (1) alluvial fan sediments (Ghost Palms member (GPM); (2) Mammoth Cove sandstone member (MCSM); (3) playa margin deposits (Inspiration Wash member); (4) playa basin deposits (Las Playas member) (LPM); and (5) fluvial floodplain/playa margin deposits (Short Wash member) (SWM). The MCSM consists of a 74 m thick megasequence of locally-derived, subaqueously deposited distal alluvial fan sediments that interfinger with extraregional lacustrine claystones (Brawley Formation, informal Colorado River group of Reynolds

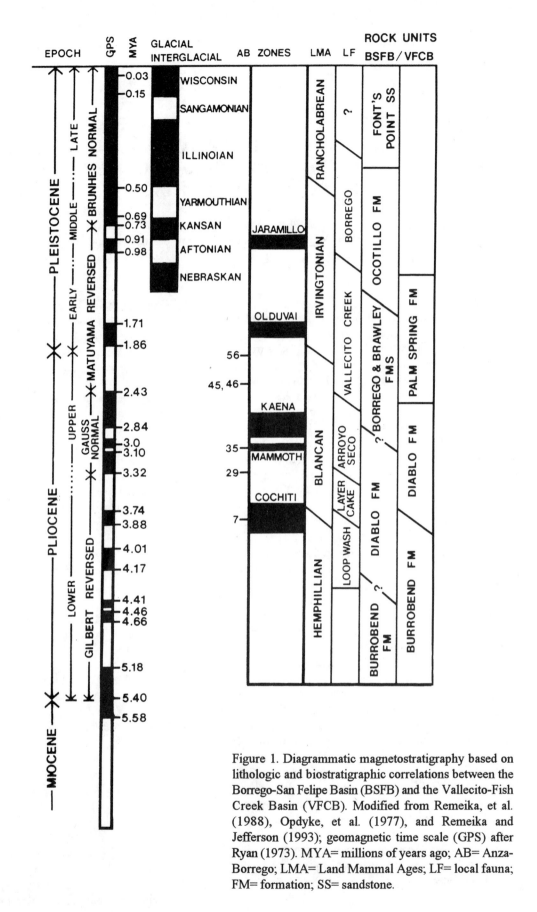

Figure 1. Diagrammatic magnetostratigraphy based on lithologic and biostratigraphic correlations between the Borrego-San Felipe Basin (BSFB) and the Vallecito-Fish Creek Basin (VFCB). Modified from Remeika, et al. (1988), Opdyke, et al. (1977), and Remeika and Jefferson (1993); geomagnetic time scale (GPS) after Ryan (1973). MYA= millions of years ago; AB= Anza-Borrego; LMA= Land Mammal Ages; LF= local fauna; FM= formation; SS= sandstone.

and Remeika, 1993). This stratigraphic relationship is exposed in the deposits beneath Font's Point in the southeastern portion of the badlands.

The 216 m thick IWM is laterally widespread and consists of interstratified fluvial-floodplain and playa-margin deposits. These grade basinward (eastward) into the LPM which is primarily composed of fine-grained sediments. The LPM grades eastward into the SWM.

Fossil localities are concentrated north of Dump Wash, with alluvial fan/fluvial deposits (MCSM) of local provenance that crop out between Tumbleweed Wash and Two-sloth Wash. Above the Ocotillo Rim, strata of the MCSM and IWM are locally folded and faulted with the section repeated in exposures across Arroyo Otro and Inspiration Wash. To the east, floodplain/playa-margin deposits (IWM) grade laterally into lacustral playa deposits (LPM and SWM) between Inspiration Wash and Font's Wash. A recent discovery of *Camelops* sp. in playa-margin sediments in Font's Wash is the first documented account from the SWM. In the BB, the Ocotillo formation is overlain by fossiliferous alluvial plain-braided fluvial deposits of the late Pleistocene Font's Point sandstone (informal name of Remeika and Pettinga, 1991). Vertebrate fossils only occur in thin, 2-3 m, sandstone interfluves within the lacustrine sediments.

Based on the preliminary magnetostratigraphic work of Scheuing, et al. (1988), and Bogen and Seeber (1986), the Ocotillo formation ranges in age from 1.25 to 0.37 Ma. (Figure 1). This period is correlative with the Irvingtonian and earliest Rancholabrean Land Mammal Ages (LMA) (Savage, 1951; Woodburne, 1987). Distal fan deposits within the basal MCSM span the Matuyama Reversed-Polarity Magnetochron, and range in age from 1.25 to about 0.90 Ma. The geomagnetic event within the lower IWM probably represents the Jaramillo Magnetosubchron which ranges from 0.98 to 0.91 Ma. The remaining IWM and LPM are magnetically normal and probably represent the Brunhes Normal-Polarity Magnetochron. The Matuyama/Brunhes geomagnetic boundary is marked by the presence of the Bishop Tuff (Izett, 1981; Sarna-Wojcicki, et al. 1984). This ashfall tuff occurs above the level of the Jaramillo event within the lower portion of the IWM. In the nearby Coyote Badlands it has been tentatively chemically correlated with the Fryant Ash member of the Bishop series dated at about 0.62 Ma (Sarna-Wojcicki, pers. comm., 1984).

## BIOSTRATIGRAPHY

The BLF (Table 1) is based primarily on fragmentary postcranial remains, however, a few articulated specimens have been recovered. Species of *Mammuthus*, *Equus*, and *Camelops* are the most abundant represented mammalian taxa. Both large and small carnivores are represented but the materials are diagnostic only to generic level. Together with large herbivores that include a relatively balanced representation of grazers, browsers and mixed feeders (Akersten, et al. 1984; Jefferson, 1988), they suggest an open, savanna-like environment with permanent water and scattered gallery forests. Aquatic vertebrates record the presence of marshy lacustrine habitats, which may have been locally ephemeral. Assemblages from each of the described, fossiliferous members of the Ocotillo formation are discussed below.

Although many of the taxa from the MCSM are shared with the Blancan-Irvingtonian age Vallecito Creek LF from the Vallecito Badlands (Downs and White, 1968) of the Vallecito-Fish Creek Basin (VFCB) (White, et al. 1991), the presence of *Mammuthus imperator*, *Equus bautistensis*, and *Camelops huerfanensis* clearly indicates an Irvingtonian Age. A partial, about 60% complete skeleton of *M. imperator* was recovered from the upper part of this unit (Miller, et al. 1991). Microvertebrates are relatively rare in this portion of the Ocotillo formation.

As well as the more common taxa, the IWM assemblage includes the humpback sucker *Xyrauchen* sp. cf. *X. texanus* and *Accipiter striatus*, the sharp-shinned hawk. A camelid trackway has been found here that is comparable to the ichnogenus *Pecoripeda* sp. cf. *Camelops* sp. from Camel Ridge in the Vallecito Badlands (Stout and Remeika, 1991). In addition to the taxa listed in Table 1, the following lower vertebrates and birds have been recovered from the IWM: *Xyrauchen* sp. cf. *X. texanus*, *Clemmys* sp., *Buteo* sp., *Aquila* sp., and *Tyto alba* (Table 1). The IWM assemblage is closely comparable to that recovered from the Coyote Badlands, 16 km northwest of the BB. Similar stratigraphic sections that include the Bishop Tuff are present in both areas. Recent field observations suggest that the Coyote Badlands are offset from Butler Canyon exposures by dip-slip movement along the Coyote Creek Fault. Both areas are coeval to, but not offset from, the Borrego Badlands. At Valle Escondido, strata high in the section (SWM) (?) are especially rich in microvertebrates. The fragmentary remains include *Lepus* sp. cf. *L. californicus*, *Thomomys* sp., *Dipodomys* sp., *Neotoma lepida*, *Ammospermophilus leucurus*, *Sigmodon hispidus*, *Peromyscus maniculatus*, and *Microtus californicus*. The arvicoline rodent *M. californicus* (Zakrzewski, pers. comm., 1988), and *S. hispidus* are evolved forms that index the Irvingtonian LMA (Repenning, et al. 1987; Kurtén and Anderson, 1980). Vertebrate remains within this unit occur in

**TABLE 2. Composite taxonomic list of the Borrego Local Fauna.**
Taxonomy follows Banks and others (1987) and Simpson (1945). Published faunal accounts include Howard (1963), Remeika (1992). Fish remains identified by M. Roeder (pers. comm., 1991) and equids by E. Scott (pers. comm., 1992). Extinct forms denoted by †.

### Class Osteichthyes
Order Teleostei
    Family Castostomidae
        *Xyrauchen* sp. cf. *X. texanus* humpback sucker)
        ? genus (minnows)

### Class Reptilia
Order Chelonia
    Family Testudinidae
        ? genus (tortoise)
    Family Emydidae
        *Clemmys* sp. (pond turtle)

### Class Aves
Order Falconiformes
    Family Accipitridae
        *Buteo* sp. (hawk)
        *Aquila* sp. (eagle)
        *Accipiter striatus* (sharp-shinned hawk)
        *A. cooperi* (Cooper's hawk)
    Family Cathartidae
        *Cathartes aura* (turkey vulture)
    Family Anseriformes
        ? genus (ducks)
Order Ciconiformes
    Family Phoenicopteridae
        *Phoenicopterus* sp. (flamingos)
    Family Strigiformes
        *Tyto alba* (barn owl)

### Class Mammalia
Order Edentata
    Family Megalonychidae
        *Megalonyx jeffersonii* † (ground sloth)
    Family Megatheriidae
        *Nothrotheriops* sp. cf. *N. shastense* † (small ground sloth)
Order Lagomorpha
    Family Leporidae
        *Sylvilagus audubonii* (desert cottontail)
        *Lepus* sp. cf. *L. californicus* (jackrabbit)
Order Rodentia
    Family Sciuridae
        *Ammospermophilus leucurus* (antelope ground squirrel)
    Family Geomyidae
        *Thomomys* sp. (pocket gopher)
        *Geomys* sp. (pocket gopher)
    Family Heteromyidae

                *Dipodomys* sp. (kangaroo rat)
        Family Cricetidae
                *Sigmodon hispidus* (cotton rat)
                *Peromyscus maniculatus* (white-footed mouse)
                *Neotoma lepida* (wood rat)
                *Microtus californicus* (California meadow vole)
Order Carnivora
        Family Mustelidae
                *Taxidea taxus* (badger)
        Family Ursidae
                *Arctodus* sp. † (short-faced bear)
                cf. *Ursus* sp. (black bear)
        Family Canidae
                *Canis dirus* † (dire wolf)
                *C. latrans* (coyote)
        Family Felidae
                *Felis concolor* (mountain lion)
                *Lynx rufus* (bobcat)
                *Smilodon gracilis* † (gracile sabertooth cat?)
                *S. fatalis* † (California sabertooth cat)
Order Proboscidea
        Family Elephantidae
                *Mammuthus columbi* † (columbian mammoth)
                *M. imperator* † (imperial mammoth)
Order Perissodactyla
        Family Equidae
                *Equus bautistensis* † (Bautista horse)
                *E. (Dolichohippus) enormis* † (giant Anza-Borrego zebra)
                *Equus (Dolichohippus)* sp. † (small zebra)
                *Equus* sp. ? *E. hemionus* † (half ass)
Order Artiodactyla
        Family Camelidae
                *Camelops* sp. ? *C. huerfanensis* † (Huerfano camel)
                *Camelops* sp. cf. *C. hesternus* † (yesterday's camel)
                *Hemiauchenia* sp. † (llama)
                *Titanotylopus* sp. † (giant camel)
        Family Antilocapridae
                *Capromeryx* sp. † (diminutive pronghorn)
                *Tetrameryx* sp. † (4-horned pronghorn)
        Family Cervidae
                *Cervus elaphus* (elk)
                *Odocoileus* sp. (deer)
        Family Bovidae
                ?*Euceratherium* sp. † (shrub oxen)

interbeds of fluviatile sandstone. Near the top of the section, these strata occur above the Bishop Tuff, and may span the Irvingtonian-Rancholabrean LMA boundary.

In summary, within ABDSP, faunal correlations between vertebrate-bearing strata in the VFCB and BB (Figure 1) record a continuous succession of vertebrate fossil assemblages ranging from the latest Hemphillian through earliest Rancholabrean ages. The stratigraphic ranges of *Mammuthus imperator* (MCSM) and *M. columbi* (MCSM-LPM) document only one example of faunal change (Downs and White, 1968; Jefferson and Remeika, 1994, and this volume) within the record.

**ACKNOWLEDGMENTS**

R. E. Reynolds and E. Scott of the San Bernardino County Museum reviewed the manuscript and provided helpful comments and suggestions. This report is an updated revision of Remeika and Jeffeson (1993).

**LITERATURE CITED**

Akersten, W. A., T. M. Foppe, and G. T. Jefferson, 1984. New source for dietary data from large extinct herbivores. Quaternary Research, 30(1):92-97.

Banks, R. C., R. W. McDiarmid, and A. L. Gardner, 1987. Checklist of Vertebrates of the United States, the U. S. Territories, and Canada. United States Department of the Interior, Fish and Wildlife Service, Resource Publication, 166:1-79.

Bogen, N. L., and L. Seeber, 1986. Neotectonics of rotating blocks within the San Jacinto Fault Zone, southern California. Abstract. EOS, 67:1200.

Downs, T., and J. A. White, 1968. A vertebrate faunal succession in superposed sediments from late Pliocene to middle Pleistocene in California. In Tertiary/Quaternary Boundary, International Global Congress, Prague, 23(10):41-47.

Howard, H. H., 1963. Fossil birds from the Anza-Borrego Desert. Los Angeles County Museum Contributions in Science, 73:1-33.

Izett, G. A., 1981. Volcanic ash beds: recorders of upper Cenozoic silic pyroclastic volcanism in the western United State. Journal of Geophysical Research, 86(B11):10200-10222.

Jefferson, G. T., 1988. Late Pleistocene large mammalian herbivores: implications for early human hunting patterns in southern California. Bulletin of the Southern California Academy of Sciences, 87(3):89-103.

Kurtén, B., and E. Anderson, 1980. Pleistocene mammals of North America. Columbia University Press, New York: 442p.

Miller, G. J., P. Remeika, J. D. Parks, B. W. Stout, and V. E. Waters, 1991. A preliminary report on half-a-million-year-old cutmarks on mammoth bones from the Anza-Borrego Desert Irvingtonian. Imperial Valley College Museum Society Occasional papers, 8:1-47.

Nicholson, C., and L. Seeber, 1989. Evidence for contemporary block rotation in strike-slip environments: examples from the San Andreas Fault system, southern California. In Paleomagnetic rotations and continental deformation, edited by C. Kissel and C. Laj. Kluwer Academic Publishers: 247-280.

Opdyke, N. D., E. H. Lindsay, N. M. Johnson, and T. Downs, 1977. The paleomagnetism and magnetic polarity stratigraphy of the mammal-bearing section of Anza-Borrego Desert State Park, California. Quaternary Research. 7:316-329.

Pettinga, J. R., 1991. Structural styles and basin margin evolution adjacent to the San Jacinto Fault Zone, southern California. Abstracts with Programs, Geological Society of America Annual Meeting, San Diego, California: A257.

Remeika, P., 1992. Preliminary report on the stratigraphy and vertebrate fauna of the middle Pleistocene Ocotillo formation, Borrego Badlands, Anza-Borrego Desert State Park, California. Abstracts of Proceedings, 6th Annual Mojave Desert Quaternary Research Symposium, San Bernardino County Museum Quarterly, 39(2):25-26.

Remeika, P., I. W. Fischbein, and S. A. Fischbein, 1988. Lower Pliocene petrified wood from the Palm Spring Formation, Anza-Borrego Desert State Park, California. Review of Palaebotany and Palynology, 56:183-198.

Remeika, P., and J. R. Pettinga, 1991. Stratigraphic revision and depositional environments of the middle to late Pleistocene Ocotillo Conglomerate, Borrego Badlands, Anza-Borrego Desert State Park, California. Symposium on the Scientific Value of the Desert, Anza-Borrego Foundation, Borrego Springs, California, Abstract 13.

Repenning, C. A., E.M. Browers, L. O. Carter, L. Marincovich, Jr., and T. A. Ager, 1987. The Beringian ancestry of *Phenacomys* (Rodentia; Cricetidae) and the beginning of the modern

Arctic Ocean borderland biota. United States Geological Survey Bulletin, 1687:1-31.

Reynolds, R. E., and R. L. Reynolds, 1990. Irvingtonian? faunas from the Pauba Formation, Temecula, Riverside County, California. In Abstracts and Proceedings, 1990 Mojave Desert Quaternary Research Symposium, edited by J. Reynolds. San Bernardino County Museum Association Quarterly, 37(3,4):37.

Reynolds, R. E., R. L. Reynolds, and A. F. Pajak III, 1991. Blancan, Irvingtonian, and Rancholabrean (?) Land Mammal Age faunas from western Riverside County, California. In Abstracts and Proceedings, 1991 Mojave Desert Quaternary Research Symposium, edited by J. Reynolds. San Bernardino County Museum Association Quarterly, 38(3,4):37-40.

Ryan, W. B. F., 1973. Paleomagnetic stratigraphy. In Initial Reports of the Deep Sea Drilling Project, LEG XIII, 2, edited by W. B. F. Ryan and K. H. Hsu. United States Government Printing Office, Washington, D. C.:1380-1386.

Sarna-Wojcicki, A. M., H. R. Boweman, C. E. Meyer, P. C. Russell, M. J. Woodward, G. McCoy, J. J. Rowe, Jr., P. A. Baedecker, F. Asaro, and H. Michel, 1984. Chemical analyses, correlations, and ages of upper Pliocene and Pleistocene ash layers of east-central and southern California. U.S. Geological Survey Professional Paper, 1293:1-40.

Savage, D.E., 1951. Late Cenozoic vertebrates of the San Francisco Bay region, California. University of California Publications Department of Geological Sciences, 28(10):215-314.

Scheuing, D. F., L. Seeber, K. W. Hudnut, and N. L. Bogen, 1988. Block rotation in the San Jacinto Fault Zone, southern California. EOS, Transactions of the American Geophysical Union 69:1456.

Shaw, C. A. 1981. The middle Pleistocene El Golfo local fauna from northwestern Sonora, Mexico. Master of Science Thesis, Department of Biology, California State University, Long Beach 141 p.

Simpson, G. G., 1945. The principles of classification and a classification of mammals. American Museum of Natural History Bulletin 85:1-350.

Stout, B.W., and P. Remeika 1991. Status report on three major camelid track sites in the lower Pliocene delta sequence, Vallecito-Fish Creek Basin, Anza-Borrego Desert State Park, California. Symposium on the Scientific Value of the Desert, Anza-Borrego Foundation, Borrego Springs: 9.

White, J. A., E. H. Lindsay, P. Remeika, B. W. Stout, T. Downs, and M. Cassiliano, 1991. Society of Vertebrate Paleontology field trip guide to the Anza-Borrego Desert. Society of Vertebrate Paleontology 51st Annual Meeting, San Diego: 23p.

Woodburne, M. O., 1987. Cenozoic mammals of North America. University of California Press, Berkeley: 336p.

# THE MID-PLESITOCENE STRATIGRAPHIC CO-OCCURRENCE OF *MAMMUTHUS COLUMBI* AND *M. IMPERATOR* IN THE OCOTILLO FORMATION, BORREGO BADLANDS, ANZA-BORREGO DESERT STATE PARK, CALIFORNIA

### George T. Jefferson and Paul Remeika
California Park Service, Colorado Desert District, Anza-Borrego Desert State Park, 200 Palm Canyon Drive, Borrego Springs, California 92004

## INTRODUCTION

The Borrego Badlands is presently the focus of intensive paleontological and geological research. Here, Irvingtonian Land Mammal Age (LMA) vertebrate fossils (Borrego Local Fauna) (Remeika, 1992; Remeika and Jefferson, 1993; and this volume) are documented from the terrestrial Ocotillo formation (informal name of Remeika, 1992). A review of *Mammuthus* cranial material recovered from this 377-m-thick unit suggests that *M. columbi* and *M. imperator* were locally contemporaneous during the middle Pleistocene. At face value, this observation counters standard models of mammoth history (Madden, 1981; Maglio, 1973), which assert that *M. imperator* and *M. columbi* exhibit an ancestor-descendent relationship. The taxonomy of Maglio (1973), which recognizes *M. columbi*, *M. imperator*, and *M. meridionalis*, is followed herein.

## DISCUSSION

A morphologic transition from *M. meridionalis* through *M. imperator* to *M. columbi* has been described (Agenbroad, 1984; Graham, 1986; Kurtén and Anderson, 1980; Madden, 1981; Maglio, 1973). This presumed evolutionary lineage is based on a variety of dental parameters, including an increase in tooth-plate numbers, increased lamellar frequency, and a decrease in enamel thickness through time. Maglio (1973) and Madden (1981) consider the transition between *M. imperator* to *M. columbi* to have occurred during the late Illinoian or early Rancholabrean, between 0.50 and 0.40 Ma. Agenbroad (1980) places this change at approximately 0.13 Ma.

## AGE AND STRATIGRAPHIC RANGE

Well-preserved specimens referable to both *M. columbi* and *M. imperator* have been recovered from one horizon within the basal sandstones of the Mammoth Cove sandstone member (MCSM) of the Ocotillo formation. Stratigraphic range of these materials is less than 5 m, and some occur within the same bed. This association is very significant because it not only documents the co-occurrence of, but also revises and constrains the timing of the transition between the two taxa.

The basal MCSM consists of a complex sequence of interstratified distal alluvial fan and fluvial-floodplain deposits (MCSM I and II) of local provenance. Basinward, the unit pinches out into the lacustrine Brawley Formation (Dibblee, 1954). Paleomagnetic calibration of Scheuing, et al. (1988) suggests the basal MCSM contact with the Brawley Formation is less than 1.66 Ma. The Jaramillo Magnetosubchron, which ranges from 0.98 to 0.91 Ma, and the Matuyama/Brunhes Magnetochron boundary have been identified (Scheuing, 1989; Scheuing, et al. 1988) in the overlaying Inspiration Wash member of the Ocotillo formation. The latter preserves tephra tentatively identified as the Bishop Tuff (K. Beratan, pers. comm., 1993; M. Rymer, 1991, pers. comm., 1993), dated at 0.74 Ma (Sarna-Wojcicki and Pringle, 1992; Sarna-Wojcicki, et al. 1984). This places the basal MCSM within the lower Matuyama Reversed-Polarity Magnetochron with an approximate age of 1 Ma.

The lower 31 m of the MCSM I consist of a coarse-grained, massive, vertebrate fossil-bearing sandstone unit. The basal contact with the underlying red clays of the Brawley Formation is sharp and well-defined. Sediments are predominantly locally-derived, siliciclastic medium- to coarse-grained arkosic sandstones that are massive in their lower parts, displaying crude normal bedding. The high-energy sandstones are light gray in color, are poorly- to moderately-sorted, horizontally bedded, and display distinct fining- and thining-upward sequences, grading up into fine-grained micaceous overbank siltstones in their upper parts. Sandstones weather a creamy yellow brown and form sub-resistant ledges. They commonly contain minor stringers of intraclastic pebbles, imbricated gravels and coarse grit sized interbeds in their lower parts. The clast mineralogy is principally sub-rounded to sub-angular granitic fragments.

Also included are four light-gray interbedded sheetflood beds that are petrographically similar to those which characterize the overlaying sandstones (MCSM II). These distinctive beds are coarse-grained and moderately sorted, ranging between 1.0-1.3 m thick in the lower part and 1.3-2.6 m in thickness in the upper part. These sandstones are composed of quartzo-feldspathic grains with a higher percentage of coarse-grained extraformational gravel clasts. The basal contact of each sandstone bed is gradational, displaying scoured, concave-up bases.

The sandstones fine upward into subordinate light brown overbank sandy siltstones and clayey siltstones that represent low-energy waning phases of the sheetfloods. The siltstones are fine-grained, frequently micaceous, and generally thin to medium-bedded (5.0-12.7 cm). Small-scale sedimentary structures are oftentimes rare although ripple drift crossbeds or planar laminations are present locally. Occasionally these are overlain by thin indurated siltstones (20.3-25.4 cm) to thick medium- to fine-grained sandstones (3.3-4.0 m) with no apparent internal structure. Within the overbanks, four anhydrite nodule beds are present, restricted to the uppermost erosional tops of the siltstones as distinct horizons. These are interpreted to represent extended subaerial, oxidizing conditions.

At outcrop scale within MC, sharply-based sandstones of MCSM I are laterally extensive for hundreds of meters along strike. In general, sand bodies were deposited as a basinward extension of an alluvial plain environment with intermittent unconfined fluvial-sheetflood conditions separated by periods of extended subaerial, oxidizing intervals. This lithofacies is significant because its uppermost horizon yields *Mammuthus imperator* and *M. columbi*. It is also noteworthy for having a lower frequency of sedimentary structures and a higher percentage of massive sandstones typical of basin-margin distal alluvial fan environments.

This sandstone unit is overlain by a 16.5 m thick, very light gray-colored sandstone (MCSM II). The basal erosional contact is also abrupt, occasionally marked by small to moderate-scale gravel-filled scour pockets with low-angle cross-stratification. These channels locally down-cut up to 3 m into the underlying strata.

MCSM II sandstones consist of predominantly moderately-sorted siliciclastic coarse-grained, arkosic sandstones that are commonly multistoried and ledge-forming, containing a host of coarse sand-sized grains and pebbles. The sandstones are interbedded with crudely-stratified, grain-supported conglomerates, especially in the lower half of the unit. Horizontal stratification is the dominant sedimentary structure. Sandstones are laterally extensive, sharply-based, and the petrography indicates that the sediment was derived from a plutonic? source terrane. Both sandstone and conglomerate beds are fossiliferous.

Sandstones fine upward (crude normal grading) into fine-grained pale yellowish-brown-colored silty sandstones and thinly-bedded micaceous siltstones with planar laminations. Small-scale crossbeds and current ripplebeds are rare in the sandstone beds, but are common in the associated overbank siltstones. In the upper half of the unit individual sequences vary from 0.6 to 1.6 m thick. Bed geometry appears to be lenticular although poorly-defined bedding planes with thick slope and pebbly drapes prevent detailed observations. At 8.3, 9.6, 11.3 and 13.6 m, 5.0-10.1 cm diameter silica-cemented sandstone concretions weather out from thin indurated sandstones. The latter forms a distinct concretionary marker horizon at the top of the unit.

The sandstones are intimately interlayered with extraformational pebbly to cobble conglomerates. The conglomerates usually display a poor to moderately imbricated fabric and crudely-developed graded bedding. Clast types present are sub-angular to sub-rounded, and include plutonic, metasedimentary and cataclastic lithologies of apparent Canebrake Conglomerate III affinity derived from a Santa Rosa Mountains source provenance.

A 10.1 cm-thick light gray-colored reworked volcanic ash occurs 2.1 m above the base of the lowermost sandstone bed. The ash is laterally continuous throughout MC, and is coeval to an ash unit that underlays the Valle Escondido *M. imperator* site (Miller, et al. 1991) above the base of the MCSM (undifferentiated) east of Vista del Malpais.

Overall, the lenticular bed geometry of the sandstones, horizontal to cross-stratification, and crudely-stratified conglomerates most likely resulted from fluviatile longitudinal bar-forming processes of outwash braided streams associated with a basinward extension of distal alluvial fans.

## MORPHOLOGY

Locally, the stratigraphic range of *Mammuthus imperator* is restricted to the basal sandstones of the MCSM, where it is represented by three measurable dental specimens (Table 1). In ABDSPV 1227/5126, a nearly complete individual, the anterior few plates on both upper and lower M/3s are worn away. Length of the M/3s exceeds 240 mm. The dentary is relatively gracile (Maximum depth and width at mid M/3, 14.0 cm, and

## Table 1. Dental parameters of *Mammuthus*.

| T | SP | TP | NP | W | LF | ET |
|---|---|---|---|---|---|---|
| C | 789 | /3 | 12+ (18+) | 94 | 6.09 (3) | 2.21 (6) |
| C | 4260 | /3 | 10+ (20+) | 85+ | 6.07 (3) | 2.17 (6) |
| I | 3981 | /3 | 10+ (12) | 104 | 4.73 (4) | 2.70 (6) |
| I | 5126 | 3/ | 12+ (15) | 110 | 5.04 (6) | 2.57 (6) |
| I | 5126 | /3 | 10+ (12) | 92 | 4.22 (4) | 2.27 (6) |

| T | PD | TP | NP | W | LF | ET |
|---|---|---|---|---|---|---|
| C | Mg | 3 | 20 - 24 | - | 5 - 7 | 2.0 - 3.0 |
| C | Ma | 3/ | 18 - 20<br>21 | 75 - 120<br>98 | 5.2 - 8.8<br>6.2 | 1.2 - 3.2<br>2.0 |
| C | Ma | /3 | 18 - 23<br>21 | 73 - 111<br>93 | 3.7 - 8.5<br>6.2 | 1.2 - 3.2<br>2.1 |
| I | Mg | 3 | 16 - 19 | - | IC | IC |
| I | Ma | 3/ | 15 - 19<br>17 | 81 - 125<br>101 | 4.0 - 8.7<br>6.0 | 1.6 - 3.4<br>2.5 |
| I | Ma | /3 | 14 - 22<br>17 | 61 - 129<br>95 | 3.6 - 7.8<br>5.5 | 1.9 - 3.6<br>2.7 |
| M | Mg | 3/ | 11 - 14<br>13 | 86 - 126<br>105 | 3.7 - 6.1<br>4.9 | 2.6 - 4.1<br>3.3 |
| M | Mg | /3 | 10 - 14<br>12 | 69 - 119<br>97 | 3.5 - 5.9<br>4.9 | 2.4 - 4.1<br>3.4 |

T = taxon; SP = ABDSP, Anza-Borrego Desert State Park specimen numbers; TP = tooth position; NP = number of plates, + indicates minimum number, ( ) estimate number based on stage of wear; W = maximum width; LF = lamellar frequency, number of plates /10 cm, ( ) number of averaged measurements; ET = enamel thickness in mm, ( ) number of averaged measurements; PD = published data; C = *M. columbi*; I = *M. imperator*, M = *M. meridionalis*; IC = intermediate condition between *M. meridionalis* and *M. columbi*; Ma = from Madden (1981), range and average; these data do not include advanced *M. columbi* which Madden places in *M. jacksoni* (= *M. jeffersonii*); Mg = from Maglio (1973), range only.

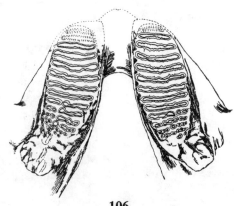

16.4 cm), conforming to the morphology of *M. imperator* (Madden, 1980). In the M/3 of ABDSPV 978/3881, a partial skull and mandible, the posterior-most plates are lightly worn and the anterior few plates are missing. Ten plates are preserved in this specimen. A fragmented cheek tooth from ABDSPV 1021/4214, is provisionally assigned to *M. imperator* based on an average enamel thickness of 2.5 mm.

Although *Mammuthus columbi* ranges throughout the Ocotillo formation, it is represented by two measurable dental specimens from the basal sandstones of the MCSM (Table 1). The M/3s in ABDSPV 1025/4260, a complete mandible comparable to the robust morphology (maximum depth and width at mid M/3, 21.5 cm, and 18.6 cm) of *M. columbi* (Madden, 1980), are in an early eruptive stage. The posterior portions of the teeth remain imbedded in the dentary, and approximately half of the crown displays ten plates in wear. The implies a total plate number of more than 20. Although the anterior one-third of ABDSPV 178/789 is missing, the posterior 6 of the 12 preserved plates in this M/3 exhibit no wear.

## CONCLUSION

Accepted diagnostic dental parameters for *Mammuthus columbi* and *M. imperator* (Table PD) exhibit an overlapping range of values. Those specimens herein considered *M. columbi* accordingly fall within the upper range of *M. imperator*, and likewise those placed in *M. imperator* fall within the lower range of *M. columbi*. But more importantly, the *M. imperator* and *M. columbi* dentaries are morphologically distinct, and the M/3s are separable into two groups with average plate numbers of 12 and 19, and average lamellar frequencies of 4.48 and 6.08 respectively. No intermediate forms have been recovered, and it is improbable that these materials sample only the end members of a morphologically variable populations. Therefore, we suggest that *M. columbi* and *M. imperator*, as presently defined, locally existed sympatrically, and that the Borrego Badlands yield the earliest recorded representatives of *M. columbi*.

\* This report is an updated revision of Jefferson and Remeika (1994).

## REFERENCES CITED

Agenbroad, L.D. 1980. Quaternary mastodon, mammoth and men in the New World. Canadian Journal of Anthropology 1(1):99-101.

----- 1984. New World mammoth distribution. In Quaternary Extinctions: A Prehistoric Revolution, edited by P. S. Martin and R. G. Klein. University of Arizona Press, Tucson. pp. 90-108.

Dibblee, T. W., Jr. 1954. Geology of the Imperial Valley region, California. California Division of Mines Bulletin 170(2):21-28.

Graham, R. 1984. Taxonomy of North American mammoths. In The Colby Mammoth Site--Taphonomy and Archaeology of a Clovis Kill in Northern Wyoming, edited by G. Frison and L. C. Todd. University of New Mexico Press, Albuquerque. Appendix 2, Part 1:165-169.

Kurtén, B., and E. Anderson 1980. Pleistocene mammals of North America. Columbia University Press, New York. 442p.

Madden, C. T. 1980. Earliest dated *Mammuthus* from North America. Quaternary Research 13:147-150.

----- 1981. Mammoths of North America. Doctoral Dissertation, Department of Anthropology, University of Colorado, Boulder. 271p.

Maglio, V. J. 1973. Origin and evolution of the Elephantidae. Transactions of the American Philosophical Society, New Series 63:1-149.

Miller, G.J., P. Remeika, J.D. Parks, B. Stout, and V. Waters 1991. A preliminary report on half-a-million year old cut marks on mammoth bones from the Anza-Borrego Desert Irvingtonian. Imperial Valley College Museum Society Occasional Paper 8:1-47.

Remeika, P. 1992. Preliminary report on the Stratigraphy and Vertebrate fauna of the middle Pleistocene Ocotillo formation, Borrego Badlands, Anza-Borrego Desert State Park, California. Abstracts of Proceedings, 6th Annual Mojave Desert Quaternary Research Symposium, San Bernardino County Museum Quarterly 39(2):25-26.

Remeika, P., and G. T. Jefferson 1993. The Borrego Local Fauna: revised basin-margin stratigraphy and paleontology of the western Borrego Badlands, Anza-Borrego Desert State Park, California. In Ashes, Faults and Basins, edited by R. E. Reynolds and J. Reynolds, San Bernardino County Museum Association Special Publication 93(1):90-93.

Rymer, M. J. 1991. The Bishop Ash bed in the Mecca Hills. In Geological excursion in southern California and Mexico, edited by M. J. Walawender and B. B. Hanna. Geological

Society of America Annual Meeting Guidebook, San Diego. pp. 388-396.

Sarna-Wojcicki, A. M., and M. S. Pringle, Jr. 1992. Laser-fusion $^{40}Ar/^{39}Ar$ ages of the Tuff of Taylor Canyon and Bishop Tuff, E. California-W. Nevada. EOS, Transactions of the American Geophysical Union, Abstracts to Meetings, p. 241.

Sarna-Wojcicki, A. M., H. R. Boweman, C. E. Meyer, P. C. Russell, M. J. Woodward, G. McCoy, J. J. Rowe, Jr., P. A. Baedecker, F. Asaro, and H. Michel 1984. Chemical analyses, correlations, and ages of upper Pliocene and Pleistocene ash layers of east-central and southern California. U. S. Geological Survey Professional Paper 1293:1-40.

Scheuing, D. F. 1989. The effect of structure on paleomagnetic data collected in complexly deformed terrains: a case study from the San Jacinto Fault Zone, southern California. Manuscript on File, Anza-Borrego Desert State Park, and Lamont-Doherty Geological Observatory. 22p.

Scheuing, D. F., L. Seeber, K. W. Hudnut, and N. L. Bogen 1988. Block rotation in the San Jacinto Fault Zone, southern California. EOS, Transactions of the American Geophysical Union, Abstracts to Meetings, 69:1456.

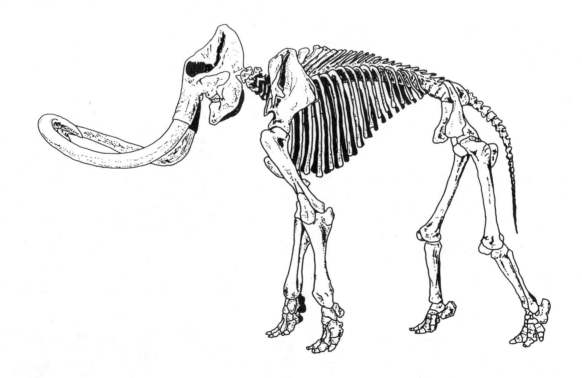

# A DIVERSE RECORD OF MICROFOSSILS AND FOSSIL PLANTS, INVERTEBRATES, AND SMALL VERTEBRATES FROM THE LATE HOLOCENE LAKE CAHUILLA BEDS, RIVERSIDE COUNTY, CALIFORNIA

**David P. Whistler, E. Bruce Lander, and Mark A. Roeder**

Paleo Environmental Associates, Inc., 2248 Winrock Ave., Altadena, California 91001, and Natural History Museum of Los Angeles County, 900 Exposition Boulevard, Los Angeles, California 90007

## ABSTRACT

As documented by Van de Kamp (1973) and Waters (1983), fresh-water lakes (Lake Cahuilla) have existed intermittently during the Holocene in the fault-bounded Salton Trough of southern California. These lakes were caused by diversions of the Colorado River westwardly into the basin. Waters (1983) provided stratigraphic and chronologic analyses of the upper 7 m of lacustrine and fluvial sedimentation based on exposures in a 6 m-long trench 8 km southwest of Indio, California.

Human occupation is well documented for at least the last four lacustrine intervals, which range in age from approximately 1,440 years before present (ybp) to approximately 470 ybp (Waters, 1993). Abundant fresh-water mollusks and some vertebrate remains were recovered previously from these intervals. These remains represent dusky snails, desert tyronia snails, California floater clams, leopard frogs, tortoises, white pelicans, cormorants, jack rabbits, white footed mice, horses, and bighorn sheep (Langenwalter, 1990).

Recent paleontologic studies are based on analyses of two test trenches, each 4 m deep and excavated as part of a paleontologic resource impact mitigation program associated with grading of a parcel in La Quinta, California. These trenches, 10 km west of the trench studied by Waters (1983), contained a succession of interbedded lacustrine and thinly bedded fluvial sediments. Lacustrine and fluvial units both contain abundant mollusk shells. The lacustrine units also contain pollen, siliceous microfossils, and ostracod valves. Bones and teeth of a diversity of small vertebrates, including fishes, lizards, snakes, birds, rabbits, and rodents, were recovered by dry screening fossiliferous sediment samples from three stratigraphic intervals/lithologic units (Whistler, et al. 1995). The diatom, ostracod, and mollusk assemblages indicate fresh or slightly saline water during the lacustrine intervals. The pollen flora was derived primarily from the adjacent Santa Rosa Mountains. The vertebrate assemblage contains small-bodied species indicative of both sandy and rocky, brush-covered desert habitat. No vertebrate species typical of a wetter or aquatic habitat was recovered. The absence of these latter vertebrate species suggests flooding of the Salton Trough occurred rapidly, leaving insufficient time for them to migrate from the Colorado River to Lake Cahuilla before the subsequent evaporation of the lake.

Charcoal was recovered from the stratigraphic intervals that yielded the fossils and from a lower interval. Radiocarbon age determinations of $1,125 \pm 80$ and $2,545 \pm 50$ ybp for the two upper intervals sampled and $5890 \pm 60$ ybp for the lower interval indicate the remains from the parcel are from stratigraphic intervals below or older than the four fluvial/lacustrine intervals defined by Waters (1983). Moreover, the radiocarbon age determination for the lower fluvial unit, which overlies a lacustrine unit in Trench II, indicates a Lake Cahuilla highstand, apparently older than any recorded previously, occurred approximately 6,000 years ago. The modern surface in the parcel is 25 m below the Lake Cahuilla highstand and the four younger fluvial/lacustrine intervals probably have been removed by subsequent eolian erosion or grading of the parcel.

## INTRODUCTION

This report presents the results of a paleontologic resource impact mitigation program conducted by Paleo Environmental Associates, Inc. (PEAI) personnel in support of grading of a parcel in La Quinta, Riverside County, California (Lander and Whistler, 1995b; Figure 1). The mitigation program was conducted in compliance with Society of Vertebrate Paleontology guidelines to reduce the adverse environmental impact of grading on the paleontologic

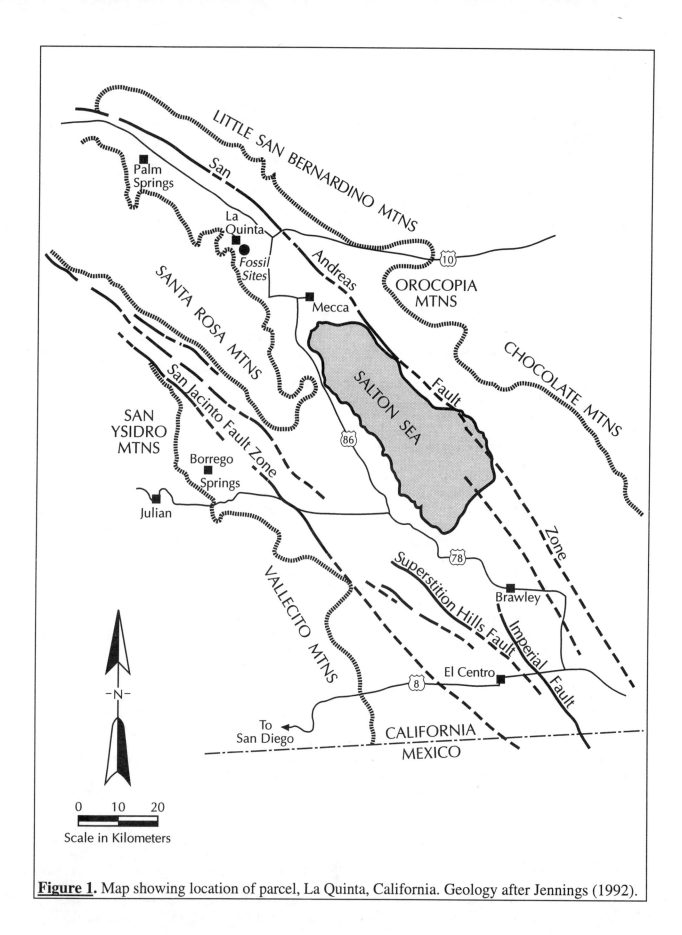

**Figure 1.** Map showing location of parcel, La Quinta, California. Geology after Jennings (1992).

resources of the parcel. The mitigation program was required of the property owner and developer by the City of La Quinta Community Development Department as part of the department's permitting process. The program was required because of the high potential for scientifically highly important fossil remains being uncovered by grading of the Lake Cahuilla beds, which underlie the entire parcel.

## GEOLOGIC SETTING

La Quinta lies in the Salton Trough a northward extension of the Gulf of California (McKibben, 1993). The Salton Trough is an active continental rift underlain by the landward extension of the East Pacific Rise. Since the beginning of the Holocene, the Colorado River delta has blocked marine water from entering the Salton Trough from the Gulf of California. Fresh-water lakes have existed intermittently in the deeper portions of the basin that developed landward of the Colorado River delta (Van de Kamp, 1973; Waters, 1983; Maloney, 1986)

Geologic mapping of the La Quinta region (Figure 1) indicates the area is immediately underlain by flat-lying lacustrine and fluvial sedimentary strata deposited below the 12-meter-elevation high shoreline of ancient Lake Cahuilla, which is believed to have existed intermittently from 400 to at least 2,300 years ago (Rogers, 1965; Van de Kamp, 1973; Waters, 1983). The lake sediments were deposited during each of at least seven lake highstands, each highstand resulting from flooding of the Salton Trough by inflow from the Colorado River (Waters, 1983). The fluvial sediments were deposited during the intervening lake lowstands, when the former lake bed was dry. The lacustrine and intervening fluvial sediments, herein referred to as the Lake Cahuilla beds, are at least 18.5 m thick in the parcel (Langenwalter, 1990), which lies near the southwestern margin of the former lake.

## METHODS

As part of the paleontologic resource impact mitigation program conducted by PEAI personnel, two test trenches (I, II) were excavated in February 1995 to document the stratigraphy underlying the parcel and to sample each stratigraphic interval or lithologic unit exposed in the trench sidewalls for fossil and carbonized plant remains (Lander and Whistler, 1995a). Each trench was excavated to a depth equal to, or greater than the maximum depth of grading planned in the parcel, thereby ensuring that the deepest stratigraphic interval or lithologic unit to be encountered by grading was exposed by trenching and made available for analysis. One trench (II) was excavated in an irrigated field, the other (I) in a dry fallow field. Trench II was water saturated below 1.5 meters in depth, making sampling difficult. Therefore, the fossil recovery effort concentrated on Trench I.

A 10-kilogram sediment sample was collected from each stratigraphic interval/lithologic unit and/or productive fossil-bearing zone exposed in the trench sidewalls. Each sample was dry screened with 1/4-, 1/8-, and 30-mesh screens to document the occurrence and allow the recovery of small invertebrate and vertebrate remains. A substantially larger sample was collected from each of four vertebrate-fossil-bearing zones. A total of approximately 3,200 kilograms of sediment was screened. Dry screening yielded a large number of small mollusks, a representative sample of which was submitted to Lindsey T. Groves of PEAI and the Natural History Museum of Los Angeles County (LACM) for identification and analysis. The remaining sediment samples were examined microscopically by the senior author (Whistler) and any small vertebrate fossil remains were removed. The fossil fish remains were sent to Kenneth W. Gobalet of California State University, Bakersfield (CSUB), for identification and analysis. The remaining vertebrate fossil remains were studied by Whistler. Four small sediment samples were submitted to Micropaleo Consultants, Inc., Encinitas, California, for identification and analysis of pollen and microfossils, including diatoms, foraminifers, and ostracods. Each piece of charcoal in excess of 2 millimeters in length was collected and four charcoal samples were submitted to Beta Analytic Inc., Miami, Florida, for radiocarbon age determinations.

Fossil mollusk and vertebrate specimens have been accessioned into the LACM collections and catalogued into the Invertebrate Paleontology Section (IP) database under LACMIP fossil site numbers 16830 (= DPW [David P. Whistler] sampling site number 2471) and 16831 (= DPW 2469/2471), and the Vertebrate Paleontology Section (VP) database under LACMVP fossil site numbers 6252 (= DPW 2471/2472), 6253 (= DPW 2470), 6255 (= DPW 2468), and LACM 6256, respectively.

## STRATIGRAPHY

The stratigraphy exposed in the Trench I and II sidewalls is presented in Figure 2. The Lake Cahuilla beds generally are composed of thinly bedded (1-to-2-centimeter-thick), cross-bedded, poorly sorted, fine-grained, immature, light grayish-brown, fluvial sandstone strata interbedded with massive, poorly sorted, bioturbated, sandy and silty, white-to-light-gray, lacustrine mudstone strata.

The sand comprising the fluvial sandstone strata is derived locally from metamorphosed crystalline

**Table 1.** Taxonomic list by sampling/fossil site, Lake Cahuilla beds, Trench I, La Quinta, Riverside County, California.

| TAXON (COMMON NAME) | DPW[a] SAMPLING SITE | | | | |
|---|---|---|---|---|---|
| | 2467[b] | 2468 | 2469 | 2470 | 2471 |
| **DIATOMS** | | | | | |
|     *Campylodiscus clypeus*? | | | X | X | |
|     *Cocconeis placentula* | | | X | | |
|     *Cyclotella kuetzingiana*? | | | X | | |
|     *Epithemia argus* | X | | X | X | |
|     *Epithemia turgida* | | | X | X | |
|     *Hantzschia taenia*? | | | X | X | |
|     *Mastogloia elliptica* | | | X | | |
|     *Navicula clementis*? | | | X | | |
|     *Navicula palpebralis* | | | X | X | |
|     *Navicula ergadensis*? | | | X | | |
|     *Nitzschia etchegoinia*? | | | X | X | |
|     *Nitzschia granulata*? | | | X | X | |
|     *Pinnularia viridis* | | | | X | X |
|     *Rhopalodia gibba* | | | X | | |
|     *Surirella striatula* | | | X | | |
|     *Synedra ulna*? | | | X | | |
|     *Terpsinoei musica* | | | X | X | |
|     *Tetracyclus lacustris* | | | X | | |
| **LAND PLANTS** | | | | | |
|     *Selaginella sinuites* (club-moss) | | | | X | |
|     Polypodiaceae (ferns) | X | | | X | |
|     *Pinus* sp. (pine) | X | | X | X | |
|     Betulaceae (alders, birches) | X | | | | |
|     *Ceanothus*? sp. (mountain lilac?) | X | | | | |
|     Chenopodiaceae (saltbushes) | X | | | X | |
|     Onagraceae (evening primroses) | | | X | X | |
|     *Quercus* sp. (oak) | | X | | | |
|     Compositae (*Ambrosia*-type) (ragweed) | | | X | | |
|     Compositae (*Helianthus*-type) (sunflower) | | | X | | |
| **PORIFERA** (sponges) | | | X | X | |
| **MOLLUSCA** (mollusks) | | | | | |
|   **BIVALVIA** (clams) | | | | | |
|     *Anodonta californiensis* (California floater) | | | X | X | X |
|     *Pisidium casertanum*? (ubiquitous pea clam) | | | X | X | X |
|   **GASTROPODA** (snails) | | | | | |
|     *Amnicola longinqua* (dusky snail) | | | X | X | X |
|     *Ferrissia walkeri*? (cloche ancylid) | | | | | X |
|     *Flumnicola* sp. (pebble snail) | | | | | X |
|     *Gyraulus parvus* (ash gyro) | | | X | X | X |
|     *Helisoma trivolvis* (rams horn) | | | X | X | X |
|     *Physella ampullacea* (paper physa) | | | X | X | X |
|     *Physella humerosa* (corkscrew physa) | | | X | X | X |
|     *Tryonia protea* (desert tryonia) | | | X | X | X |

| Taxon | DPW 2467 | DPW 2468 | DPW 2469 | DPW 2470 | DPW 2471 |
|---|---|---|---|---|---|
| **CRUSTACEA (crustaceans)** | | | | | |
|   OSTRACODA (microscopic bivalved crustaceans) | | | | | |
|     *Cypridopsis vidua* | | | X | X | |
|     *Cyprinotus torosa* | | | X | | |
|     *Limnocythere ceriotuberosa* | | | X | X | |
| **OSTEICHTHYES (bony fish)** | | | | | |
|   CYPRINIDAE (minnows) | | | | | |
|     ?*Cyprinodon macularius* (desert pupfish) | | | | | X |
|     *Gila elegans* (bonytail) | | X | | X | X |
|   CATOSTOMIDAE (suckers) | | | | | |
|     *Xyrauchen texanus* (razorback sucker) | | | | | X |
| **REPTILIA (reptiles)** | | | | | |
|   SQUAMATA (lizards, snakes) | | | | | |
|     IGUANIDAE (iguanid lizards) | | | | | |
|       *Phrynosoma platyrhinos* (desert horned lizard) | | | | | X |
|       *Sceloporus magister* (desert spiny lizard) | | | | | X |
|       *Uma inornata* (Coachella Valley fringe-toed lizard) | | X | | | X |
|       *Urosaurus graciosus* (long-tailed brush lizard) | | | | | X |
|     COLUBRIDAE (colubrid snakes) | | | | | |
|       *Chionactis occipitalis* (western shovel-nosed snake) | | | | | X |
|       *Hypsiglena torquata* (night snake) | | | | | X |
|       *Pituophis melanoleucus* (gopher snake) | | | | X | X |
|       *Sonora semiannulata* (western ground snake) | | | | | X |
|     CROTALIDAE (rattlesnakes) | | | | | |
|       *Crotalus cerastes* (sidewinder) | | | | | X |
|       *Crotalus* sp. (large rattlesnake) | | | | | X |
| **AVES (birds)** | | | | | |
|   PASSERIFORMES (advanced land birds) | | | | | X |
| **MAMMALIA (mammals)** | | | | | |
|   LAGOMORPHA (rabbits, pikas) | | | | | |
|     LEPORIDAE (rabbits) | | | | | |
|       *Sylvilagus* sp. (cottontail) | | | | | X |
|   RODENTIA (rodents) | | | | | |
|     SCIURIDAE (squirrels) | | | | | |
|       *Ammospermophilus leucurus* (antelope ground squirrel) | | | X | | X |
|     HETEROMYIDAE (pocket mice, kangaroo rats) | | | | | |
|       *Perognathus longimenbris* (pocket mouse) | | | X | | X |
|       *Dipodomys*? sp. (kangaroo rat) | | | | | X |
|     CRICETIDAE (wood rats, white-footed and harvest mice) | | | | | |
|       *Neotoma lepida* (desert wood rat) | | | | | X |
|       *Peromyscus* sp. (white-footed mouse) | | | | | X |

[a]DPW: David P. Whistler.
[b]DPW 2467 = middle lacustrine? unit;
 DPW 2468 = LACMVP 6255, middle fluvial unit;
 DPW 2469 = LACMIP 16830, upper lacustrine unit;
 DPW 2470 = LACMIP 16830, VP 6253, upper lacustrine unit;
 DPW 2471 = LACMIP 16831, VP 6252, upper fluvial unit.

basement appearing to be a diorite schist. The sand, being immature, contains considerable amounts of biotite and other mafic minerals that generally are rapidly weathered, and the sand grains are highly angular. Bedding planes between the strata are coated with clay and orange limonite? staining, which imparts a brownish hue to the otherwise gray sediments. Clay rip-up clasts, cross bedding, channeling, and load casts are prevalent. The lacustrine mudstone strata display very little internal structure. The upper lacustrine unit is highly bioturbated and has numerous vertical *Anodonta* (clam) burrows, some still containing the shells of the individuals that excavated the burrows (Figure 2). The upper surface of the upper lacustrine unit is channeled locally, and the burrows are filled with cleaner sand from the overlying upper fluvial unit, these phenomena suggesting a period of fluvial deposition across the former lake margin as the shoreline retreated and the lake bed emerged due to the loss of water inflow from the Colorado River. The middle lacustrine? unit in Trench I contains a clay-ball breccia, suggesting local redeposition of the clay by subsequent fluvial action during deposition of the overlying middle fluvial unit. The lower lacustrine unit, which, like the underlying lowermost fluvial unit, is not exposed in Trench I, is well developed in Trench II, but bioturbation of its upper surface is not apparent as at the top of the upper lacustrine unit.

The fossilized shells of fresh-water mollusks are abundant in nearly every stratigraphic interval exposed in the Trench I and II sidewalls (Figure 2). Most of the surface over thousands of square meters surrounding the Trench I site is littered with shells. Shells are particularly abundant in the upper lacustrine unit (DPW 2469/2470) and the overlying upper fluvial unit (includes DPW 2471/2472) in the Trench I sidewall. One fossiliferous zone in the upper fluvial unit contains a particularly high concentration of small snail shells that comprise more than 50 percent of the sediment. This fossiliferous zone also is present in the Trench II sidewall.

Charcoal was observed at two stratigraphic levels (DPW 2468, 2472 = 2471) in the Trench I sidewall, and at two stratigraphic levels (DPW 2473, 2474) in the Trench II sidewall (Figure 2).

## PALEONTOLOGY

Fossil remains representing a diversity of fresh-water diatoms, land plants, sponges, ostracods, mollusks, fish, and small terrestrial vertebrates were recovered from the Lake Cahuilla beds at La Quinta. Taxonomic lists by stratigraphic level are presented in Table 1, and sampling sites are depicted stratigraphically in Figure 2.

Pollen, diatoms, and sponge spicules were recovered from the lacustrine units. Ostracods, generally

**Table 2.** Radiocarbon age determinations, Lake Cahuilla beds, Trenches I and II, La Quinta, Riverside County, California.

| TRENCH | STRATIGRAPHIC INTERVAL | DPW[a] SAMPLE | AGE (ybp[b]) |
|---|---|---|---|
| I | upper fluvial unit | 2472 | 1,080 ± 80 |
| I | middle fluvial unit | 2468 | 2,500 ± 50 |
| II | upper fluvial unit | 2474 | -- |
| II | lower fluvial unit | 2473 | 5,890 ± 60 |

[a]DPW: David P. Whistler.
[b]ybp: years before present.

rare, were recovered only from the upper lacustrine unit. Taxonomic composition of the mollusk samples did not differ significantly between the fluvial and lacustrine units or by stratigraphic interval. Fossil fish were recovered from both fluvial and lacustrine units, but were most common in the upper fluvial unit. Most of the remaining small vertebrate fossil remains were recovered from the base of the upper fluvial unit.

### Pollen

Pollen was found in three sediment samples from the upper lacustrine unit (DPW 2469, 2470) and the underlying middle lacustrine? unit (DPW 2467) in Trench I. As reported by Hideyo Haga of Micropaleo Consultants, Inc., the predominance of pine pollen, particularly in the lower two samples, indicates the primary source of the pollen was in the nearby Santa Rosa Mountains west of the parcel. Other pollen species represent a mix of lowland desert shrubs and annuals with some additional montane species.

### Siliceous microfossils

As reported by Stanley A. Kling of Micropaleo Consultants, Inc., diatom tests and siliceous sponge spicules were found in sediment samples from two levels in the upper lacustrine unit (DPW 2469, 2470), and from the underlying lacustrine unit (DPW 2467) in Trench I. The upper samples contained abundant remains representing a diversity of diatom species that suggest fresh-water to slightly brackish-water (low-salinity) environments. The sample from the lower half of the upper lacustrine unit (DPW 2469), in particular, indicates the presence of a large persistent lake with a low sedimentation rate. The higher sample (DPW 2470), which contains fewer diatom species, probably indicates waning of the lake and an increased sedimentation rate. Sponge spicules also were recovered from the upper samples. Though common to abundant in the upper lacustrine unit, siliceous microfossil remains are rare in the lower lacustrine unit in Trench I (DPW 2467), possibly because a lacustrine environment was poorly developed compared to the depositional environment represented by the upper two samples.

### Mollusks

The fossilized shells of fresh-water snails and clams are abundant in nearly every stratigraphic interval exposed in the Trench I and II sidewalls, particularly in the upper fluvial unit (DPW 2471 = LACMIP fossil site 16831) and the underlying lacustrine unit (DPW 2469/2470 = LACMIP fossil site 16830) in Trench I. As reported by Lindsay T. Groves of PEAI and the LACM, the remains represent species that still exist in the Colorado Desert of southeastern California and western Arizona. These species occupy fresh-water habitats with muddy to sandy substrates in comparatively shallow (less-than-2-meter-deep), relatively permanent bodies of water with abundant rooted vegetation and debris.

### Ostracods

As reported by Kenneth L. Finger of Micropaleo Consultants, Inc., ostracod valves representing three species were found in only two sediment samples (DPW 2469, 2470) from the upper lacustrine unit in Trench I. *Cypridopsis vidua* and *Limnocythere ceriotuberosa* inhabit fresh-water lakes, although species of *Limnocythere* also inhabit inland salt-water lakes. The lack of sorting of valves by size and the presence of juvenile valves indicate a low-energy environment. The predominance of disarticulated valves indicates a low sedimentation rate. In general, the ostracod assemblages suggest fresh-water to slightly brackish-water (low-salinity) environments.

### Fish

As reported by Kenneth W. Gobalet of CSUB, remains representing three species of fresh-water fish were recovered from three stratigraphic intervals (DPW 2468, 2470, 2471 = LACMVP fossil sites 6255, 6253, 6252, respectively) in Trench I. The razorback sucker and bonytail have been reported previously from the Lake Cahuilla beds (Gobalet, 1992, 1994). However, a damaged vertebra, if correctly identified, represents the oldest record of a desert pupfish from the Salton Trough (Wilke, 1979), although the modern distribution of the desert pupfish includes springs surrounding the Salton Trough (Schoenherr, 1993).

### Terrestrial Vertebrates

With one exception (see below), all of the fossil terrestrial vertebrate remains represent small animals, the largest being the cottontail rabbit. These latter remains consist of bones and teeth and, except for the bird remains, represent small reptile and mammal species. Table 1 summarizes the assemblages by stratigraphic level. Of the three samples recovered from Trench I, only one (DPW 2471 = LACMVP fossil site 6255) yielded remains representing a diversity of species.

The terrestrial vertebrates comprise an assemblage very similar to the fauna currently inhabiting the brush-covered Salton Trough floor. Three species, including *Sceloporus magister* (desert spiny lizard), *Ammospermophilus leucurus* (antelope ground squirrel), and *Neotoma lepida* (desert wood rat) prefer habitats where scattered or prominent boulders are present, but also are found less commonly in open desert habitats. *Phrynosoma platyrhinos* (desert horned lizard) and *Uma inornata* (Coachella Valley fringe-toed lizard) frequent

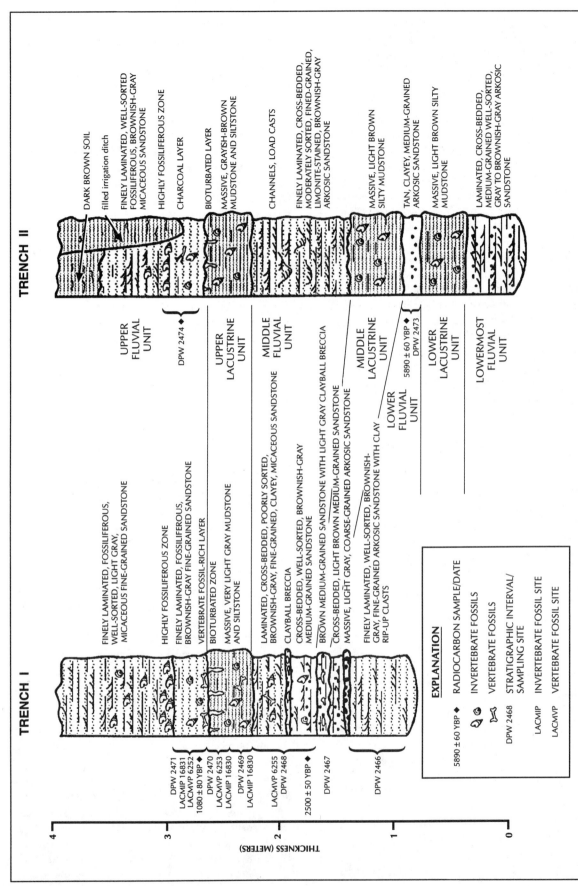

**Figure 2.** Stratigraphic columnar sections and sampling/fossil sites, Lake Cahuilla beds, La Quinta, Riverside County, California.

areas with loose sand. Otherwise, the species present are generalists in terms of their habitat preferences.

The nearest shoreline during a Lake Cahuilla highstand was less than 2 km west of the parcel at the base of the Santa Rosa Mountains. The mountains rise abruptly from the basin floor, and rocky outcrops and scattered boulders dominate the slopes. The close proximity of the parcel and the rocky mountain slopes probably explains the recovery of species preferring rocky habitats.

The absence of species, such as frogs, toads, aquatic turtles, watersnakes, waterfowl, shrews, gophers, cotton rats, meadow voles, and muskrats, that frequent wetter or aquatic habitats suggests Lake Cahuilla was not a large persistent body of water. Apparently, these species did not have sufficient time to migrate around the lake margin from the Colorado River after the river had overflowed into the Salton Trough to form Lake Cahuilla and before the lake had evaporated. Desert species already present in the basin as it filled with overflow from the Colorado River simply were displaced with the margin of the expanding lake, their isolated bones and teeth subsequently accumulating as fossils within the lacustrine and fluvial sediments.

A single tooth-bearing jaw of a large sheep, presumably a bighorn sheep, was recovered from LACMVP fossil site 6256 during paleontologic monitoring of grading in the parcel after completion of the test trenches.

## RADIOCARBON AGE DETERMINATIONS

Four charcoal samples, two each from Trenches I and II, were submitted to Beta Analytic Inc. for radiocarbon age determinations. Preliminary treatment indicated only three samples contained sufficient carbon to yield reliable age determinations. Two samples (DPW 2471, 2473) were analyzed with the more precise accelerator-mass-spectrometer (AMS) method, and the third (DPW 2468) was analyzed with an extended counting time. Using these techniques, the three samples yielded very precise age determinations with small statistical errors. The final age determinations are presented in Table 2 as years before present (ybp), by convention, "present" being 1950 AD. Sampling sites are depicted stratigraphically in Figure 2.

The youngest age determination ($1,080 \pm 80$ ybp) from the parcel suggests the upper fluvial unit in Trench I is the same as the fluvial unit at the top of fluvial/lacustrine interval 1 (fourth interval below surface), as recognized by Waters (1983) in a trench 10 km east of the parcel and 8 km southwest of Indio, and that the upper three fluvial/lacustrine intervals (2 to 4) recognized by Waters (1983) are not present in the parcel. The two older age determinations from the parcel ($2,500 \pm 50$, $5,890 \pm 60$ ybp) are older than the oldest age determination ($2,300 \pm 120$ ybp) reported by Waters (1983) for the sixth lacustrine unit below the surface, and suggest the middle fluvial unit in the parcel is the same as the seventh fluvial unit recognized below the surface by Waters (1983). However, the oldest age determination from the parcel suggests the lower fluvial unit in Trench II is a correlative of the eighth fluvial unit recognized by Waters (1983) below the surface, or is a correlative of an even older fluvial unit than any exposed in the trench studied by Waters (1983). The lowermost fluvial unit in Trench II is even older.

## SUMMARY

The recovered fossil remains and the radiocarbon age determinations for samples from the La Quinta area demonstrate the occurrence of fossil vertebrate assemblages similar to the fauna now inhabiting the Salton Trough existing in the basin during the later Holocene. Lacustrine conditions did not persist for periods long enough to allow aquatic species to migrate from the Colorado River delta around the shoreline of Lake Cahuilla. The radiometric age determination on the sample (DPW 2473) from the lower fluvial unit, which overlies a lacustrine unit in Trench II, indicates that a lake highstand, apparently older than any recorded previously by Waters (1983), occurred approximately 6,000 years ago. The other younger age determinations suggest the three youngest lake highstands are not recorded by the Lake Cahuilla beds in the parcel, and could have been removed by eolian erosion or earlier grading of the parcel.

The species represented by the diatom, sponge, ostracod, and mollusk remains suggest Lake Cahuilla consisted of fresh to slightly brackish water. The mollusks indicate that, at least in the parcel, the lake did not exceed 2 m in depth for any significant period of time. Though presumably rare, desert pupfish inhabited the lake. Most of the mollusk and terrestrial vertebrate species were not reported previously from the Lake Cahuilla beds by Langenwalter (1990) and, like the pupfish, at least some of the recovered remains probably represent the first prehistoric records of their respective species from the Salton Trough.

## ACKNOWLEDGMENTS

The authors are indebted to KSL Recreation Corporation for access to the PGA West Tom Weiskopf Signature Course and for supporting this study. We especially wish to thank Steven Walser and Steven Auckland of KSL Recreation Corporation for their assistance. Gino Calvano of PEAI assisted in collecting and field processing the fossiliferous sediment samples. Micropaleo Consultants, Inc., of Encinitas, California, provided identifications and analyses of the pollen, siliceous microfossils, and ostracods. Lindsey T. Groves of PEAI and the LACM provided identifications and analyses of the mollusks. Kenneth W. Gobalet of CSUB provided identifications and analyses of the fossil fish. Samuel A. McLeod of the LACM reviewed the manuscript. Figures 1 and 2 were drafted by Lisa Yoshisato.

## LITERATURE CITED

Gobalet, K.W. 1992. Colorado River fishes of Lake Cahuilla, Salton Basin, southern California: a cautionary tale for zoo-archaeologists. Bulletin of the Southern California Academy of Sciences 91(2):70-83.

----- 1994. Additional archeological evidence for Colorado River fishes in the Salton Basin of southern California. Bulletin of the Southern California Academy of Sciences 93(1):38-41.

Jennings, W.W., compiler. 1992. Preliminary fault activity map of California. California Division of Mines and Geology, 92-03.

Lander, E.B., and D.P. Whistler 1995a. Preliminary technical report of findings, paleontologic resource impact mitigation program, PGA West, Tom Weiskopf Signature Course, La Quinta, Riverside County, California-- results of paleontologic testing of two trenches. Paleo Environmental Associates, Inc., project 93-5. Prepared for KSL Recreation Corporation.

----- 1995b. Paleontologic resource impact mitigation program final report, PGA West, Tom Weiskopf Signature Course, La Quinta, Riverside County, California. Paleo Environmental Associates, Inc., project 93-5. Prepared for KSL Recreation Corporation.

Langenwalter, P.E., II. 1990. A paleontological survey and assessment of the PGA West 5th golf course property near La Quinta, Riverside County, California. Heritage Resource Consultants project 198. Prepared for Douglas Wood & Associates.

Maloney, N.J. 1986. Coastal landforms of Holocene Lake Cahuilla, northeastern Salton Basin, California. In Geology of the Imperial Valley, California, edited by P.D. Guptil, E.M. Gath, and R.W. Ruff, South Coast Geological Society, Annual Field Trip Guidebook, Santa Ana, California 14:151-158.

McKibben, M.A. 1993. The Salton Trough rift. In Ashes, Faults and Basins, edited by R.E. Reynolds and J. Reynolds. San Bernardino County Museum Association Special Publication 93-1:76-80.

Rogers, T.H. 1965. Geologic map of California, Santa Ana sheet. California Division of Mines and Geology. Scale 1:250,000.

Schoenherr, A.A. 1993. Rise and fall of the desert pupfish *Cyprinodon macularius* at the Salton Sea. In Ashes, Faults and Basins, edited by R.E. Reynolds and J. Reynolds. San Bernardino County Museum Association Special Publication 93-1:67-70.

Van de Kamp, P.C. 1973. Holocene continental sedimentation in the Salton Basin, California: a reconnaissance. Geological Society of America Bulletin 84:827-848.

Waters, M.R. 1983. Late Holocene lacustrine chronology and archaeology of ancient Lake Cahuilla, California. Quaternary Research 19:373-387.

Whistler, D.P., E.B. Lander, and M.A. Roeder 1995. First diverse record of small vertebrates from late Holocene sediments of Lake Cahuilla, Riverside County, California. Abstracts of Proceedings. 9th Annual Mojave Desert Quaternary Research Symposium, San Bernardino County Museum Quarterly 42(2):46.

Wilke, P.J. 1980. Prehistoric wier fishing on recessional shorelines of Lake Cahuilla, Salton Basin, southwestern California. Proceedings of the Desert Fishes Council 11:101-102.

# STROMATOLITES: LIVING FOSSILS IN ANZA-BORREGO DESERT STATE PARK

## H. Paul Buchheim
Department of Natural Sciences, Section of Paleontology, Loma Linda University, Loma Linda, California 92350

### ABSTRACT

The investigator has discovered the only known occurrence of modern algal stromatolites north of Mexico living in an ephemeral stream in Carrizo Gorge and San Felipe Creek. Stromatolites are living fossils, and are highly significant paleoenvironmental indicators. They are also the frequent subject of scientific discussions about the origin of life on earth.

The Anza-Borrego stromatolites exhibit the criteria for identifying them as stromatolites, including the presence of cyanobacteria, biotically enhanced cementation by calcite, laterally linked hemispheroids, and calcite lamination. This research documented their occurrence and nature as well as the physical and chemical characteristics of their aqueous environment. The research sought to answer questions of (1) why are the stromatolites growing in some streams of the Anza-Borrego Desert while not others? and (2) is there something about the water chemistry, water temperature, or other physical or biological aspects of the streams that allow stromatolites to grow?

The association of stromatolites with concentrated water (greater than 4000 ppm dissolved solids) and their absence from fresher water leads to the conclusion that water chemistry is the major factor favoring stromatolite growth in Anza-Borrego's streams.

## INTRODUCTION

Stromatolites were only known as fossils until Logan (1961) reported his discovery of stramatolites in the hypersaline intertidal waters of Hamelin Pool at Shark Bay, Western Australia. A significant amount of controversy has arisen concerning their paleoenvironmental meaning, and most recently this controversy has been renewed by the discovery of giant subtidal algal stromatolites (Dill, et al. 1986) forming in normal salinity waters in current-swept channels between the Exuma Islands on the eastern Bahama Bank. Until the discovery of the Bahama stromatolites the only subtidal marine examples known to be living while undergoing lithification were in the hypersaline intertidal waters of Shark Bay, Western Australia. Since then, stromatolites have been described from numerous fresh and marine environments, including abundant stromatolites in streams and lakes of Cuatro Cienegas, Mexico (Winsborough, 1990). Previously, the occurrence of ancient stramatolites has been interpreted as indicative of hypersaline and intertidal paleoenvironmental conditions.

Stromatolites in both their domal and oncolitic form are reported and described from perennial streams, rivers, and ponds in northern Mexico (Winsborough, 1990). Modern stromatolites are not reported from ephemeral stream environments north of Mexico. This study investigated newly discovered stromatolites from remote Carrizo Gorge, 70 km east of San Diego, California (Figure 1). Although the stromatolites are only up to two cm thick, they nevertheless form large structures encrusting granitic cobbles and boulders. The stromatolites build by the rapid and abundant growth of filamentous blue-green algae (cyanobacteria) which induces the

Figure 1. Stromatolites from Carrizo Creek coat cobbles and boulders up to highest water line. Stream desiccates in zone of stromatolite buildup in late spring..

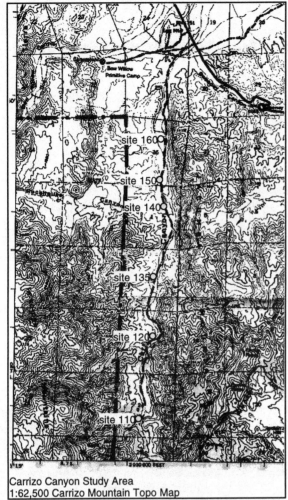

**Figure 2.** Carrizo Creek Study Area

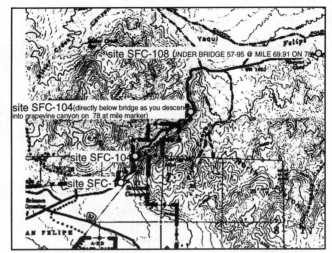

**Figure 3.** San Felipe Creek Study Area

**Figure 4.** Clean cobbles in Carrizo Creek, not coated as yet by stromatolites.

which fine sediment particles stick. This process results in periodic laminae composed of calcite, detrital stream particles, and an organic component composed of filamentous algae. These structures fulfill the criteria for identifying stromatolites: a knobby surface of digitate heads, lamination, internal calcification, and accretion due to the growth of mucilaginous cyanobacteria.

Modern stromatolites are not common, and are subject to intense study because of their implications to our understanding of ancient paleoenvironments and the origin of life. They are the dominant fossil in the Proterozoic. It has been hypothesized that their dramatic drop in abundance thereafter was due to the evolution of grazing metazoans. Hypersalinity and high-velocity currents have been cited (Logan, 1961; Dill, 1986) as factors limiting grazing on modern stromatolites. The stromatolites of Anza-Borrego Desert State Park's (ABDSP) streams provide an additional location to study the prerequisites for their abundant growth and factors to be considered in making interpretations about ancient stromatolite paleoenvironments.

**METHODS**

This research was conducted within Carrizo Gorge (figure 2) and San Felipe Creek (Figure 3) in Sentenac Canyon (adjacent to State Highway 78). Because of severe flooding in ABDSP during the spring of 1993, and significant modifications in some aspects of the research had to be made. Unfortunately, the stromatolites in Carrizo Gorge were entirely washed out. Stream flood velocities were high enough to totally rework boulders as large as 1 m in diameter. The stream was totally scoured and the original stromatolite-covered boulders either scoured clean or were transported away. Fresh, un-incrusted boulders, cobbles, and sand were deposited (Figure 4). However, this was not all bad news, but rather provided an opportunity to study the stromatolites as they reestablished themselves and continued to accumulate.

The methods in Carrizo Gorge included observation of stream bottom conditions, collection of water samples, and observation of stromatolite growth at approximately two week intervals from February through May, 1993. Less frequent checks were made through the spring of 1994 and continue at this time. Coated cobbles and water samples were collected at sites indicated on the map in Figure 2. Other environmental factors including water chemistry and pH, temperature, velocity and depth were determined. As a control, San Felipe Creek (Figure 3) was studied as well. It does not contain stromatolites in the upper part of the canyon, but is alkaline enough to yield precipitate travertine.

deep pools (Figure 6). Some stromatolites were up to 60 cm in length and 25 cm high. Where the stromatolites occurred, the cobbles were solidly cemented in place by calcite. Dark green algae coated most of the boulders and cobbles. The cyanobacteria are composed of sub-millimeter filaments coated with mucilaginous sheaths, and are tentatively identified as a species of *Schisothrix*. The algae apparently goes into a dormant state when the stream desiccates.

Stromatolite lamina (Figure 7) are 1-2 mm thick and are distinguished by textural variations and/or a dark brown to green algal-rich upper surface. Lamina are composed of couplets of alternating porous and dense calcite. The basal lamina are parallel to the substrate surface and become increasingly more digitate upward forming small heads 1-2 cm wide. Individual lamina consist of palisades of vertical

**Figure 5**. Carrizo Creek during the spring of 1990 in an area containing abundant stromatolites.

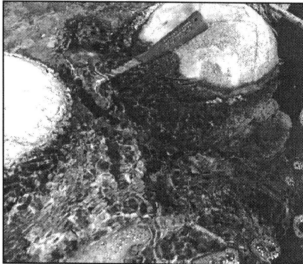

**Figure 6**. Stromatolites coat cobbles and boulders in a well cemented section of Carrizo Creek.

## RESULTS

Before the major flooding during the spring of 1993, dark green cyanobacteria were abundant in a 1.5 km stretch of Carrizo Gorge where water flowed for only a few months in the spring of each year (Figure 5). The stream varied in depth from .10 to 1.5 m and in width from 1-4 m. The stream bottom was sandy and contained abundant granitic cobbles and boulders that formed shallow pools up to 60 cm deep.

The best developed stromatolites coated cobbles and boulders in a 100 m stretch of 20-30 cm

calcite tubes as can be readily seen in the scanning electron micrograph (figure 8). Thin-section study revealed long algal filaments and masses of spore-like structures at the top of each laminae. Vertical algal filaments are coated with calcite microspar. Porosity decreased deeper into the stromatolite, due to in-filling by calcite cement. Also noted is the diatom, *Phopalodia gibba*, which is indicative of "stressful" conditions; in this case hypersalinity (pers. comm., Barbara Winsborough, 1995).

During the spring of 1993, water samples

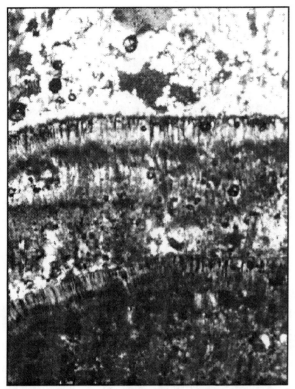

**Figure 7**. Thin section of stromatolites revealing micro structure of vertical algal filaments coated with calcite. Note the brown upper surface of the laminae which are composed of apparently spore cases. Individual lamina are about 1.2 mm thick.

**Figure 8**. Scanning electron photograph of stromatolites microstructure showing vertical calcite tubes. Also note the diatom in lower right corner, *Phopalodia gibba*, which is indicative of "stressful" conditions; in this case hypersalinity (personal communication, Dr. Barbara Winsborough). Magnification approximately 800x.

**Figure 9**. By the end of the 1993 season most of the cobbles and boulders were coated with half-hemispheroids 1-3 mm in diameter.

**Figure 10**. In the spring of 1994 cobbles were coated with a dark green mat that accumulated up to 3 mm of stromatolitic coating. Note cut out square in lower center

**Table 1.** Carrizo Creek water analysis in ppm.

| Location | 2245 (300 m. east of cc 110) | cc110 | cc110 | cc120 | cc140 | cc110 |
|---|---|---|---|---|---|---|
| Date | 5/3/90 | 2/19/93 | 4/1/93 | 4/22/93 | 5/6/93 | 5/27/93 |
| Cl | 1372 | 224 | 209 | 287 | 312 | 396 |
| SO4 | 2565 | 682 | 218 | 330 | 359 | 476 |
| HC0 | 272 | 400 | 387 | 387 | 312 | 296 |
| Na | 865 | 246 | 207 | 207 | 215 | 281 |
| K | 25 | 15 | 7 | 8 | 10 | 11 |
| Ca | 601 | 169 | 106 | 125 | 129 | 141 |
| Mg | 304 | 83 | 45 |  | 58 | 77 |
| Total | 5074 | 1819 | 1179 | 1344 | 1395 | 1678 |
| pH | 8.5 | 8.5 | 8.5 | 8.7 | 8.4 | 8.5 |

were collected from both Carrizo Gorge and San Felipe Creek at approximately two week intervals. During this period no stromatolites rew in Carrizo Gorge and the stromatolites at the terminal end of San Felipe Creek showed no growth. Fresh granitic cobbles (Figure 4) deposited in Carrizo Gorge accumulated a thin film (.5-1 mm) of calcium carbonate. Green algal forms clogged Carrizo Gorge for the most part, unlike previous seasons when the blue-greens dominated. It is suspected that fresher water conditions precluded the growth of cyanobacteria (blue-green algae). By the end of the 1993 season when the stream desiccated, most of the cobbles and boulders were coated with calcium carbonate and isolated half-hemispheroids 1-3 mm in diameter (Figure 9). Carrizo Gorge was revisited during the spring of 1994 when conditions returned to normal, although stream volume was at least 2-3 times as high as that observed in several years previous to 1993. The cyanobacteria had returned and coabbles were coasted with a dark green mat that accumulated up to 3 mm of stromatolitic coating (Figure 10).

## CONCLUSIONS

Stromatolites are known to thrive in harsh conditions, particularly where the disovled solutes are high. The high concentrations of solutes observed during the spring of 1990 (Table 1 and Figure 11) when stromatolite growth was at its highest is consistent with this. This also supports the observation that stromatolites are most abundant in the lower reaches of both Carrizo Gorge and San Felipe Creek where maximum evaporation takes place and where solutes were observed to be the highest. It is concluded, therefore, that the major factor controlling stromatolite growth in these streams is the high solute concentration as well as calcite supersaturation. The presence of calcite coating cobbles and cementing the stream bottom into a concrete substrate indicates supersaturation with respect to calcite. In the upper reaches of Sentenac Canyon tufa was observed, which is a porous calcium carbonate deposit (easily distinguished from stromatolites, which are laminated). This also indicates supersaturation with calcite, however, tufa is plentiful in many desert alkaline streams. Some

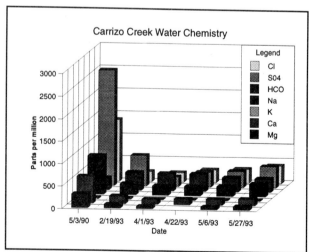

**Figure 11.** The comparitively high disolved solutes present in Carrizo Creek in 1990 coincide with active stromatolite growth. The dilute waters of 1993 were not favorable to stromatolite growth.

stromatolites have been observed in this area, but they appear to be non-active. This suggests that higher solute concentrations are required for active cyanobacteria growth. A drinkable fresh water can be saturated with calcite and not be saline.

The 1993 spring floods nearly destroyed all of the stromatolites in Carrizo Gorge, however the San Felipe stromatolites suffered much less damage. Unfortunately, the San Felipe stromatolites are less well developed. The observation of several millimeters of new stromatolite growth in Carrizo Gorge during the spring of 1994 suggests that within 3-5 more seasons the stromatolites will be back to the pre-1993 size and appearance, barring no new devastating flood events.

The application of this study to ancient stromatolite paleoenvironments is significant. It suggests that not all paleo-stromatolites indicate perennial water conditions, such as that found in lakes and oceans. One must consider the possibility of fluvial (stream) environments, which should also be obvious from the entombing sediments and facies. It further substantiates that extreme conditions are suggested by the presence of stromatolites (high salinities, ephemerality). Moreover, the presence of abundant gastropods associated with the Carrizo Gorge stromatolites calls into question the long held belief that abundant stromatolite growth is restricted or inhibited by grazing metazoans.

## REFERENCES CITED

Dill, R.F., and E.A. Shinn, et al. 1986. Giant subtidal stromatolites forming in normal salinity waters. Nature 324(6092):55-58.

Logan, B.W. 1961. *Cryptozoon* and associate stromatolites from the Recent of Shark Bay, Western Australia. Journal of Geology 69:517-533.

Winsborough, B. 1990. Some ecological aspects of modern freshwater stromatolites in lakes and streams of the Cuatro Cienegas Basin, Coahuila, Mexico. Ph.D. Dissertation, University of Texas. Microfilms, Ann Arbor, MI, 362 p.

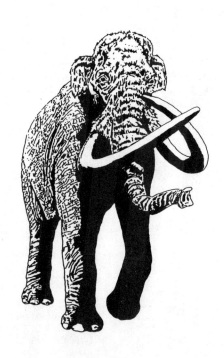

# INTERAGENCY COOPERATIVE AGREEMENT BETWEEN CALIFORNIA PARK SERVICE, COLORADO DESERT DISTRICT, ANZA-BORREGO DESERT STATE PARK AND UNITED STATES DEPARTMENT OF THE INTERIOR, BUREAU OF LAND MANAGEMENT, CALIFORNIA DESERT CONSERVATION DISTRICT

## Paul Remeika

California Park Service, Colorado Desert District, Anza-Borrego Desert State Park, 200 Palm Canyon Drive, Borrego Springs, California 92004

## I. PURPOSE:

The purpose of this Interagency Cooperative Agreement (ICA) is to enhance the conservation and management of paleontological resources on public lands administered by the Bureau of Land Management (BLM), California Desert Conservation District (CDCD), and the California Park Service (CPS), Anza-Borrego Desert State Park (ABDSP) within Imperial, Riverside, and San Diego counties, California. Furthermore, it is to establish stewardship for the long-term procurement of paleontological resources of common concern recovered on public lands under their jurisdiction directly adjourning one another.

## II. DEFINITION OF PALEONTOLOGICAL RESOURCES:

Henceforth, paleontological resources referred to in this document are defined as all non-renewable fossil evidences of prehistoric fauna and flora, including but not limited to, vertebrate (mammalian, reptilian, avian) bones, molluscan (marine and non-marine invertebrate) shells, botanical (leaf impressions and petrified wood), natural casts and molds, and trace fossils (including vertebrate footprints) of prehistoric fauna and flora naturally occurring within geological formations exposed on CPS- and BLM-administered public lands situated as an integral part of the geomorphic Salton Trough-Gulf of California structural depression. Paleontological resources shall not include cultural artifacts, remains, or any archeological resources which are related to human activity.

Paleontological resources are part of the public heritage and belong in professionally curated and permanently maintained collections where they must be accessioned, curated, and remain available to all public land management agencies and legitimate interested parties.

Vertebrate fossil resources require consideration as a separate category relative to invertebrate and other fossil resources. The Government Liaison Committee of the Society of Vertebrate Paleontology recommends the following position statement concerning effective management of fossil vertebrate resources on federal and state lands:

**1.** Vertebrate fossils are a non-renewable resource on federal, state and private lands.

**2.** Vertebrate fossils that are currently being exposed by erosion on federal and state lands cannot be allowed to disintegrate and be destroyed by weathering processes; periodic survey of such lands by competent professional vertebrate paleontologists is necessary to salvage such fossils.

**3.** Permits to collect vertebrate fossils on federal and state lands should be required of all individuals. Fossils acquired through these permits should be available for study in universities, museums and other repositories as part of the public heritage.

**4.** Scientific values on federal and state lands must take precedence over recreational values. Once destroyed, the scientific resources are lost forever, and the public heritage cannot be rcovered through reliance on a recreational or commercial ethic for federal and state lands.

In summary, the premise of the ICA is that paleontological resources on public lands are an important public trust to be wisely managed and conserved by federal and state agencies who act as stewards of that trust. Permission to recover paleontological resources should be limited to qualified and trained persons, and fossils must be placed in public accredited repository collections where there is established research and educational programs in paleontology. Fossils so collected must be accessible/available to qualified persons for scientific study and education when housed in such repositories. Current professional practices whereby scientific data is gathered on fossils at the time of collection in the field must be followed.

## ARTICLE I. BACKGROUND AND OBJECTIVES

WHEREAS in the California Desert Conservation District (CDCD) and Anza-Borrego Desert State Park (ABDSP) there occurs a scientifically-important series of biochronostratigraphic geologic formations with documented, nationally and internationally recognized paleontological vertebrate, invertebrate and paleobotanical resources protected and preserved in public trust. ABDSP's paleontological resources are included in the National Registry of Natural Landmarks. Furthermore, ABDSP is a component of the Mojave/Colorado Desert Biosphere Reserve.

WHEREAS the California Park Service (CPS) of the Resources Agency and the Bureau of Land Management (BLM) of the Department of the Interior manage natural-heritage lands with exposures containing these significant geologic formations and paleontological resources.

WHEREAS there is a need for the proper and responsible management, including inventory, field salvage recovery, identification, preparation, and cataloging, of these resources.

WHEREAS there is an obligation for the proper accountability and disposition for scientific specimens collected from the public lands.

WHEREAS the ability of both agencies to manage these resources **(1)** will be enhanced by gathering and sharing information provided by on-going and future scientific research efforts; and **(2)** will encourage mutual cooperative paleontologic/geologic resource management efforts.

WHEREAS ABDSP **(1)** is equipped with the mandate, technical staff, and equipment to provide scientific research, salvage and curatorial services; and **(2)** is dedicated to the responsible exploration for recovery, study, and research of paleontological resources for the purpose of educating the public and its members, and increasing knowledge of paleontological resources.

WHEREAS the BLM desires ABDSP's assistance in identifying and preserving scientifically-significant paleontological resources on public lands.

WHEREAS the expertise, established fossil repository collections, and facilities required for the appropriate conservation of these resources exists at ABDSP.

## ARTICLE II. RESOLUTION AND TERMS OF THE AGREEMENT

This agreement is subject to the provisions of the Antiquities Act of 1906, its regulations (43 CFR 3), and the following special resolutions:

**ANZA-BORREGO DESERT STATE PARK agrees to:**

**A.** ABDSP's initiation of field work or other activities under the authority of this ICA permit signifies ABDSP's acceptance of the terms and conditions of this ICA.

**1.** Before undertaking any work on lands administered by the BLM, authorization will be obtained from the Area Manager of the resource area concerned.

**2.** Disturbance shall be kept to the minimum area consistent with the nature and purpose of the paleontological field work.

**3.** Vehicular activity shall be restricted to existing roads and trails unless otherwise provided by the Area Manager.

**4.** All excavated areas shall be restored by filling in the excavations and otherwise leaving the area in as near to original condition as is practicable.

**5.** ABDSP shall conduct all operations in such a manner as to prevent the erosion of the land, pollution of the water resources, and damage to the watershed, and to do all things necessary to prevent or reduce to the fullest extent the scarring of the lands.

**6.** ABDSP shall take precautions to protect livestock, wildlife, the public, or other users of the public lands from accidental injury in any excavation unit.

**7.** Living trees or plants shall not be cut or otherwise damaged, unless authorized by the Area Manager.

**8.** Within ninety (90) days of completion of field work, ABDSP shall submit two copies of a preliminary report.

**9.** The following information shall be submitted to the appropriate Area Manager as part of the preliminary report, or as part of a separate report, upon completion of the field work. See # 8 above.

    **a.** A statement of what work has been accomplished under the authority of this ICA including a map of areas surveyed and/or excavated;

    **b.** The significance of the identified paleontological resources and their potential for contributing to paleontological data;

    **c.** A completed site inventory (fossil

fauna/flora list) for each site found with appropriate geological and stratigraphic map indicating the location of each paleontological locality/information investigated and/or excavated.

10. ABDSP shall submit one copy of all published journal articles and other published or unpublished reports, papers, and manuscripts resulting from the paleontological work to the Area Manager of the resource area involved.

11. ABDSP shall deposit all samples and collections, as applicable, and copies of all records, data, photographs, and other documents resulting from work conducted under this ICA within their accredited curatorial facility. The final report submitted to the BLM shall include a catalog of all materials deposited.

12. Collections of paleontological specimens acquired under the provisions of this ICA permit remain the property of the United States Government and may be recalled at any time for scientific use and study of the Department of the Interior or other agencies of the Federal Government.

13. The person in direct charge of field work, or a qualified designee, shall be on site at all times when work is in progress.

**B.** Provide input and technical expertise in the management of paleontological/geological resources occurring on BLM lands adjacent to ABDSP.

1. Assist with paleontological surveys, site evaluation, surface-salvage and resource inventories.
2. Assist in determining significance of fossil localities and specimens.
3. Establish stratigraphic control and contextual information for the geologic formations yielding the fossils. Supply taxonomic and osteological identifications of paleobotanical, invertebrate and vertebrate paleontological material.
4. Supply BLM copies of ABDSP's Paleontology Collection Managment Policy (PCMP), Paleontology Resource Management Plan (PRMP), comprehensive bibliographic listing for ABDSP (including the Salton Trough-Gulf of California and Peninsular Ranges), and data compiled during the Inventory of Natural Features for ABDSP's General Plan upon request.

**C.** Provide curatorial facilities and accountable museum storage of retrieved paleontological specimens.

1. Specimens must be compatible with ABDSP's PCMP and PRMP.
2. Specimens stored in the ABDSP repository collection will receive ABDSP catalog numbers.

**D.** Provide facilities and materials as available, for paleontological related work such as site documentations, research, lab and fossil preparation, curation, computer database records, exhibitry and collection storage.

**E.** Cooperate with the BLM to provide the BLM with technical expertise, scientific guidance, advice and consultation to render judgment on paleontological queries that come before the agency in-charge, depending upon circumstances and needs. Other assistance of certified paleo-volunteers will be made available, at the invitation of the Park Paleontologist, on perceived paleontological projects where fossil vertebrates are concerned, which may be suggested by the BLM.

**F.** Assist with the BLM to develop criteria for the identification, permitting, exploration for, recovery, study and research of scientifically-significant paleontological resources on public lands. Permitting is necessary in order to: **(1)** inform land managers of the nature of the fossil resources under their jurisdiction and administrative control; **(2)** protect fossil resources from unscrupulous and untrained persons; **(3)** provide a record of persons who have had access to fossil resources on public lands; and **(4)** professional review of these permit requests by qualified investigators knowledgeable and experienced in the paleontology and geology of the area in question.

**G.** Notify the BLM if a specimen of unique scientific significance is found.

**H.** In every case, ABDSP will be responsible for following all applicable federal and state laws and regulations. If questions arise, ABDSP will contact the BLM for information.

**THE BUREAU OF LAND MANAGEMENT agrees to:**

**A.** Permit ABDSP staff with authorization from the appropriate Area Manager, or his/her designee, to collect scientifically-significant paleontological specimens. Activities involving collection of specimens will be contingent upon prior approval by the BLM.

**B.** Inform ABDSP of paleontological projects whereby the BLM can utilize assistance from ABDSP.

fauna/flora list) for each site found with appropriate geological and stratigraphic map indicating the location of each paleontological locality/information investigated and/or excavated.

10. ABDSP shall submit one copy of all published journal articles and other published or unpublished reports, papers, and manuscripts resulting from the paleontological work to the Area Manager of the resource area involved.

11. ABDSP shall deposit all samples and collections, as applicable, and copies of all records, data, photographs, and other documents resulting from work conducted under this ICA within their accredited curatorial facility. The final report submitted to the BLM shall include a catalog of all materials deposited.

12. Collections of paleontological specimens acquired under the provisions of this ICA permit remain the property of the United States Government and may be recalled at any time for scientific use and study of the Department of the Interior or other agencies of the Federal Government.

13. The person in direct charge of field work, or a qualified designee, shall be on site at all times when work is in progress.

**B.** Provide input and technical expertise in the management of paleontological/geological resources occurring on BLM lands adjacent to ABDSP.

1. Assist with paleontological surveys, site evaluation, surface-salvage and resource inventories.

2. Assist in determining significance of fossil localities and specimens.

3. Establish stratigraphic control and contextual information for the geologic formations yielding the fossils. Supply taxonomic and osteological identifications of paleobotanical, invertebrate and vertebrate paleontological material.

4. Supply BLM copies of ABDSP's Paleontology Collection Managment Policy (PCMP), Paleontology Resource Management Plan (PRMP), comprehensive bibliographic listing for ABDSP (including the Salton Trough-Gulf of California and Peninsular Ranges), and data compiled during the Inventory of Natural Features for ABDSP's General Plan upon request.

**C.** Provide curatorial facilities and accountable museum storage of retrieved paleontological specimens.

**1.** Specimens must be compatible with ABDSP's PCMP and PRMP.

**2.** Specimens stored in the ABDSP repository collection will receive ABDSP catalog numbers.

**D.** Provide facilities and materials as available, for paleontological related work such as site documentations, research, lab and fossil preparation, curation, computer database records, exhibitry and collection storage.

**E.** Cooperate with the BLM to provide the BLM with technical expertise, scientific guidance, advice and consultation to render judgment on paleontological queries that come before the agency in-charge, depending upon circumstances and needs. Other assistance of certified paleo-volunteers will be made available, at the invitation of the Park Paleontologist, on perceived paleontological projects where fossil vertebrates are concerned, which may be suggested by the BLM.

**F.** Assist with the BLM to develop criteria for the identification, permitting, exploration for, recovery, study and research of scientifically-significant paleontological resources on public lands. Permitting is necessary in order to: **(1)** inform land managers of the nature of the fossil resources under their jurisdiction and administrative control; **(2)** protect fossil resources from unscrupulous and untrained persons; **(3)** provide a record of persons who have had access to fossil resources on public lands; and **(4)** professional review of these permit requests by qualified investigators knowledgeable and experienced in the paleontology and geology of the area in question.

**G.** Notify the BLM if a specimen of unique scientific significance is found.

**H.** In every case, ABDSP will be responsible for following all applicable federal and state laws and regulations. If questions arise, ABDSP will contact the BLM for information.

**THE BUREAU OF LAND MANAGEMENT agrees to:**

**A.** Permit ABDSP staff with authorization from the appropriate Area Manager, or his/her designee, to collect scientifically-significant paleontological specimens. Activities involving collection of specimens will be contingent upon prior approval by the BLM.

**B.** Inform ABDSP of paleontological projects whereby the BLM can utilize assistance from ABDSP.

C. Inform ABDSP of those BLM-administered lands which are designated as research, or where exploration, recovery of paleontological resources is restricted or prohibited. Such areas with fossils that require critical managment attention can include Areas of Critical Environmental Concern, Wilderness Study Areas, Research Natural Areas, National Natural Landmarks, and other specially designated management areas.

D. Coordinate with ABDSP on activities or actions which could potentially affect paleontologic/geologic resources (e.g., land use plans, activity plans, research proposals).

E. Provide limited support and logistical assistance, dependant upon budgetary and man-power capabilities. All monies must be identified through the BLM's annual work plan process. Approved monetary assistance may be used for defraying the costs of storage, preparation, cataloging, and conservation of objects collected from BLM lands and stored within the ABDSP repository collection.

F. Agree to recognize the value to the scientific community of keeping the collection intact, however, the BLM may use other repositories.

## ARTICLE III. STATUTORY AUTHORITY FOR MANAGEMENT OF FOSSIL RESOURCES (ANZA-BORREGO DESERT STATE PARK):

The California Park Service is organized within the Resources Agency (Government Code 12805 and Public Resources Code 501), whose objective is to further the conservation of California's resources.

1). **Public Resources Code (Division 5, Chapter 1, Article 2) 5003.** The California Park Service shall administer, protect, develop, and interpret the property under its jurisdiction for the use and enjoyment of the public.

2). **Public Resources Code (Division 5, Chapter 1, Article 1) 5019.53**. The purpose of state parks shall be to preserve outstanding natural, scenic, and cultural values, indigenous aquatic and terrestrial fauna and flora, and the most significant examples of such ecological regions of California.... Each state park shall be managed as a composite whole in order to restore, protect, and maintain its native environmental complexes to the extent compatible with the primary purpose for which the park was established.

3). **Public Resources Code (Division 5, Chapter 1.7) 5097.5.** No person shall knowingly and willingly excavate upon, or remove, destroy, injure, or deface, any historic or prehistoric ruins, burial grounds, archeological, or vertebrate paleontological site, including fossilized footprints, inscriptions made by human agency, or any other archeological, paleontological or historic feature, situated on public lands, except with the express permission of the public agency having jurisdiction over such lands.

4). **Commision Policy Number 2**. Land acquired for the California Park Service shall be dedicated to public use and protected against exploitation in accordance with its classifications, with its adopted resource management directives, and as outlined in approved resource elements of general plans.

5). **Resource Management Directive (Department Operations Manual), Section 1831, Number 3. Geology.** Geologic resources constitute another group of environmental elements of great importance and considerable variety. In some instances, geological values may form the primary feature of interest within a given unit of the California Park Service, for scenic interest or for scientific values.

6). **Resource Management Directive (Department Operations Manual), Section 1831, Number 4. Paleontological Values.** The occurrence of fossil plants or animals in any rock or soil formation is more often of interest to the scientist than to the layman. However, there are numerous instances where fossil remains of prehistoric plants or animals are striking in their occurrence, and are of great interest.

(40) Paleontological resources in the California Park Service are to be reported when observed by any employee of the Service. The resources shall be investigated, evaluated, and recorded by qualified personnel designated by the Director, who will then determine the necessary protective measures to be taken.

(41). Paleontological resources in the California Park Service shall be protected against damaging influences, including deterioration or adverse modification of their environment. Sites proposed for development will be evaluated for paleontological resources in the preliminary planning stages. Stabilization of paleontological resources may be required to prevent loss, but will be done in ways that protect the integrity of the sites.

7). **Department Operations Manual, Section 1621.4**. Defines geological resources as those features of the earth's crust which are conspicuous features in the landscape. A primary concern is to protect the integrity of rock features such as cliffs, buttes, and crags, by protecting them against excessive erosion and other forms of natural and human-induced damage.

**8). Department Operations Manual, Section 1621.5.** Defines paleontological features as the records in stone of plant and/or animal species from past geologic ages. They represent the only source of knowledge concerning life on earth in the geological past. They are irreplaceable, and if destroyed through carelessness or greed, are lost forever.

**9). Department Operations Manual, Section 1831, Number 31.** It is the objective of the California Park Service to apply creative and effective techniques of environmental resource management found by scinetific analysis to be required to achieve the protection and perpetuation of the values around which the units are built.

In instances where paleontological resources happen to occur in an archeological context, the fossils fall under the enhanced guardianship offered to the latter discipline through the Archeological Resources Protection Act of 1979. Section 3 of that Act, as amended, specifically excludes nonfossilized and fossilized paleontological specimens from the Act's authorities unless those specimens are found in an archeological context. This exclusion is reaffirmed in Section .3 (A) 4 of the Act's uniform regulations (43 CFR 7; 36 CFR 296; 18 CFR 1312; 32 CFR 229) which states that paleontological remains **"shall not be considered of archeologic interest, and shall not be considered to be archeological resources...unless found in a direct physical relationship with archeological resources...."** Archeological resources are defined in Section .3 (A) of the uniform regulations to mean **"any material remains of human life or activities which are at least 100 years of age, and which are of archeological interest."**

**(BUREAU OF LAND MANAGEMENT):**

The following amended BLM regulations were adopted during the December 4-6, 1989 meetings at Boulder, Colorado, on Negotiated Rule Making for Collection of Fossils on Federal Lands. BLM and the USFS have integrated these into their internal working procedures. For the BLM, providing for the proper management, use, and protection of fossils is accomplished using the regular planning and environmental assessment processes required by the Federal Land Policy and Management Act of 1976, and the National Environmental Act.

**1). National Environmental Policy Act, Section 101 [a] 4.** Fossil vertebrates are protected as natural aspects of our national heritage and should be preserved.

**2). Bureau of Land Management Instruction Memorandum 79-111.** All vertebrate fossils have been categorized as being of significant scientific value.

**3). National Environmental Policy Act, Section 102.** Policies, regulations, and public laws of the United States shall be interpreted and administered in accordance with the policies set forth in this Act.

**4).** The Bureau of Land Management has statutory and regulatory authority to provide for the management and disposition of fossils (1-17-86 BLM.ER.0354 memorandum to the Director, BLM, from Associate Solicitor, Energy and Resources).

**5). Federal Land Policy and Management Act of 1976, Section 302 [b].** The Bureau of Land Management has the charge to manage and regulate the use of public lands by permitting procedures.

**6). Federal Land Policy and Management Act, Section 102 [8].** Public lands should be managed in a manner that will protect the quality of scientific values.

**7). Federal Land Policy and Management Act, Section 102 [a].2.** The national interest will be best realized if the public lands and their resources are periodically and systematically inventoried and their present and future use is projected through a land use planning process coordinated with other federal and state planning efforts.

**ARTICLE IV. TERMS OF AGREEMENT**
**BOTH AGENCIES agree to:**

**A.** Share paleontological/geological research results.
**B.** Share paleontology/geology geographic information databases.
**C.** Ensure the timely consideration of applications for research permits potentially involving the land bases of both agencies.

    **1.** Designated sites or areas may be set aside for specific periods of time to allow for scientific study and research.

**D.** Meet at least once every fiscal year to up-date and/or coordinate relevant activities.

    **1.** Based on information assembled and analyzed, planning decisions may be made for the best management, use, and protection of fossils on public lands.

    **2.** If fossil resources are identified as being an issue in scoping a specific area, a literature review and further consultation may be undertaken. A limited practical reconnaissance inventory of fossil resources, where such resources are of scientific significance, or where the strong probability of such resources exists, should accompany this effort.

    **3.** Convene an advisory board to identify and evaluate potential paleontological localities of national significance (on both public and private lands) for possible designation as National Natural Landmarks, pursuant to the existing National Natural Landmark Program administered by the National Park Service.

**E.** This ICA in no way restricts the BLM from participating with other public or private agencies, organizations, and individuals, or from accepting contributions and/or gifts for the management and conservation of paleontological resources on the public lands.

**F.** No part of this ICA shall entitle ABDSP to any share or interest in BLM avtivities other than that provided by applicable laws and regulations.

**G.** Nothing in this ICA shall be construed as obligating the BLM to expend money or as involving the United States Government, Department of the Interior, in any contract of other obligations for the present or future payment of money except mutually agreed upon appropriations authorized by law and administratively allocated for activities and projects initiated pursuant to this ICA.

**H.** Nothing in this ICA shall be construed as giving ABDSP any type of exclusive arrangement with the BLM to the exclusions or detriment of other interest groups or organizations.

**I.** This ICA may be revised as necessary by the issuance of a written amendment that is consented to, signed, and dated by both parties. This ICA is renewable upon mutual agreement of the BLM and ABDSP.

**J.** This ICA shall be reviewed for modifications, termination, or renewal, at least every five years.

**K.** Nothing in this ICA will be construed as binding upon the BLM or ABDSP to perform beyond their respective authorities.

## ARTICLE V. TERMINATION

This agreement may be terminated by either party by providing 90 days written notice.

## ARTICLE VI. REQUIRED CLAUSES

During the performannce of this agreement, the participants agree to abide by the terms of Executive Order 11246 on nondiscrimination and will not discriminate against any person because of race, color, religion, sex, physical handicap or national origin. The participants will take affirmative action to ensure that applicants are employed without regard to their race, color, religion, sex or national origin.

No member or delegate to Congress, or resident Commissioner, shall be admitted to any share or part of this ICA, or to any benefit that may arise therefrom, but this provision shall not be construed to extend to this agreement if made with a corporation for its general benefit.

In Witness Whereof, each party has caused this ICA in paleontologic stewardship to be executed by an authorized official on the day of the year set forth by their signatures.

## ARTICLE VII. KEY OFFICIALS
**REPRESENTING ANZA-BORREGO DESERT STATE PARK:**

Mr. David H. Van Cleve, District Superintendent, Colorado Desert District, 200 Palm Canyon Drive, Borrego Springs, California 92004 619-767-5311 FAX 619-767-3427

Mr. Paul Remeika, Park Paleontologist/State Park Ranger I, Colorado Desert District, Anza-Borrego Desert State Park, 200 Palm Canyon Drive, Borrego Springs, California 92004 619-767-5311 FAX 619-767-3427

Mr. George T. Jefferson, District Archeologist, Colorado Desert District, 200 Palm Canyon Drive, Borrego Springs, California 92004 619-767-4974 FAX 619-767-3427

**REPRESENTING BUREAU OF LAND MANAGEMENT:**

Mr. G. Ben Koski, Jr., Area Manager, El Centro Resource Area, California Desert District, Bureau of Land Management, 1661 South Fourth Street, El Centro, California 92243 619-353-1060 FAX 619-353-7461

Mrs. Julia Dougan, Area Manager, Palm Springs--South Coast Resource Area, Bureau of Land Management, 63-500 Garnet Avenue, P.O. Box 2000, North Palm Springs, California 92258-2000 619-251-0812 FAX 619-251-0248

# NEOGENE STRATIGRAPHY OF THE BORREGO MOUNTAIN AREA, ANZA-BORREGO DESERT STATE PARK, CALIFORNIA

C. Herzig, A. Carrasco, T. Schar, G. Murray, D. Rightmer, J. Lawrence, Q. Milton, T. Wirths
Department of Geological Sciences, San Diego State University,
San Diego, CA, 92182-1020

## ABSTRACT

Neogene strata exposed at Borrego Mountain record the progradation of the Colorado River delta into this area. Sediments were deposited unconformably across Cretaceous crystalline basement. An influx of basin-margin fanglomerate, reflecting tectonic uplift and denudation in the source area, buried the delta. Deltaic sedimentation resumed following cessation of fanglomerate deposition. The deltaic sediments are in turn overlain by lacustrine facies.

Mudstones and siltstones resting unconformably on crystalline basement may be the base of the Pliocene Palm Spring Formation, or deltaic transition facies of the Camels Head member of the Imperial Formation. Basin-margin fanglomerate is part of the Canebrake Conglomerate. Overlying deltaic facies correspond to the Diablo and Olla members of the Palm Spring Formation. The lacustrine rocks are tentatively assigned to the Pliocene Borrego Formation.

## INTRODUCTION

Previous studies of Neogene strata along the western margin of the Salton Trough have focused on differentiating deltaic rocks from basin-margin materials (Dibblee, 1954; Merriam and Bandy, 1965; Muffler and Doe, 1968; Winker, 1987). We continue this work in the Borrego Mountain area. This paper presents the preliminary results of our investigation of the Neogene strata at Borrego Mountain. Our study seeks new insights into the relationship between tectonics and sedimentation within this rapidly evolving extensional basin.

## BORREGO MOUNTAIN AREA

The transtensional plate boundary between the North American and Pacific plates within the Salton Trough is a diffuse zone of complex wrench tectonics (Atwater, 1970; Elders et al., 1972; Crowell, 1974; Lonsdale, 1989). This geology is manifested along the western margin of the trough by right-lateral strike-slip faults of the San Jacinto and Elsinore fault zones, uplifted Paleozoic and Mesozoic basement underlying tthe major mountain ranges, and extensive outcrops of Neogene strata that have been folded and disrupted by strike-slip and normal faults.

Cretaceous plutonic rocks of the Peninsular Ranges batholith are the dominant crystalline basement materials in the region. East and West Buttes of Borrego Mountain, and other smaller basement exposures in the study area, are composed of pegmatitic granitoids. Contacts between the basement rocks and overlying Neogene strata are disrupted by strike-slip or normal faults, however an unconformable depositional contact is preserved in Hawk Canyon (Fig. 1).

Neogene sedimentary rocks crop out from Hawk Canyon westward towards "The Slot," and northward to San Felipe Wash (Fig. 1). This gently west/southwest dipping sequence (20 - 30° dip) comprises the south limb of a northwest-plunging anticline. The strata are locally offset by northeast-striking normal faults with meter-scale displacements. Strata east of West Butte are chopped into horsts and grabens by northeast-striking normal faults making this area difficult for stratigraphic investigation.

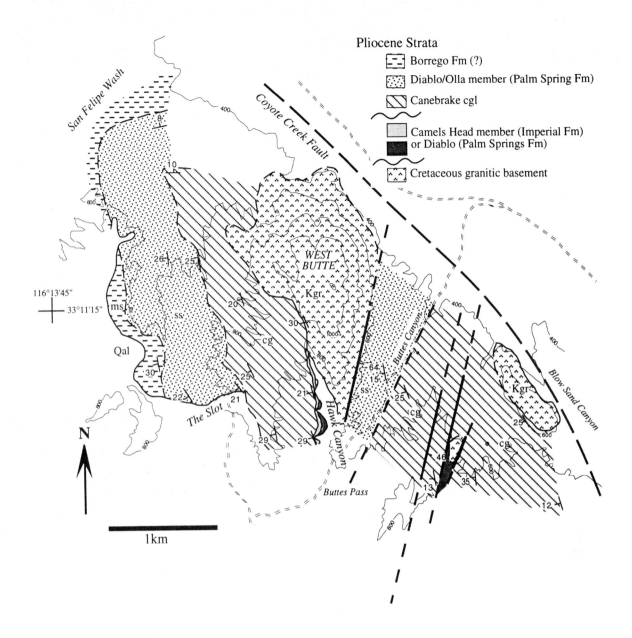

Fig. 1. Simplified geologic map for the West Butte area of Borrego Mountain. Qal = Quaternary alluvium.

### Stratigraphy west of West Butte

The base of the Neogene strata is exposed in Hawk Canyon (Fig. 1). Previous mapping shows conglomerate resting on the Cretaceous plutonic basement in this area (Weber, 1959; Winker, 1987). This contact is a nonconformity north of the 800' contour in Hawk Canyon, where conglomerate laps directly onto crystalline basement (Fig. 1). However, south of the 800' contour in Hawk Canyon, and stratigraphically downsection beneath the conglomerate, sandstones, mudstones, and grus, were deposited on the granitic rocks before the conglomerate.

The unconformable contact between the sedimentary rocks and the granitic basement in Hawk Canyon is an irregular topographic surface with poorly-sorted, granitic grus filling paleo-depressions in the basement. Clasts of the granitic basement occur in the grus. Very coarse-grained sandstone and pebble conglomerate cover the grus and lap onto the basement rocks, filling paleo-depressions in the crystalline basement.

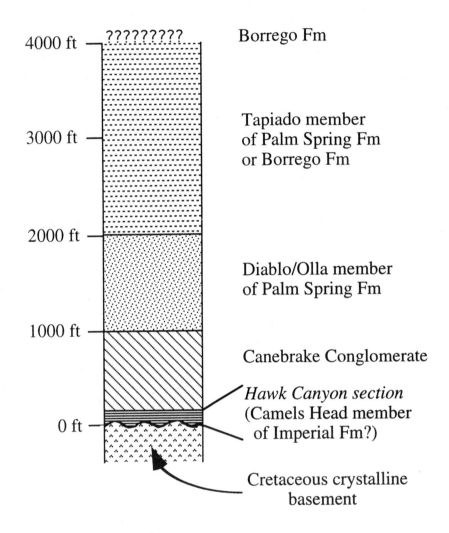

Fig. 2. Simplified stratigraphic column for Neogen strata at Borrego Mountain.

Granitic outcrops "pop-up" through this sandstone unit, exhuming paleotopography.

The next stratigraphically higher unit consists of green-brown mudstones and siltstones with thinly bedded quartz-rich sandstones. Biotite grains are very common in these sedimentary rocks. This mudstone-siltstone interval grades upsection into 0.75 meter-thick beds of red-brown sandstones with low-angle planar bedding, interbedded red mudstones and siltstones, and channel-lag deposits. Subrounded to rounded granitic and gneissic pebbles in the channel lag deposits range in size from 1 to 5 cm.

The red-brown sandstones are unconformably overlain by poorly sorted conglomerate with interbedded coarse-grained sandstones This contact is a very low-angle unconformity. Subrounded to rounded clasts of plutonic and gneissic rocks contained within the conglomerate range in size from 1 to 30 cm; rare clasts are 1 to 2 meters in diameter.

The plutonic and metamorphic rocks do not resemble the underlying granitic basement exposed at West and East Buttes, suggesting other sources. Conversely, clasts of the underlying granitic basement were not found within the conglomerate where it laps onto plutonic basement north of the 800' contour in Hawk Canyon.

The conglomerate grades into overlying brown to red, fine- to coarse-grained sandstones with interbedded red/brown siltstones and mudstones, and channel lag deposits. Cross-bedding, scour and fill, and mudstone rip-up clasts are common. Some sandstone beds comprise channel fill up to 10-m-thick. The frequency of siltstone and mudstone laminae increases upsection. All strata in this unit are biotite-rich, with abundant muscovite. Concretions are very common in the sandstones.

The stratigraphically highest unit consists of red/brown and green mudstones with interbedded fine- to medium-grained sandstones. Cross-bedding in the sandstones is common. Biotite and muscovite grains are abundant. Tan-colored sandstones were changed to red-brown colors by diagenesis. Concretions are also common. Although not shown in Fig. 1, the mudstone unit continues to crop out westward in the canyons opening into San Felipe Wash. Here, brown to green mudstone and siltstone beds are up to several meters thick. Fine- to medium-grained sandstones are uncommon, and increase in abundance stratigraphically upward within this unit.

The Neogene strata at Borrego Mountain coarsen upward from the basal contact with the underlying crystalline basement to the conglomerate in Hawk Canyon (Fig. 2). The conglomerate and overlying strata in turn have a fining upward trend toward "The Slot" and San Felipe Wash.

## Stratigraphic Nomenclature and Depositional Environment

Strata at Borrego Mountain record the progradation of the Colorado River delta into this region during the Pliocene. This was temporarily interrupted by the influx of basin-margin fanglomerate. This basic sedimentary pattern is recorded by other exposures of Neogene strata located elsewhere in the western Salton Trough (Winker, 1987). Stratigraphic nomenclature for these rocks was standardized by Dibblee (1954; 1984), and will be followed here, with modifications from Woodard (1963) and Winker (1987).

Outcrops in Hawk Canyon at the base of the sedimentary section (Fig. 2) are interpreted as deltaic facies of the Colorado River lapping onto exposed crystalline basement. These rocks may record the transition from the uppermost part of the Camels Head member of the Miocene Imperial Formation to the lowermost facies of the Diablo member of the Pliocene Palm Spring Formation (Winker, 1987). It is difficult however, to distinguish the sandstone bodies in Hawk Canyon from those in the overlying Diablo member of the Palm Spring Formation; therefore it is possible that these sandstones may be part of the Diablo member.

Canebrake Conglomerate was emplaced across the underlying deltaic facies with an angular unconformity. This unit, most likely a gravelly braided-stream system, records the influx of basin-margin detritus into the area. A basin-margin source is indicated by the plutonic and metamorphic clasts that do not resemble local plutonic basement exposures. This braided-stream system records one of many pulses of rapid uplift and denudation along the western margin of the Salton Trough.

Waning and cessation of conglomerate deposition allowed renewed progradation of Colorado River deltaic facies. Sandstones and mudstones are assigned to the Olla member of the Palm Spring Formation. The Olla member is part of a gradational facies tract between the Canebrake Conglomerate and the Diablo member of the Palm Spring Formation. The Olla and Diablo members are laterally continuous and interfinger with each other (Winker, 1987).

The Olla/Diablo members are characteristic lithologies of the Palm Spring Formation. They represent an overall fining-upward deltaic sequence in the Borrego Mountain area. These rocks are in turn depositionally overlain by mudstones, siltstones and sandstones tentatively assigned to either the lacustrine Tapiado member of the Palm Spring Formation, or basal units of the lacustrine Borrego Formation. The Tapiado member of the Palm Spring Formation is a restricted local unit in Arroyo Tapiado of the Vallecito Basin (Winker, 1987), so a better correlation for the lacustrine rocks in the Borrego Mountain area may be with the Borrego Formation. The lacustrine rocks in the Borrego Mountain area and those northward in the Borrego Badlands have since been disrupted by strike-slip activity along the Coyote Creek fault.

## ACKNOWLEDGMENTS

We thank Paul Remeika of Anza-Borrego Desert State Park for assistance with this project.

## LITERATURE CITED

Atwater, T., 1970, Implications of plate tectonics for the Cenozoic tectonic evolution of western North America: Geological Society of America Bulletin, v. 81, p. 3513-3536.

Crowell, J.C., 1974, Origin of late Cenozoic basins in southern California, in W.R. Dickinson (ed.), Tectonics and Sedimentation: Special Publication Society of Economic Paleontologists and Mineralogists, v. 22, p. 190-204.

Dibblee, T.W., 1954, Geology of the Imperial Valley region, California, in Jahns, R.H., (ed.), Geology of California, Chapter 2: Geology of the natural provinces: California Division of Mines Bulletin, 170, p. 21-28.

Dibblee, T.W., 1984, Stratigraphy and tectonics of the San Felipe Hills, Borrego Badlands, Superstition Hills, and vicinity, in Rigsby, C.A., (ed.), The Imperial Basin - Tectonics, Sedimentation, and Thermal Aspects, Publication 40: Pacific Section, Society of Economic Paleontologists and Mineralogists, Bakersfield, California, p. 31-44.

Elders, W.A., Rex, R.W., Meidev, T., Robinson P.T., and Biehler, S., 1972, Crustal spreading in southern California: Science, v. 178, p. 15-24.

Lonsdale, P., 1989, Geology and tectonic history of the Gulf of California, in Winterer, E.L., et al., (eds.), The eastern Pacific Ocean and Hawaii: Boulder, Colorado, Geological Society of America, Geology of North America, v. N, p. 499-521.

Merriam, R., and Bandy, O.L., 1965, Source of upper Cenozoic sediments in Colorado delta region: Journal of Sedimentary Petrology, v. 35, p. 911-916.

Muffler, L.J.P., and Doe, B.R., 1968, Composition and mean age of detritus of the Colorado River delta in the Salton Trough, southeastern California: Journal of Sedimentary Petrology, v. 38, p. 384-399.

Weber, F.H., Jr., 1959, Geology and mineral resources of San Diego County, California: California Division of Mines and Geology, Scale 1:125,000.

Winker, C.D., 1987, Neogene stratigraphy of the Fish Creek-Vallecito section, southern California: implications for early history of the northern Gulf of California and Colorado delta: Ph.D. dissertation, University of Arizona, Tucson, 494 p.

Woodard, G.D., 1963, The Cenozoic succession of the west Colorado Desert, San Diego and Imperial Counties, southern California: Ph.D. dissertation, University of California, Berkeley, 173 p.

**THE END**

# SELECTED BIBLIOGRAPHY OF THE WESTERN SALTON TROUGH DETACHMENT AND RELATED SUBJECTS, ANZA-BORREGO DESERT STATE PARK, CALIFORNIA

FIELD TRIP GUIDEBOOK AND VOLUME FOR THE 1995 SAN DIEGO ASSOCIATION OF GEOLOGIST'S FIELD TRIP TO ANZA-BORREGO DESERT STATE PARK

VOLUME II

Edited By
**PAUL REMEIKA**
Park Paleontologist/State Park Ranger I
Anza-Borrego Desert State Park
and
**ANNE STURZ**
Vice-chairperson/San Diego Association of Geologists
University of San Diego/Marine/Environmental Studies

# BIBLIOGRAPHY

Abbott, P.L. 1980. Provenance of Salton Dunes southwest of Salton Sea, California. In Geology and Mineral Wealth of the California Desert, edited by D.L. Fife, South Coast Geological Society, Santa Ana, California p. 77-70.

Abbott, P.L., R.P. Kies, and D.R. Kerr 1984. Right-slip offset of the Eocene Ballena River Valley across the Elsinore Fault Zone, southern California. Ciencias Marinas 9:87-94.

Abbott, P.L., R.P. Kies, D. Krummenacher, and D. Martin 1983. Potassium-argon ages of rhyolitic bedrock and conglomerate clasts in Eocene strata, northwestern Mexico and southern California. In Tectonics and Sedimentation Along Faults of the San Andreas System, edited by D.W. Anderson and M.J. Rymer, Society of Economic Paleontologists and Mineralogists, Pacific Section p. 59-66.

Abbott, P.L., and T.E. Smith 1978. Trace-element comparison of clasts in Eocene conglomerates, southwestern California and northwestern Mexico. Journal of Geology 86:753-762.

----- 1979. Itinerary for field trip to the Peninsular Ranges Batholith, San Diego County. In Mesozoic Crystalline Rocks, Peninsular Ranges Batholith and Pegmatites, Point Sal Ophiolite, edited by P.L. Abbott and V.R. Todd, Department of Geological Sciences, San Diego State University, California p. 233-286.

Abbott, P.L., and V.R. Todd 1979. Mesozoic crystalline rocks: Peninsular Ranges Batholith and pegmatites, Point Sal ophiolite. Department of Geological Sciences, San Diego State University, California 286 p.

Abbott, P.L., and J.K. Victoria 1977. Geologic hazards in San Diego; earthquakes, landslides and floods. San Diego Society of Natural History, California 96 p.

Abbott, W.O. 1969. Salton Basin, a model for Pacific rim diastrophism. Geological Society of America Special Paper 121:1.

Abe, K. 1984. The present condition of geothermal well drilling in Cerro Prieto and Imperial Valley. Chinetsu, Journal of Japan Geothermal Energy Association 21(5[85]):24-30.

Adrian, B.M., D.S. Smith, D.B. Detra, and T.A. Romer 1984. Analytical results and sample locality map of stream-sediment, heavy-mineral-concentrate, and rock samples from the Indian Pass (CDCA-355) and Picacho Peak (CDCA-355A) Wilderness Study Areas, Imperial County, California. U.S. Geological Survey Open-File Report 84-779:1-39.

Adsit, R.J. 1983. Heavy minerals in quartzites of the Julian Schist and other metasedimentary rocks in southern California. Undergraduate Research Reports, Department of Geology, San Diego State University, California 42(1):1-19.

Agenbroad, L.D., J.I. Mead, and R.E. Reynolds 1992. Mammoths in the Colorado River corridor. In Old Routes to the Colorado, edited by R.E. Reynolds and J. Reynolds, San Bernardino County Museum Special Publication 92(2):104.

Agnew, D.C. 1984. The 1852 Fort Yuma Earthquake; two additional accounts. Seismological Society of America Bulletin 68:1761-1762.

Agnew, D.C., M. Legg, and C.L. Strand 1979. Earthquake history of San Diego. In Earthquakes and Other Perils -- San Diego Region, edited by P.L. Abbott and W.J. Elliott, Geological Society of America Field Trip, San Diego Association of Geologists p. 123-138.

Agnew, D.C., and F.K. Wyatt 1989. The 1987 Superstition Hills Earthquake sequence: strains and tilts at Pinon Flat Observatory. Seismological Society of America Bulletin 79:480-492.

Ague, J.J., and G.H. Brimhall 1987. Granites of the batholiths of California: products of local assimilation and regional-scale crustal contamination. Geology 15:63-66.

----- 1988a. Magmatic arc asymmetry and distribution of anomalous plutonic belts in the batholiths of California: effects of assimilation, crustal thickness, and depths of crystallization. Geological Society of America Bulletin 100:912-927.

----- 1988b. Regional variations in bulk chemistry, mineralogy, and the compositions of mafic and accessory minerals in the batholiths of California. Geological Society of America Bulletin 100:891-911.

Albright, L.B., and M.O. Woodburne 1993. Refined chronologic resolution of the San Timoteo Badlands, Riverside County, California, and tectonic implications: a prospectus. In Ashes, Faults and Basins, edited by R.E. Reynolds and J. Reynolds, San Bernardino County Museum Association Special Publication 93-1:104-105.

Alcock, E.D. 1974. Comments on "Comparison of earthquake and micrometer ground motions in El Centro, California" by F.E. Udwadia and M.D. Trifunac. Seismological Society of America Bulletin 64:495.

Allen, C.R. 1954. Geology of the north side of San Gorgonio Pass, Riverside County. In Geology of Southern California, edited by R.H. Jahns, California Division of Mines Bulletin 170:map sheet 20.

----- 1957. San Andreas Fault Zone in San Gorgonio Pass, southern California. Geological Society of America Bulletin 68:315-349.

----- 1981. The modern San Andreas Fault. In The Geotectonic Development of California, edited by W.G. Ernst, Prentice-Hall, Ruby Volume 1:511-534.

Allen, C.R., P. St. Amend, C.F. Richter, and J.M. Nordquist 1965. Relationship between seismicity and geologic structure in the southern California region. Seismological Society of America Bulletin 21(7):103-106.

Allen, C.R., A. Grantz, J.N. Brune, M.M. Clarke, R.V. Sharp, T.G. Theodore, E.W. Wolfe, and M. Wyss 1968a. The Borrego Mountain Earthquake, April 9, 1968. Mineral Information Service 21(7):103-106.

----- 1968b. The Borrego Mountain Earthquake, April 9, 1968, a preliminary report. Seismological Society of America Bulletin 58:1183-1186.

Allen, C.R., and D.V. Helmberger 1973. Search for the temporal change in seismic velocities using large explosions in southern California. In Conference of Tectonic Problems of the San Andreas Fault System Proceedings, edited by R.L. Kovach and A. Nur, School of Earth Sciences, Stanford University, California p. 436-445.

Allen, C.R., and J.M. Nordquist 1972. Foreshocks, main shock, and larger aftershocks of the Borrego Mountain Earthquake. In The Borrego Mountain Earthquake of April 9, 1968, U.S. Geological Survey Professional Paper 787:16-23.

Allen, C.R. M. Wyss, J.N. Brune, A. Grantz, and R.E. Wallace 1972. Displacements on the Imperial, Superstition Hills, and San Andreas Faults triggered by the Borrego Mountain Earthquake. In The Borrego Mountain Earthquake of April 9, 1968, U.S. Geological Survey Professional Paper 787:87-104.

Allison, M.L. 1974a. Geologic and geophysical reconnaissance of the Elsinore-Chariot Canyon Fault systems. In Recent Geologic and Hydrologic Studies, Eastern San Diego County and Adjacent Areas, edited by M.W. Hart and R.W. Dowlen, San Diego Association of Geologists Guidebook, California p. 21-35.

----- 1974b. Geophysical studies along the southern portion of the Elsinore Fault. Master of Science Thesis, San Diego State University, California 229 p.

----- 1990. Remote detection of active faults using borehole breakouts in the Heber geothermal field, Imperial Valley, California. Transactions of the Geothermal Resources Council 14(1-2):1359-1364.

Allison, M.L., and J.H. Whitcomb 1975. Elsinore Fault seismicity; the September 13, 1973, Agua Caliente Springs, California earthquake series. EOS, Transactions of the American Geophysical Union 57:153.

----- 1976. Elsinore Fault seismicity; the September 13, 1973, Agua Caliente Springs, California earthquake series. EOS, Transactions of the American Geophysical Union 56:398.

Alvarez, L.J., and J.F. Hermance 1980. Magnetotelluric sounding and magnetic variation profiling in the Imperial Valley of southern California. EOS, Transactions of the American Geophysical Union 61:225.

Ambriano, J. 1973. Geology of San Diego County, a bibliography of the holdings in Malcolm A. Love Library San Diego State University. Sciences and Engineering Library, San Diego State University, California 94 p.

American Association of Petroleum Geologists, Pacific Section 1962. California fault bibliography and cross index. Los Angeles, California 57 p.

----- 1973. Imperial Valley regional geology and geothermal exploration. Guidebook 30, Los Angeles, California 56 p.

American Geographical Society 1906. The Salton Sink, geographical record. American Geographical Society Bulletin 38:1-561.

Anderson, D.L. 1971. The San Andreas Fault. Scientific American 225(5):52-68.

Anderson, D.N. 1974. First- and second-order geothermal project survey, Imperial County, California, October 1973-April 1974. Report PB-261 511/OWN, 7 p.

----- 1976. Update of geothermal resources and development in California and northern Mexico. American Association of Petroleum Geologists Bulletin 60:2174.

Anderson, D.N., and S. Willard 1978. Direct utilization of geothermal energy in the Imperial Valley, California. Geothermal Resources Council, Davis, California p. (not numbered).

Anderson, J.G. 1985. Two observations about low-frequency signals on accelerograms from the October 15, 1979 Imperial Valley, California earthquake. Earthquake Engineering & Structural

Dynamics 13:97-108.

Anderson, J.G., and P.G. Silver 1984. Accelerogram evidence for southward rupture propagation on the Imperial Fault during the October 15, 1979 earthquake (M 6.5). EOS, Transactions of the American Geophysical Union 65:1004.

----- 1985. Accelerogram evidence for southward rupture propagation on the Imperial Fault during the October 15, 1979 earthquake. Geophysical Research Letters 12:349-352.

Anderson, J.G., and T.H. Heaton 1982. Aftershock accelerograms recorded on a temporary array. In The Imperial Valley, California, Earthquake of October 15, 1979. U.S. Geological Survey Professional Paper 1254:443-451.

Anderson, J.G., and P. Bodin 1987. Earthquake recurrence models and historic seismicity in the Mexicali-Imperial Valley. Seismological Society of America Bulletin 77:562-578.

Anderson, J.R. 1983. Petrology of a portion of the eastern Peninsular Ranges mylonite zone, southern California. Contributions to mineralogy and Petrology 84:253-271.

Andes, J.P., Jr. 1987. Mineralogic and fluid inclusion study of ore-mineralized fractures in drillhole State 2-14, Salton Sea Scientific Drilling Project, California, U.S.A. Master's Thesis, Department of Geological Sciences, University of California, Riverside 125 p.

Andrews, M.C., C. Deitel, T. Noce, E. Sembra, and J.D. Bicknell 1989. Digital seismograms of the Superstition Hill, California, aftershock sequence; November 24 to December 8, 1987. U.S. Geological Survey Open-File Report 88-700:1-154.

Angelier, J., B. Colletta, J. Chorowicz, L. Ortlieb, and C. Rangin 1981. Fault tectonics of the Baja California peninsula and the opening of the Sea of Cortez, Mexico. Journal of Structural Geology 3:347-357.

Anonymous 1964. From the earth a new source of energy. Western Engineer 48(3):6-7.

----- 1976. Geothermal project summaries; geothermal energy research, development and demonstration program. Energy Research and Development Administration, Division of Geothermal Energy, Washington D.C. ERDA-76/53/1 p. (variously numbered).

----- 1978. Southern Elsinore Fault Zone in California. U.S. Geological Survey Professional Paper 1100:267.

----- 1982. The 1979 Imperial Valley, California, earthquake. Earthquake Information Bulletin 14:173-179.

----- 1986. Salton Sea Scientific Drilling Program monitor. Geothermal Resources Council Bulletin 15:15-18.

----- 1988. Special section on results of the Salton Sea Scientific Drilling Project, California. Journal of Geophysical Research, B, Solid Earth and Planets, 93:12,953-13,186.

Arabasz, W.J., J.N. Brune, and G.R. Engen 1970. Locations of small earthquakes near the trifurcation of the San Jacinto Fault southeast of Anza, California. Seismological Society of America Bulletin 60(2);617-627.

Archuleta, R.J. 1982a. Analysis of near source static and dynamic measurements from the 1979 Imperial Valley Earthquake. Seismological Society of America Bulletin 72:1927-1956

----- 1982b. An in-depth look at the 1979 Imperial Valley Earthquake. Earthquake News 54:62

----- 1982c. Faulting model for the 1979 Imperial Valley Earthquake based on synthetic seismograms EOS, Transactions of the American Geophysical Union 63:1037-1038

----- 1984a. A faulting model for the 1979 Imperial Valley Earthquake. Journal of Geophysical Research 89:4559-4585

----- 1984b. Modeling strong motion data from the 1979 Imperial Valley Earthquake. In Ground Motion and Seismicity (Earthquake Engineering Research Institute), Proceedings of the World Conference on Earthquake Engineering 8:369-376

Archuleta, R.J., and M. Bouchon 1984. A study of Q in the Imperial Valley, California, earthquake. Earthquake Notes 55:24.38, Terra Cognita 2:165-166

Archuleta, R.J., and R.B. Sharp 1980. Source parameters of the Oct. 15, 1979 Imperial Valley Earthquake from near-field observations. EOS, Transactions of the American Geophysical Union 61:297

Archuleta, J.H., and P. Spudich 1981. An explanation for the large amplitude vertical accelerations generated by the 1979 Imperial Valley, California, earthquake. EOS, Transactions of the American Geophysical Union 62:323.

----- 1982. Analysis of near-source static and dynamic measurements from the 1979 Imperial Valley Earthquake. In The Dynamic Characteristics of Faulting Inferred from Recordings of Strong Ground Motion, edited by J. Boatwright, U.S. Geological Survey Open-File Report 82-591:784-838.

Armitage, A.L. 1989. Composition of Holocene Colorado River sand; an example of mixed-provenance sand derived from multiple tectonic elements of the Cordilleran continental margin. Master of Science Thesis, San Diego State University, California 169 p.

Armstrong, R.L. 1968. Sevier orogenic belt in Nevada and Utah. Geological Society of America Bulletin 79:429-458.

Armstrong, R.L., and J. Suppe 1973. Potassium-argon

geochronometry of Mesozoic igneous rocks in Nevada, Utah, and southern California. Geological Society of America Bulletin 84:1375-1392.

Arnal, R.E. 1954. Preliminary reports on the sediments and foraminifera from the Salton Sea, southern California. Geological Society of America Bulletin 65:1227-1228.

----- 1958. Rhizopoda from the Salton Sea, California. Cushman Foundation for Foraminiferal Research Contributions 9:36-45.

----- 1961. Limnology, sedimentation, and microorganisms of the Salton Sea, California. Geological Society of America Bulletin 72:427-478.

Arnold, R.E. 1904. The faunal relations of the Carrizo Creek beds of California. Science 19:1-503.

----- 1906. The Tertiary and Quaternary pectens of California. U.S. Geological Survey Professional Paper 47:84-85.

----- 1909a. Paleontology of the Coalinga District, Fresno and Kings Counties, California. U.S. Geological Survey Bulletin 396: 1-173.

----- 1909b. Environment of the Tertiary faunas of the Pacific Coast of the United States. Journal of Geology 17:509-533.

Arnold, R.E., and R. Anderson 1910. Geology and oil resources of the Coalinga District, California. U.S. Geological Survey Bulletin 398:1-354.

Arthur, M.A. 1974. Stratigraphy and sedimentation of lower Miocene non-marine strata of the Orocopia Mountains; constraints from late Tertiary slip on the San Andreas Fault system in southern California. Master's Thesis. University of California, Riverside.

Aschmann, H.H. 1959. The evolution of a wild landscape, and its persistence, in southern California. Annals of the Association of American Geography 49:34-56.

----- 1966. The head of the Colorado delta. In Geography as Human Ecology, Methodology by Example, edited by S.R. Eyre and R.J. Jones, Edward Arnold Limited, London p. 231-263.

Atwater, T. 1970. Implications of plate tectonics for the Cenozoic tectonic evolution of western North America. Geological Society of America Bulletin 81:3513-3536.

----- 1992. Constraints from reconstructions for Cenozoic tectonic regimes of southern and eastern California. In Deformation Associated with the Neogene Eastern California Shear Zone, Southeastern California and Southwestern Arizona, edited by S.M. Richard, San Bernardino County Museum Special Publication 92-1:1-2.

Augur, I.V. 1920. Resume of oil-well operations in Imperial Valley. California State Mining Bureau Summary of Operations, California Oil Fields, Fifth Annual Report State Oil and Gas Supervisor 5:5-9

Axelrod, D.I. 1937. A Pliocene flora from the Mount Eden beds, southern California. Carnegie Institute of Washington Publication 476:125-183.

----- 1944a. The Mulholland flora. Carnegie Institute of Washington Publication 553(5):103-146.

----- 1944b. The Sonoma flora. Carnegie Institute of Washington Publication 553(7):167-206.

----- 1948. Climate and evolution in western North America during middle Pliocene time. Evolution 2:127-144.

----- 1950a. Further studies of the Mount Eden flora, southern California. Carnegie Institute of Washington Publication 590:73-117.

----- 1950b. The evolution of desert vegetation in western North America. Carnegie Institute of Washington Publication 590:215-306.

----- 1957. Late Tertiary floras and the Sierra Nevadan uplift. Geological Society of America Bulletin 68:19-46.

----- 1966. The Pleistocene Soboba flora of southern California. University of California Publications in the Geological Sciences 60:1-79.

----- 1979. Age and origin of the Sonoran Desert vegetation. Occasional Papers of the California Academy of Sciences 132:1-74.

Babcock, E.A. 1968. On the high density core of the Durmid Dome, Imperial Valley. American Geophysical Union Transactions 49:759

----- 1969. Structural geology and geophysics of the Durmid area, Imperial Valley, California. Doctoral Dissertation, Department of Geological Sciences, University of California, Riverside. 145p.

----- 1970. Basement structure and faulting along the northeast margin of the Salton Sea. Geological Society of America Abstracts 2:68

----- 1971a. Detection of active faulting using oblique infrared aerial photography in the Imperial Valley, California. Geological Society of America Abstracts 3:77

----- 1971b. Detection of active faulting using oblique infrared aerial photography in the Imperial Valley, California. Geological Society of America Bulletin 82:3189-3196.

------ 1974. Geology of the northern margin of the Salton Trough, Salton Sea, California. Geological Society of America Bulletin 85:321-332.

Babcock, J.N. 1961. Geology of a portion of the Pinyon Well quadrangle, Riverside County, California. Master's Thesis, University of California, Los Angeles.

Bailey, G.E. 1902. The saline deposits of California. California Mining Bureau Bulletin 24:1-216.

----- 1924. California, a geologic wonderland. The Times-Mirror Press, Los Angeles 119 p.

Bailey, L. Thomas, and R.H. Johns 1954. Geology of the

Transverse Range province, Southern California. In Geology of Southern California, edited by R.H. Jahns, California Division of Mines Bulletin 170(2):83-106.

Bailey, T.P. 1977. A hydrogeological and subsurface study of Imperial Valley geothermal anomalies, Imperial Valley, California. Master's Thesis, University of Colorado, Boulder.

Baird, A.K., K.W. Baird, and E.E. Welday 1979. Batholithic rocks of the northern Peninsular and Transverse Ranges, southern California; chemical composition and variation. In Mesozoic Crystalline Rocks, Peninsular Ranges Batholith and Pegmatites, Point Sal Ophiolite, edited by P.L. Abbott and V.R. Todd, Department of Geological Sciences, San Diego State University, California p. 111-132.

Baird, A.K., and Meisch, A.T. 1984. Batholithic rocks of southern California - a model for the petrochemical nature of their source materials. U.S. Geological Survey Professional Paper 1284:1-42.

Balderman, M.A., C.A. Johnson, D.G. Miller, and D.L. Schmidt 1978a. The 1852 Fort Yuma Earthquake. Seismological Society of America Bulletin 68:1717-1729.

----- 1978b. The 1852 Fort Yuma Earthquake. Seismological Society of America Bulletin 68:699-710.

Baldwin, J.E. 1986. Martinez Mountain rock avalanche. In Geology of the Imperial Valley, California, edited by P.D. Guptil, E.M. Gath and R.W. Ruff, South Coast Geological Society Annual Field Trip Guidebook, Santa Ana, California 14:37-48.

Ballard, A. 1971. The paleoecology of a marine deposit in the northwest Indio Hills. Geology 111B, Department of Geological Sciences, University of California, Riverside 12 p.

Ballog, A.P., Jr., and W.R. Moyle Jr. 1980. Water resources and geology of the Los Coyotes Indian Reservation and vicinity, San Diego County, California. U.S. Geological Survey Open- File Report 1978, 80-0960:1-29.

Banks, P.O., and L.T. Silver 1968. U-Pb isotope analyses of zircons from Cretaceous plutons of the Peninsular and Transverse Ranges, southern California. Geological Society of America Special Papers 121:17.

Barker, C.E., B.L. Crysdale, and M.J. Pawlewicz 1986. Relationship between vitrinite reflectance, metamorphic grade, and temperature in the Cerro Prieto, Salton Sea, and East Mesa geothermal systems, Salton Trough, United States and Mexico. In Studies in Diagenesis, edited by F.A. Mumpton, U.S. Geological Survey Bulletin 1578:83-95.

Barker, T.G. 1976. Quasi-static motions near the San Andreas Fault Zone. Royal Astronomical Society Geophysical Journal 45:689-705.

Barker, T.G., and J.L. Stevens 1983. Shallow shear wave velocity and Q structures at the El Centro strong motion accelerograph array. Geophysical Research Letters 10:853-856.

----- 1984. Shallow shear wave velocity and Q structures at the El Centro strong motion accelerograph array. Earthquake Notes 55:25.

Barkman, J.H., D.A., Smith, J.L. 1976. East Mesa; geology, reservoir properties and an approach to reserve determination. In Second Workshop on Geothermal Reservoir Engineering; Summaries, edited by Kruger, Stanford University, California p. 116-125.

Barmore, R.L. 1980. Soil survey of Yuma-Wellton area; parts of Yuma County, Arizona, and Imperial County, California. Washington D.C., U.S. Department of Agriculture, Soil and Conservation Service 104 p.

Barnard, F.L. 1968. Structural geology of the Sierra de los Cucapas, northeastern Baja California, Mexico, and Imperial County, California. Doctoral Dissertation, University of Colorado, Boulder 157 p.

Barnes, N.E. 1983. Geology of Yaqui Ridge and The Narrows, eastern San Diego County, California. Undergraduate Research Reports, Department of Geology, San Diego State University, California 42(1):1-23.

Barrows, D.P. 1900. The Colorado Desert. National Geographic Magazine 11:337-351.

Bartel, J. 1981. Stratigraphic report of the northeast flank of Superstition Mountain, Imperial Valley. Undergraduate Research Reports, Department of Geology, San Diego State University, California 38:19.

Barth, A.P., R.M. Tosdal, and J.L. Wooden 1990. A petrologic comparison of Triassic plutonism in the San Gabriel and Mule Mountains, southern California. Journal of Geophysical Research, B, Solid Earth and Planets 95(12):20,075-20,096.

Bartholomew, M.J. 1968. Geology of the northern portion of Seventeen Palms and Font's Point quadrangles, Imperial and San Diego Counties, California. Master of Science Thesis, University of Southern California, Los Angeles 60 p.

----- 1970. San Jacinto Fault Zone in the northern Imperial Valley. Geological Society of America Bulletin 81:3161-3166.

Bartling, J.W. 1976. Rainfall and runoff characteristics of the Cuyamaca watershed. Undergraduate Research Reports, Department of Geology, San Diego State University, California 28(11):1-11.

Bartow, J.A., A.M. Sarna-Wojcicki, and J.C. Matti 1991. Cenozoic stratigraphic correlation for the southern

California areal mapping project: a progress report. Geological Society of America Abstracts with Programs p. A475.

Bateman, A.M. 1938. [Review of] The Colorado delta, by G.G. Sykes, 1937. Economic Geology, 33:119-120.

Batzle, M.L., and G. Simmons 1975. Microfractures in rocks from two geothermal areas. Geological Society of America Abstracts with Programs 7:993.

----- 1976. Microfractures in rocks from two geothermal areas. Earth and Planetary Science Letters 30:71-93.

Bauersfeld, A.G. 1971. Paleoecology of the Tertiary fauna, Willis Palms Formation, Thousand Palms quadrangle, California. Geology 111B, Department of Geological Sciences, University of California, Riverside 19 p.

Beal, C.H. 1915. The earthquake in the Imperial Valley, June 22, 1915. Seismological Society of America Bulletin 5:130-149.

Becker, J.J., and J.A. White 1981. Late Cenozoic geomyids (Mammalia; Rodentia) from the Anza-Borrego Desert, southern California. Journal of Vertebrate Paleontology 1:211-218.

Beckett, G.D. 1991. Pinyon and Vallecito Mountains detachment fault, general geometry and morphology. Undergraduate Research Report, Department of Geology, San Diego State University, California, 69 p.

Begland, Lt. E. 1876. Preliminary report upon the operations of party no. 3, California sections, seasons of 1875-1876, with a view to determine the feasibility of diverting the Colorado River for the purposes of irrigation. Report of the Secretary of War 2(2):334-336, appendix B.

Begnoche, D.J. 1961. Geology and ground waters of Henshaw Basin. Undergraduate Research Reports, Department of Geology, San Diego State University, California 5(1):1-37.

Bell, C.J., and J.I. Mead 1993. Fossil lizards from the Elsinore Fault Zone, Riverside County, California, with comments on the Neogene fossil history of the legless lizard, *Anniella*. Abstracts of Proceedings for the 7th annual Mojave Desert Quaternary Research Symposium and the Desert Studies Consortium Symposium. San Bernardino County Museum Association Quarterly 40(20):20-21.

Bell, P.J. 1980. Environments of deposition, Pliocene Imperial Formation, southern Coyote Mountains, Imperial County, California. Master of Science Thesis, San Diego State University, California 72 p.

Bell-Countryman, P.J. 1984. Environments of deposition, Pliocene Imperial Formation, Coyote Mountains, southwest Salton Trough. In The Imperial Basin, Tectonics, Sedimentation and Thermal Aspects, edited by C.A. Rigsby, Society of Economic Paleontologists and Mineralogists 40:45-70.

Bellemin, G.S., and R. Merriam 1958. Petrology and origin of the Poway Conglomerates, San Diego County, California. Geological Society of America Bulletin 69:199-220.

Bennet, G.E. 1966. Geology of a portion of Clark Lake quadrangle. Undergraduate Research Reports, Department of Geology, San Diego State University, California 10(1):1-31.

Bennet, M.J. 1984. Influence of depositional environment on liquefaction susceptibility in the Imperial Valley, California. Society of Economic Paleontologists and Mineralogists, Abstract 1:11.

----- 1988. Sand boils and their source beds; November 24, 1987, Superstition Hills Earthquake, Imperial Valley, California. In Exploration Methods and Applications, New and Revised, Association of Engineering Geologists, National Meeting 31:38.

Bennet, M.J., T.L. Youd, E.L. Harp, and G.F. Wieczorek 1981. Subsurface investigation of liquefaction, Imperial Valley Earthquake, California, October 15, 1979. U.S. Geological Survey Open-File Report 81-0502:1-87.

Bent, A., and D.V. Helmberger 1989. A re-examination of historic earthquakes in the western Imperial Valley, California. Seismological Research Letters 60:23.

Bent, A., D.V. Helmberger, R.J. Stead, and P. Ho-Liu 1989. Waveform modeling of the November 1987 Superstition Hills Earthquakes. Seismological Society of America Bulletin 79:500-514.

Bent, A., P. Ho-Liu, and D.V. Helmberger 1988. The November 1987 Superstition Hills Earthquake and comparisons with previous neighboring events. Seismological Research Letters 59:49.

Beratan, K.K. 1992. Age constraints on the end of detachment faulting, Colorado River extensional corridor, southeastern California-western Arizona. In Deformation Associated with the Neogene Eastern California Shear Zone, Southeastern California and Southwestern Arizona, edited by S.M. Richard, San Bernardino County Museum Special Publication 92-1:3-5.

Berg, L. 1982. Mid-Tertiary detachment faulting in the Midway Mountains, Imperial County, California. Undergraduate Research Reports, Department of Geology, San Diego State University, California 40:38

Berg, L., G. Levielle, and P. Geis 1982. Mid-Tertiary detachment faulting and manganese mineralization in the Midway Mountains, Imperial County, California. In Mesozoic-Cenozoic Tectonic Evolution of the Colorado River Region, California, Arizona, and Nevada, edited by E.G. Frost and D.L. Martin, Cordilleran Publication, San Diego,

California p. 298-311.

Berger, J., L. Baker, J. Brune, J. Fletcher, T. Hanks, and F. Vernon 1984. The Anza array: a high dynamic range, broadband, digitally radiotelemetered seismic array. Seismological Society of America Bulletin 74:1469-1481.

Berggren, R.G. 1976. Petrology, structure, and metamorphic history of a metasedimentary roof pendant in the Peninsular Ranges Batholith, San Diego County, California. Master of Science Thesis. San Diego State University, California 77 p.

Berggren, W.A., D.V. Kent, J.J. Flyn, and J.A. VanCouvering 1985. Cenozoic geochronology. Geological Society of America Bulletin 96:1407-1418.

Berry, F.A.F. 1966a. Proposed origin of subsurface thermal brines, Imperial Valley, California. American Association of Petroleum Geologists Bulletin 50:644-645.

----- 1966b. Role of membrane hyperfiltration on the origin of the thermal brines, Imperial Valley, California. In The 11th Pacific Science Congress Proceedings, Tokyo, Scientific Council Japan, Division Meeting Solid Earth Physics II 3:43.

----- 1967. Geothermal brines. In Natural Gas, Coal, Ground Water - Exploring New Methods and Techniques in Resources Research, 8th Western Resources Conference, Colorado School of Mines, University of Colorado Press p. 155-169.

Berry, L. 1981. Chemical analysis of thermal water in the Salton Sea area. Undergraduate Research Reports, Department of Geology, San Diego State University, California 38:36.

Biehler, S. 1963. Geophysical investigations in the Salton Trough, southern California. American Geophysical Union Transactions 44:105-106.

----- 1964. Geophysical study of the Salton Trough of southern California. Doctoral Dissertation, California Institute of Technology, Pasadena 145 p.

----- 1965. Geophysical study of the Salton Trough of southern California. Dissertation Abstracts 25:4648.

----- 1971a. Gravity models of the crustal structure of the Salton Trough. Geological Society of America Abstracts 3:82-83.

----- 1971b. Gravity models of the crustal structure of the Salton Trough. Geological Society of America Abstracts 3:506.

Biehler, S., R.L. Kovach, and V.R. Allen 1964. Geophysical framework of northern end of Gulf of California structural province. In Marine Geology of the Gulf of California - A Symposium, American Association of Petroleum Geologists Memoir 3:126-143.

Biehler, S., and R.W. Rex 1971. Structural geology and tectonics of the Salton Trough, southern California. In Geological Excursions in Southern California, edited by W.A. Elders, University of California, Riverside Campus Museum Contributions 1:30-42.

Bilham, R. 1989. Surface slip subsequent to the 24 November 1987 Superstition Hills, California, earthquake monitored by digital creepmeters. Seismological Society of America Bulletin 79:424-450.

Bilham, R., and P. Williams 1985. Sawtooth segmentation and deformation processes on the southern San Andreas Fault, California. Geophysical Research Letters 12:557-560.

Birchard, G.F., and W.F. Libby (no date). Soil Radon concentration changes preceding and following four M 4.2- M 4.7 earthquakes on the San Jacinto Fault in southern California. Institute of Geophysics and Planetary Physics, Manuscript on File Anza-Borrego Desert State Park, California 17 p.

Bird, D.K. 1975. Geology and geochemistry of the Dunes hydrothermal system, Imperial Valley of California. Master of Science Thesis, Department of Geological Sciences, University of California, Riverside 123 p.

Bird, D.K., M. Cho, C.J. Janik, J.G. Liou, and L.J. Caruso 1988. Compositional, order-disorder, and stable isotope characteristics of Al-Fe epidote, State 2-14, drill hole, Salton Sea geothermal system. Journal of Geophysical Research, B, Solid Earth and Planets 93:13,135-13,144.

Bird, D.K., and D. Norton 1979. Correlation of the spatial variations in the composition of geothermal reservoir fluids from petrochemical observations and thermodynamic calculations in the Salton Sea geothermal system. Geological Society of America Abstracts with Programs 11:388.

Bird, D.K., and W.A. Elders 1973. Petrology of silicified cap rocks in the Dunes geothermal anomaly, Imperial Valley of California. EOS, Transactions of the American Geophysical Union 54:1214.

----- 1975a. Hydrothermal alteration and mass transfer in the discharge portion of the Dunes geothermal system, Imperial Valley of California, U.S.A. United Nations Symposium on the Development and Use of Geothermal Resources Abstracts, No. 2 (unpaginated).

----- 1975b. Investigations of the Dunes geothermal anomaly, Imperial Valley. Institute of Geophysics and Planetary Physics, UCR/IGPP Report 75/14:1-20.

Bird, P., and R.W. Rosenstock 1984. Kinematics of present crust and mantle flow in southern California. Geological Society of America Bulletin 95:946-957.

Bishop, H.K. 1978. Sampling and analysis of geothermal brines from Niland field, California. Proceedings of the Second Workshop on Sampling Geothermal Effluents, Report EPA-600/7:78-121.

Black, W.E. 1974. A geophysical investigation of Yuha Desert, Imperial County, California. Master's Thesis, University of California, Riverside.

Black, W.E., J.S. Nelson, and J.Combs 1973. Thermal and electrical resistivity investigation of the Dunes Geothermal Anomaly, Imperial Valley, California. EOS, Transactions of the American Geophysical Union 54:1214.

Blackford, L.C. 1995. Postcranial skeletal analysis of the Pleistocene deer *Navahoceros fricki* (Cervidae). Master of Science Thesis, Northern Arizona University, 144p.

Blackwelder, E. 1934. Origin of the Colorado River. Geological Society of America Bulletin 45:551-566.

Blair, W.N. 1978. Gulf of California in Lake Mead area of Arizona and Nevada during late Miocene time. American Association of Petroleum Geologists Bulletin 62(7):1159-1170.

Blair, M.C., and J.P. Bradbury 1979. Gulf of California in Lake Mead area during late Miocene time: reply. American Association of Petroleum Geologists Bulletin 63:1140-1142.

Blake, M.C., Jr., R.H. Campbell, T.W. Dibblee, D.C. Howell, T.H. Nilsen, W.R. Normark, J.C. Veder, and E. A. Silver 1978. Neogene basin formation in relation to plate-tectonics of San Andreas Fault system, California. American Association of Petroleum Geologists Bulletin 62:344-372.

Blake, R.L. 1974. Extracting minerals from geothermal brines; a literature study. U.S. Bureau of Mines Information Circular 8638:1-25.

Blake, W.P. 1854. Ancient lake of the Colorado Desert. American Journal of Science 17:435-438.

----- 1855. Preliminary geological report of the expedition under command of Lieutenant R.S. Williamson, United States Topographical Engineers. U.S. 33rd Congress, 1st Session, House Executive Document 129:1-80.

----- 1857. (geological report). In 1855 Reports of Exploration and Surveys to Ascertain the Most Practicable and Economic Route for a Railroad from the Mississippi River to the Pacific Ocean, under command of Lieutenant R.S. Williamson, United States Topographical Engineers, U.S. 33rd Congress, 2nd Session, Senate Executive Document 78, 5(2):1-310.

----- 1907. Lake Cahuilla, the ancient lake of the Colorado Desert. National Geographic Magazine 18:830.

----- 1908. Destruction of the salt works in the Colorado Desert by the Salton Sea. American Institute of Mining Engineers Bulletin 19:81-82.

----- 1914. The Cahuilla basin and desert of the Colorado. In The Salton Sea, edited by D.T. MacDougal, Carnegie Institution of Washington Publication 193:1-12.

----- 1915. Sketch of the region at the head of the Gulf of California. In The Imperial Valley and the Salton Sink, edited by H.T. Cory, San Francisco, California p. 1-35.

Blom, R.G., R.E. Crippen, and E.G. frost 1988. Geometry and role of major east dipping detachment faults in the initiation of the Salton Trough and localization of the San Andreas Fault system. Geological Society of America, Abstracts with Programs 20:A381.

Blom, R., R. Crippen, and T. Rockwell 1990. Use of SPOT and LANDSAT thematic mapper images for detection and study of northeast trending faults between the Elsinore and San Jacinto Faults southern California. Geological Society of America Abstracts with Programs 22(3):9.

Boatwright, J., K.E. Budding, and R.V. Sharp 1989. Inverting measurements of surface slip on the Superstition Hills Fault. Seismological Society of America Bulletin 79:411-423.

Boehm, M.C. 1984. An overview of the lithostratigraphy, biostratigraphy, and paleoenvironments of the late Neogene San Felipe marine sequence, Baja California, Mexico. In Geology of the Baja California peninsula, edited by V.A. Frizzell. Society of Economic Paleontologists and Mineralogists p. 253-266.

Bogen, N.L., and L. Seeber 1986a. Neotectonics of rotating blocks within the San Jacinto Fault Zone, southern California. EOS, Transactions of the American Geophysical Union 67:1200.

----- 1986b. Late Quaternary block rotations within the San Jacinto Fault Zone, southern California. Geological Society of America Special Paper 18(2):88.

Boley, J.L., J.P. Stimac, J.R. Weldon, II, and M.J. Rymer 1994. Stratigraphy and paleomagnetism of the Mecca Hills and Indio Hills, southern California. In Geological Invstigations of an Active Margin, edited by S.F. McGill and T.M. Ross, U.S. Geological Survey, Cordilleran Section Guidebook Trip 15, pp. 336-344.

Boore, D.M., and R.L. Porcella 1980. Peak horizontal ground motions from the 1979 Imperial Valley earthquake; comparison with data from previous earthquakes. In Selected Papers on the Imperial Valley, California, Earthquake of October 15, 1979, edited by C. Rojhan, U.S. Geological Survey Open-File Report 80-1094:17-27.

----- 1982. Peak horizontal ground motions from the main shock, comparison with data from previous earthquakes. In The Imperial Valley, California, Earthquake of October 15, 1979, U.S. Geological

Survey Professional Paper 1254:439-442.

Boore, D.M., and J.B. Fletcher 1980. A preliminary study of selected aftershocks of the 1979 Imperial Valley earthquake from digital acceleration and velocity recordings. In Selected Papers on the Imperial Valley, California, Earthquake of October 15, 1979, edited by C. Rojhan, U.S. Geological Survey Open-File Report 80-1094:43-67.

----- 1982. Preliminary study of selected aftershocks from digital acceleration and velocity recordings. In The Imperial Valley, California, Earthquake of October 15, 1979, U.S. Geological Survey Professional Paper 1254:109-118.

Borden, J. 1967. Geology of the San Felipe Hills, Imperial Valley, California. Undergraduate Research Reports, Department of Geology, San Diego State University, California 11:15.

Bowden, J.K., and E. Scott 1992. New record of *Smilodon fatalis* (Leidy), 1868 (Mammalia; Carnivora; Felidae) from Riverside County, California. Mojave Desert Quaternary Research Symposium, Abstracts of Proceedings, San Bernardino County Museum Association Quarterly 39(2):27.

Bowers, S. 1901. Reconnaissance of the Colorado Desert mining district. California State Mining Bureau 19 p.

Bowersox, R.J. 1972a. Molluscan paleontology and paleoecology of Holocene Lake Cahuilla. Southern California Academy of Sciences, Abstracts (unpaginated).

----- 1972b Molluscan paleontology and paleoecology of Holocene Lake Cahuilla. Undergraduate Research Reports, Geology Department, San Diego State University, California 21:1-22.

Bowie, W. 1931. Geodetic work lays the basis for study of earth movements. American Geophysical Union Transactions 12:65-66.

----- 1973. Geodetic work lays the basis for study of earth movements. In Reports on Geodetic Measurements of Crustal Movement, 1906-71, National Oceanic, and Atmospheric Administration, National Ocean Survey-National Geodetic Survey 2 p.

Bozkurt, U. 1989. A fluid inclusion study of selected boreholes, Salton Sea geothermal system, Imperial Valley, California. Master's Thesis, University of California, Riverside.

Bradshaw, G.B., W.W. Donnan, and H.F. Blaney 1951. Preliminary progress report on cooperative investigation in Imperial Valley, California for year 1950. Unpublished manuscript, U.S. Soil Conservation Service 105 p.

Brady, A.G., V. Perez, and P.N. Mork 1980. Seismic engineering data report; the Imperial Valley earthquake, October 15, 1979; digitalization and processing of accelerograph records. U.S. Geological Survey Open-File Report 80-703:1-311.

----- 1982. Digitization and processing of main-shock groundmotion data from the U.S. Geological Survey accelerograph network. In The Imperial Valley, California, Earthquake of October 15, 1979, U.S. Geological Survey Professional Paper 1254:385-406.

Brady, A.G., and Mork, P.N. 1989. Long-period signal content in the accelerations at the Imperial Wildlife liquefaction array during the Superstition Hills, California, earthquakes, 24 Letters 60:4.

Bramkamp, R.A. 1935. Stratigraphy and molluscan fauna of the Imperial Formation of San Gorgonio Pass, California. Doctoral Dissertation, University of California, Berkeley 371 p.

Branham, A.D. 1988. Gold mineralization in low angle faults, American Girl Valley, Cargo Muchacho Mountains, California. Master's Thesis, Washington State University, Pullman.

Breau, S.F. 1994. Water wells as strain meters along the San Jacinto Fault, southern California, 1978-1985. Master of Science Thesis, Department of Geology, Duke University 107 p.

Breed, W.J., and E. Roat 1976. Geology of the Grand Canyon. Museum of Northern Arizona and Grand Canyon Natural History Association, Flagstaff 186 p.

Bringhurst, K.N. 1987. Major element chemistry and mineralogy in well Fee #5, in the Salton Sea geothermal field, California. Master's Thesis, University of California, Riverside 135 p.

Brinkman, D.J. 1968. The geology of the southern portion of the Santa Rosa Mountains, California. Undergraduate Research Reports, Department of Geology, San Diego State University, California 12(1):1-24.

Brook, C.A., and C.W. Mase 1981. The hydrothermal system at the east Brawley KGRA, Imperial Valley, California. In Geothermal Energy, the International Success Story, Transactions of the Geothermal Resources Council 5:157-160.

Brooks, B., and R. Ellis 1954. Geology of the Jacumba area, San Diego and Imperial Counties. In Geology of Southern California, edited by R.H. Jahns, California Division of Mines and Geology Bulletin 170:map sheet 23.

Brooks, K.R. 1981. A geographical survey across the Coyote Creek Fault, San Jacinto Fault Zone, Ocotillo Wells, California. Undergraduate Research Reports, Department of Geology, San Diego State University, California 38(1):1-56.

Brown, A.J. 1983. Channel changes in arid badlands, Borrego Springs, California. Physical Geography

4:82-102.

Brown, A.R. 1968. Geology of a portion of the southeastern San Jacinto Mountains, Riverside County, California. Master of Science Thesis, Department of Geological Sciences, University of California, Riverside 95 p.

----- 1981. Structural history of the metamorphic, granitic and cataclastic rocks in the southeastern San Jacinto Mountains. In Geology of the San Jacinto Mountains, edited by A.R. Brown and R.W. Ruff, South Coast Geological Society, Annual Field Trip Guidebook, Santa Ana, California 9:100-138.

Brown, E.T., Edomnd, J.M., Campbell, A.C., Measures, C.I., Bowers, T.S., and Palmer, M.R. 1986. Preliminary report on metal chemistry of Salton Sea drilling project hydrothermal fluids. EOS, Transactions of the American Geophysical Union 67:1256.

Brown, J.S. 1920. Routes to desert watering places in the Salton Sea region, California. U.S. Geological Survey Water-Supply Paper 490-A:1-86.

----- 1921. Fault features of Salton Basin, California. Washington Academy of Sciences Journal 11:423.

----- 1922. Fault features of the Salton Basin, California. Journal of Geology 30:217-226.

----- 1923. The Salton Sea region, California. U.S. Geological Survey Water Supply Paper 497:1-292.

Brown, J., and J. Boyd 1922. History of San Bernardino and Riverside Counties. The Western Historical Association 1:403-409.

Brown, N.N. 1989. A structural and seismotectonic analysis of the Ocotillo Badlands, southern California. Master's Thesis, University of California, Santa Barbara.

Brown, N.N., M.D. Fuller, and R.H. Sibson 1991. Paleomagnetism of the Ocotillo Badlands, southern California, and implications for slip transfer through an antidilational fault jog. Earth and Planetary Science Letters 102:277-288.

Brown, N.N., and R.H. Sibson 1987. Deformation at a compressional fault jog; a study of the Ocotillo Badlands, Southern California. EOS, Transactions of the American Geophysical Union 68:1507.

----- 1988. Deformation at a compressional fault jog; a study of the Ocotillo Badlands, Southern California. Geological Society of America Abstracts with Programs 20:108.

----- 1989. Structural geology of the Ocotillo Badlands antidilational fault jog, southern California. U.S. Geological Survey Open-File Report 89-315:89-315.

Brown, N.N., M.D. Fuller, and R.H. Sibson 1991. Paleomagnetism of the Ocotillo Badlands, southern California, and implications for slip transfer through an antidilational fault jog. Earth and Planetary Science Letters 102:277-288.

Browne, P.R.L., and Elders, W.A. 1976. Hydrothermal alteration of diabase, Heber geothermal field, Imperial Valley, California. Geological Society of America Abstracts with Programs 8:793.

Brusca, R.C. 1980. Common intertidal invertebrates of the Gulf of California. University of Arizona Press, Tucson 513 p.

Brune, J.N., and C.R. Allen 1967. A low stress drop, low-magnitude earthquake with surface faulting from the Imperial, California, earthquake of March, 4 1966. Seismological Society of America Bulletin 57:501-514.

----- 1968. Low stress drop low-magnitude earthquake with surface faulting from the Imperial, California, earthquake of March 4, 1966. Geological Society of America Special Paper 115:314.

Brune, J.N., and A. Anooshehpoor 1988. Soil-structure interaction and source mechanism for the El Centro 1940 earthquake. Academy Scinica, National Tiwan University, Taipei, and University of Missouri, Rolla p. 41-61.

Brune, J.N., W. Arabasz, and G.R. Engen 1969. Locations of small earthquakes near the trifurcation of the San Jacinto Fault southeast of Anza, California. Earthquake Notes 40:25.

Brune, J.N., R.S. Simons, F. Vernon, L. Canales, and A. Reyes 1980. Digital seismic event recorders, description and examples from the San Jacinto Fault, the Imperial Fault, the Cerro Prieto Fault, and the Oaxaca, Mexico subduction fault. Seismological Society of America Bulletin 70:1395-1408.

Bryan, K. 1931. Physiographic study in the Salton Sea region. Geographica Review 21:153.

Buchheim, H.P. 1990. Discovery of fresh water stromatolites in Carrizo Creek, California. Geological Society of America Abstracts with Programs 22(7):A358-A359.

Buckley, C.P., and C.W. Kohlenberger 1980. The Imperial Valley earthquake; inertial displacement measurements by tiltmeters in the Los Angeles Basin. EOS, Transactions of the American Geophysical Union 61:368.

Buckley, C.P., M.E. Magee, and N.A. Hayden 1977. Recent fractures along the northeast margin of the Salton Trough, Imperial County, California. California Geology 30:58-60.

Buckley, C.P., N.J. Maloney, G.L. Cooper, C. Kirkpatrick, and C.W. Kohlenberger 1975. Short-term anomalous tilting prior to earthquakes in southern California. Geological Society of America Abstracts with Programs 7:399.

Budding, K.E., J. Boatwright, R.V. Sharp, and J.L. Saxton 1989. Compilation and analysis of displacement measurements obtained on the Superstition Hills Fault zone and nearby faults in Imperial Valley, California, following the earthquakes of November 24, 1987. U.S. Geological Survey Open-File Report 89-140:1-90.

Budding, K.E., and R.V. Sharp 1988. Surface faulting associated with the Elmore Desert Ranch and Superstition Hills, California, earthquakes of 24 November 1987. Seismological Research Letters 59:49.

Buising, A.V. 1986. The Bouse Formation and bracketing units, southeast California and west Arizona: investigations into the depositional and tectonic evolution of the proto-Gulf of California. Geological Society of America Abstracts with Programs 18(2):190.

----- 1988a. Depositional and tectonic evolution of the northern proto-Gulf of California and lower Colorado River, as documented in the mid-Pliocene Bouse Formation and bracketing units, southeastern California and western Arizona. Doctoral Dissertation, University of California, Santa Barbara.

----- 1988b. Contrasting subsidence histories, northern and southern proto-Gulf of California and lower Colorado River, as documented in the Mio-Pliocene Bouse Formation, lower Colorado River area, southeastern California western Arizona. In Conglomerates in Basin Analysis, a Symposium Dedicated to A.G. Woodford, edited by I.P. Colburn, P.L. Abbott, and J. Minch, Society of Economic Paleontologists and Mineralogists, Pacific Section 62:53-72.

----- 1990. The Bouse Formation and bracketing units, southeastern California and western Arizona: implications for the evolution of the proto-Gulf of California and the lower Colorado River. Journal of Geophysical Research, 95(B12):20, 111-20, 132.

----- 1991. Stratigraphic evidence for five states in the evolution of the lower Colorado River region (SE CA - W AZ) between ≈18 and ≈4 ma. Mojave Desert Quaternary Research Symposium, Abstracts of Proceedings, San Bernardino Museum Association Quarterly 38(2):45.

----- 1992a. Small-scale late Cenozoic faulting in the Colorado River extensional corridor, western Arizona and southeastern California. In Deformation Associated with the Neogene Eastern California Shear Zone, Southeastern California and Southwestern Arizona, edited by S.M. Richard. San Bernardino County Museum Special Publication 92-1:11-15.

----- 1992b. The Bouse Formation and bracketing units, southeastern California and western Arizona. In Old Routes to the Colorado, edited by R.E. Reynolds and J. Reynolds, San Bernardino County Museum Special Publication 92(2):103.

Bull, W.B. 1984. Alluvial fans and pediments of southern Arizona. In Landscapes of Arizona, the Geological Story, edited by T.L. Smiley. University Press of America Incorporated p. 229-252.

----- 1988. Stream terraces - are they the result of base-level fall, climate-change induced variations in discharge of water in sediment, or are they a complex response? Geosciences, University of Arizona, Tucson, Manuscript on File Anza-Borrego Desert State Park, California 4 p.

Bull, W.E. 1979. A gravity survey of the Elsinore Fault along the southwest flank of the Coyote Mountains. Undergraduate Research Reports, Department of Geology, San Diego State University, California 35:44.

Burdick, L.J. 1975. A determination of the Borrego Mountain earthquake source mechanism using a generalized linear inverse technique. EOS, Transactions of the American Geophysical Union 56:400.

Burdick, L.J., and G.R. Mellman 1976. Inversion of the body waves from the Borrego Mountain earthquake to the source mechanism. Seismological Society of America Bulletin 66:1485-1499.

Burford, R.O. 1972. Continued slip on the Coyote Creek Fault after the Borrego Mountain Earthquake. In The Borrego Mountain Earthquake of April 9, 1968, edited by R.V. Sharp, U.S. Geological Survey Professional Paper 787:105-111.

Burns, H. 1963. Salton Sea Story. Desert Printers Incorporated, Palm Desert, California 36 p.

Burham, W.L. 1954. Data on water wells in Borrego, Ocotillo, San Felipe and Vallecito Valley areas, eastern San Diego County, California. U.S. Geological Open-File Report 54: 1-60.

Burke, A.E. 1948. The Coachella Valley, a geographical survey. Master of Arts Thesis, University of California, Los Angeles.

Bushee, J., J. Holden, B. Geyer, and G. Gastil 1954. Lead-alpha dates for some basement rocks of southwestern California. Geological Society of America Bulletin 74:803-806.

Butler, E.W., and J.B. Pick 1982. Geothermal energy development; problems and prospects in the Imperial Valley of California. Plenum Press, New York 361 p.

Butler, R. 1982. Surface wave analysis of the 9 April, 1968, Borrego Mountain Earthquake. Seismological Society of America Bulletin 73:879-883.

Buttram, G.N. 1962. Geology of the Agua Caliente quadrangle, San Diego County, California. Master of Science Thesis, University of Southern California, Los Angeles 96 p.

Buwalda, J.P. 1930. Geological events in the history of the Indio Hills and the Salton Basin, southern California. Science, n.s., 71:104-106.

Buwalda, J.P., and C.F. Richter 1941. Imperial Valley Earthquake of May 18, 1940. Geological Society of America Bulletin 52:1944-1945.

Buwalda, J.P., and W.L. Stanton 1930. Geological events in the history of the Indio Hills and the Salton Basin, southern California. Science 71:104-106.

Bycroft, G.N. 1980. El Centro, California, differential ground motion array. U.S. Geological Survey Open-File Report 80-919:1-15.

----- 1982a. Anomalous record of October 15, 1979, Imperial Valley, California, earthquake from Coachella Canal Engine House No. 4. U.S. Geological Survey Open-File Report 82-317:1-14.

----- 1982b. El Centro differential ground motion array. In The Imperial Valley, California, earthquake of October 15, 1979. U.S. Geological Survey Professional Paper 1254:351-356.

Bycroft, G.N., and P.N Mork 1987a. Differential displacements and spectra for the April 26, 1981 Westmorland and the January 26, 1986 Hollister earthquakes. U.S. Geological Survey Open-File Report 87-62:1-52.

----- 1987b. Differential spectra for the 1979 El Centro and the 1984 Morgan Hill, California, earthquakes. U.S. Geological Survey Open-File Report 87-94:1-33.

Byseeda, J.J., and J.D. Hunter 1985. Metal recovery from Imperial Valley hypersaline brine. Geothermal Resources Council Transactions 9:2

Cagnetti, V., and C. Bufe 1980. Seismicity patterns preceding the 1979 Imperial Valley Earthquake. EOS, Transactions of the American Geophysical Union 61:108-109.

Caine, R.L. 1951. Legendary and geological history of lost desert gold. Desert Magazine Press, Palm Desert, California 71 p.

California Department of Water Resources 1952. Groundwater basins in California. Quality of Water Investigations Report 3:1-44.

----- 1954. Groundwater occurrence and quality, Colorado River basin region. Qaulity of Water Investigations Report 4.

----- 1968. Water wells and springs in Borrego, Carrizo and San Felipe Valley areas, San Diego and Imperial Counties, California. Bulletin 91-15:1-16.

----- 1970. Geothermal wastes and the water resources of the Salton Sea area. California Department of Water Resources Bulletin 118(2):153 p.

Callian, J.T. 1984. A paleomagnetic study of Miocene volcanics from the Colorado River and mainland Mexico regions. Master of Science Thesis, San Diego State University, California 114 p.

Cameron, J.L. 1980. The Lucky Five pluton in the southern California batholith; a history of emplacement and solidification under stress. Master of Science Thesis, University of California, Los Angeles 145 p.

----- 1989. Anomalous velocities in Salton Sea area. American Association of Petroleum Geologists 73:535

Campbell, D.A. 1983. Plans for drilling and testing the Fee No. 7 well in the Salton Sea geothermal field. EOS, Transactions of the American Geophysical Union 64:864.

Campbell, H.W. 1986. Mineral resources of the Fish Creek Mountains Wilderness Study Area, Imperial County, California. U.S. Bureau of Mines Report MLA 23-86:1-22.

Campbell, K.W., and D.M. Hampson 1989. Processed digital recordings for selected aftershocks of the October 15, 1979 Imperial Valley, California, earthquake. U.S. Geological Survey Open-File Report 89-437:1-330.

Carey, D.L. 1976. Forms and processes in the pseudokarst topography of Arroyo Tapiado, Anza-Borrego Desert State Park, San Diego County, California. Master of Science Thesis, University of California, Los Angeles 126 p.

Caruso, L.J., D.K. Bird, M. Cho, and J.G. Liou 1988. Epidote-bearing veins in the State 2-14 drill hole. Implications for hydrothermal fluid composition. Journal of Geophysical Research, B, Solid Earth and Planets 93:13,123-13,134.

Cassiliano, M.L. 1994. Paleoecology and taphonomy of vertebrate faunas from the Anza-Borrego Desert of California. Doctoral Dissertation, Department of Geosciences, University of Arizona, Tucson 421p.

Castle, R.O., and T.L. Youd 1972. Engineering geology. In The Borrego Mountain Earthquake of April 9, 1968, edited by R.V. Sharp, U.S. Geological Survey Professional Paper 787:158-174.

Chan, M.A., and J.D. Tewhey 1977. Subsurface structure of the southern portion of the Salton Sea geothermal field. Lawerence Livermore Laboratory, UCRL, 52354:1-13.

Chang, S.R., C. Allen, R. Allen, and J. Kirshvink 1987. Stratigraphy and a test for block rotation of sedimentary rocks within the San Andreas Fault Zone, Mecca Hills, southeastern California. Quaternary Research 27:30-40.

Charles, R.W., D.R. Janecky, F. Goff, and M.A. McKibben 1988. Chemographic and thermodynamic analysis of the paragenesis of the major phases in the vicinity of the 6120-foot (1866m) flow zone, California State Well 2-14. Journal of Geophysical

Research, B, Solid earth and Planets 93:13,145-13,158.

Chase, C.G., H.W. Menard, R.L. Larsen, G.F. Sharman, S.M. Smith 1970. History of sea-floor spreading west of Baja California. Geological Society of America Bulletin 81:491-498.

Chasteen, A.J. 1974. Geothermal steam condensate reinjection. In Conference on Research for the Development of Geothermal energy Resources, California Institute of Technology, Jet Propulsion Laboratory, Pasadena, California p. 340-344.

Chavez, D., J. Gonzales, A. Reyes, M. Medina, C. Duarte, J.N. Brune, F.L. Vernon III, R. Simons, L.K. Hutton, P.T. German, and C.E. Johnson 1982. Main-shock location and magnitude determination using combined U.S. and Mexican data. In the Imperial Valley, California, Earthquake of October 15, 1979. U.S. Geological Survey Professional Paper 1254:51-54.

Chen, A.T.F. 1982. Site characterization for stations 6 and 7, El Centro strong motion array, Imperial Valley, California. U.S. Geological Survey Open-File Report 82-1040:1-40.

----- 1983. A study of seismic response at stations 6 and 7, El Centro strong motion array, Imperial Valley, California. Report PB-84 119 924.

Chesterman, C.W. 1957. Fluorspar. In Mineral Commodities of California, California Division of Mines Bulletin 176:202-203.

Chew, R.T., III, and D.R. Antrim 1982. Data releases on the Salton Sea quadrangle, California and Arizona. Geologic maps 1:500,000; site location maps 1:500,000 (Report No.GJBX-190(82):1-6.

Childers, W.M. 1964. Sandstone chimneys of the Imperial Formation. Pacific Discovery 17:29-31.

Christensen, A.D. 1957. Part of the geology of the Coyote Mountain area, Imperial County, California. Master of Arts Thesis, University of California, Los Angeles 188 p.

Christensen, R.J. 1973. Petrographic and textural analysis of a barchan dune southwest of the Salton Sea, Imperial County, California. Master's Thesis, California State University, San Diego 120 p.

Christensen, R.J., and P.L. Abbott 1980. Provenance of Salton dunes, southwest of the Salton Sea. In Geology and Mineral Wealth of the California Desert, edited by D.L. Fife and A.R. Brown, South Coast Geological Society, Santa Ana, California p. 409-413.

Christie, J.M., A. Ord 1980. Flow stress from microstructures of mylonites: example and current assessment. Journal of Geophysical Research 85(B11):6253-6262.

Chou, C.K. 1989. A study of phyllosilicates of the Fee #5 well, Salton Sea geothermal field, California. Master's Thesis, University of California, Riverside.

Cho, M., L.C. Liou, and D.K. Bird 1988. Prograde phase relations in the State 2-14 Well metasandstones, Salton Sea geothermal field, California. Journal of Geophysical Research, B, Solid Earth and Planets 93:13,081-13,103.

Clardy, B.I. 1967. Geology of a portion of the Palo Verde Mountains Imperial County, California. Undergraduate Research Reports, Department of Geology, San Diego State University, California 11:45.

Clark, M.M. 1972a. Collapse fissures along the Coyote Creek Fault. In The Borrego Mountain Earthquake of April 9, 1968, edited by R.V. Sharp, U.S. Geological Survey Professional Paper 787:190-207.

----- 1972b. Intensity of shaking estimated from displaced stones. In The Borrego Mountain Earthquake of April 9, 1968, edited by R.V. Sharp, U.S. Geological Survey Professional Paper 787:175-182.

----- 1972c. Surface rupture along the Coyote Creek Fault. In The Borrego Mountain Earthquake of April 9, 1968, edited by R.V. Sharp, U.S. Geological Survey Professional Paper 787:55-86.

----- 1974. Character and distribution of Recent movement along the southeastern part of the Elsinore Fault Zone, southern California. In Recent Geologic and Hydrologic Studies, Eastern San Diego County and Adjacent Areas, edited by M.W. Hart and R.J. Dowlen, San Diego Association of Geologists, Field Trip Guidebook p. 57.

----- 1975. Character and distribution of recent movement along the southeastern part of the Elsinore Fault Zone, southern California. Geological Society of America, Abstracts with Programs 7:303.

----- 1983. Map showing recently active breaks along the Elsinore and associated faults, California, between Lake Henshaw and Mexico. U.S. Geological Survey Map I-1329.

----- 1984. Map showing recently active breaks along the San Andreas Fault and associated faults, between Salton Sea and Whitewater River-Mission Creek, California. U.S. Geological Survey Miscellaneous Investigations Series I-1483:1-6.

Clark, M.M., and A. Grantz 1969. Geologic effects of the Borrego Mountain Earthquake, southern California, of 9 April 1968. Geological Society of America Abstracts with Programs p. 11.

----- 1970. Geologic effects of the Borrego Mountain, southern California, of 9 April 1968. In Proceedings of the International Symposium on Recent Crustal Movements and Associated Seismicity, edited by N.Z. Willington, Royal Society of New Zealand p. 14-15.

----- 1971. Late Holocene activity along the San Jacinto Fault Zone in lower Borrego Valley, southern California. Geological Society of America Abstracts with Programs 3:97.

Clark, M.M., A. Grantz, and M. Rubin 1972. Holocene activity of the Coyote Creek Fault as recorded in sediments of Lake Cahuilla. In The Borrego Mountain Earthquake of April 9, 1968, edited by R.V. Sharp, U.S. Geological Survey Professional Paper 787:112-130.

Clarke, A.O., and C.L. Hansen 1988. Geomorphology of Upper Palm Wash, Anza-Borrego Desert, California, San Diego and Imperial Counties. California Geology 41:111-116.

Clarke, S.H., and T.H. Nilsen 1973. Displacement of Eocene strata and implications for the history of offset along the San Andreas Fault, central and northern California. In Proceedings of the Conference on Tectonic Problems of the San Andreas Fault System, edited by R.L. Kovach and A. Nur, Stanford University Publication, Geological Sciences 13:358-367.

Clayton, R.N., L.J.P. Muffler, and D.E. White 1968. Oxygen isotope study of calcite and silicates of the River Ranch No. 1 well, Salton Sea geothermal field, California. American Journal of Science 266:968-979.

Clinkenbeard, J.P. 1987. The mineralogy, geochemistry and geochronology of the La Posta pluton, San Diego and Imperial Counties, California. Master of Science Thesis, San Diego State University, California 215 p.

Clinkenbeard, J.P., M.J. Walawender, K.E Parrish, and M.S. Wardlaw 1986. Geochemical and isotopic composition of the La Posta Granodiorite, San Diego County, California. Geological Society of America Abstracts with Programs 18(2):95.

Cloud, W.K., and N.H. Scott 1968. The Borrego Mountain, California, earthquake of 9 April 1968; a preliminary engineering seismology report. Seismological Society of America Bulletin 58:1187-1191.

Coble, B.A. 1978. Intertidal environment in the Imperial Formation of southern California. Undergraduate Research Reports, Department of Geology, San Diego State University, California 33:16

Cockerell, T.D. 1945. The Colorado Desert of California: its origin and biota. Kansas Academy of Science Transactions 48(1):1-39.

----- 1946. The age of Lake Cahuilla. Science 103:235.

Cohn, S.N., C.R. Allen, R. Gilman, and N.R. Goulty 1982. Pre-earthquake and post-earthquake creep on the Imperial Fault and the Brawley Fault Zone. In The Imperial Valley, California, Earthquake of October 15, 1979, U.S. Geological Survey Professional Paper 1254: 161-168.

Cole, K.L., and T.R. Van Devender 1984. Seasonality and the summer monsoon; packrat midden sequences from the lower Colorado Valley. American Quaternary Association Program and Abstracts 8:24

Cole, M. 1980. The lower Colorado River Valley. Quaternary Research 25:392-400.

Coleman, G.A. 1929. A biological survey of the Salton Sea. California Fish and Game 15(2):218-227.

Colletta, B., and L. Ortlieb 1984. Deformations of middle and late Pleistocene deltaic deposits at the mouth of the Rio Colorado, northwestern Gulf of California. In Neotectonics and Sea Level Variations in the Gulf of California Area, a Symposium, edited by V. Malpica-Cruz, S. Celis-Gutiérrez, J. Guerrero Garcia, and L. Ortlieb, Instituto de Geologia, Universidad Autònoma de México, México, D.F. p. 31-53.

Combs, J. 1977. Seismic refraction and basement temperature investigation of East Mesa KGRA, southern California. Transactions of the Geothermal Resources Council, Davis, California 1:45-47.

Combs, J., H.K. Gupta, and C.E. Helsley 1976. Lateral variations of the subsurface velocity, East Mesa geothermal field, Imperial Valley, California. EOS, Trans. of the American Geophysical Union 57:153.

Combs, J., H.K. Gupta, and D.C. Jarabek 1976. Seismic travel time delays and attenuation anomaly associated with the East Mesa geothermal field, Imperial Valley, California. EOS, Transactions of the American Geophysical Union 57:594.

Combs, J., and D.M. Hadley 1973. Microearthquake investigation of the Mesa Geothermal Anomaly, Imperial Valley, California. EOS, Transactions of the American Geophysical Union 54:1213-1214.

----- 1977a. Microearthquake investigation of the Mesa geothermal anomaly, Imperial Valley, California. Geophysics 42:17-33.

----- 1977b. Microearthquake investigation of the Mesa geothermal anomaly, Imperial Valley, California. University of Texas Institute of Geosciences Contribution 280:1-17.

Combs, J., and D. Jarzabek 1978. Seismic wave attenuation anomalies in the East Mesa Geothermal Field, Imperial Valley, California; preliminary results. Transactions of the Geothermal Resources Council, Davis, California 2:109-112.

Combs, J., and D. Langenkamp 1973. A microearthquake study of the Elsinore Fault, southern California. EOS, Transactions of the American Geophysical Union 54:376.

Combs, J., and R.W. Rex 1971. Geothermal investigations in the Imperial Valley of California. Geological

Society of America Abstracts 3:101-102.

Conrad, T.A. 1854. Descriptions of new fossil shells of the United States. Journal of the Academy of Natural Sciences Philadelphia Series 2, 2(4):299-300.

----- 1855. Report on the fossil shells collected by W.P. Blake, geologist to the expedition under the command of Lieutenant R.S. Williamson, United States Topographical Engineers, 1852. In W.P. Blake, Preliminary Geological Report, U.S. Pacific Railroad Exploration, U.S. 33rd Congress, 1st session, House Executive Document 129:5-21, appendix.

Cooper, J.G. 1894. Catalogue of California fossils. California Mining Bureau Bulletin 4:1-65.

Copenhaver, G.C., Jr. 1963a. Investigation of the geology of the Hidden Treasure Mine, Julian mining district, California. Undergraduate Research Reports, Department of Geology, San Diego State University, California 7(2):1-9.

----- 1963b. Structural and economic geology of the Golden Gem No. 1 (Gardiner) Mine, Julian mining district, San Diego County, California. Undergraduate Research Reports, Department of Geology, San Diego State University, California 7(2):1-13.

----- 1963c. Structural and mining geology of a part of the Julian Schist Formation, Julian, San Diego County, California. Undergraduate Research Reports, Department of Geology, San Diego State University, California 7(2):1-21.

----- 1964. A geochemical orientation survey of nickel in soils over Cuyamaca gabbro. Undergraduate Research Reports, Department of Geology, San Diego State University, California 8(2):1-10.

----- 1966. Relationship between nickel mineralization and overlying soil composition in the Cuyamaca gabbro, San Diego County, California. Geological Society of America Special Paper 87:199.

----- 1970. Geochemical prospecting for nickel in the Julian-Cuyamaca area, California. Master of Science Thesis, San Diego State University, California 160 p.

Coplen, T.B. 1973a. Exploration for geothermal systems in the Imperial Valley Area, California, using the Na-K-Ca technique. EOS, Transaction of the American Geophysical Union 54:1213.

----- 1973b. Isotopic composition of calcite and water from the Dunes-DWR #1 Geothermal Test Corehole, Imperial Valley, California. EOS, Transactions of the American Geophysical Union 54:488.

----- 1974a. Investigations of the hydrology of the Imperial Valley geothermal area using stable isotopes. Geological Society of America Abstracts 6:694.

----- 1974b. The origin of ground water from various sources in the Imperial Valley, California. International symposium on water-rock interaction, Czechoslovakia Geological Survey, Prague p. 9-10.

----- 1976. Cooperative geochemical resource assessment of the Mesa geothermal system. Report IGPP-UCR-76-1:1-97.

----- 1977. Oxygen, hydrogen and carbon isotope studies of the Mesa geothermal system, California. Geological Society of America Abstracts with Programs 9:935.

Coplen, T.B., Combs, J., Elders, W.A. 1973. Preliminary findings of an investigation of the Dunes thermal anomaly, Imperial Valley, California. Report PB-261 220/8WE:1-25.

Coplen, T.B., and P. Kolesar 1974. Investigations of the Dunes geothermal anomaly, Imperial Valley, California; Part I, Geochemistry of geothermal fluids. Report PB-261 221/6WE:1-25.

Corbett, E.J. 1978. Changes in stress indicated by "pre-shocks" to the 1968 Borrego Mountain Earthquake. EOS, Transactions of the American Geophysical Union 59:1127.

Corey, H.T. 1915. Imperial Valley and Salton Sink. J.J. Newbegin, San Francisco, California 91 p.

Cornell, W.C. 1979. Gulf of California in Lake Mead area during late Miocene time: discussion. American Association of Petroleum Geologists Bulletin 63:1139-1140.

Cornett, J.W. 1989. Desert Palm Oasis. Palm Springs Desert Museum, California 48 p.

Corona, F.V., and F.F. Sabins, Jr. 1993a. The San Andreas Fault of the Salton Trough region, California, as expressed on remote sensing data. Abstracts of Proceedings for the 7th annual Mojave Desert Quaternary Research Symposium and the Desert Studies Consortium Symposium. San Bernardino County Museum Association Quarterly 40(2):24

----- 1993b. The San Andreas Fault of the Salton Trough region, California, as expressed on remote sensing data. In Ashes, Faults and Basins, edited by R.E. Reynolds and J. Reynolds. San Bernardino County Museum Association Special Publication 93-1:69-75.

Corones, J. 1982. Geological reconnaissance of the Paymaster Quarry site northeastern Imperial County, California. Undergraduate Research Reports, Department of Geology, San Diego State University, California 40(1):38.

Costello, S.C. 1985. A paleomagnetic investigation of Mid-Tertiary volcanic rocks in the lower Colorado River area, Arizona and California. Master of Science Thesis, San Diego State University, California 109 p.

Craig, H. 1966. Isotopic composition and origin of the Red Sea and Salton Sea geothermal brines. Science 154:1544-1548

----- 1969. Discussion [of paper by H.C. Helgeson, 1968] - Source fluids for the Salton Sea geothermal system. American Journal of Science 267:249-255.

Cramer, M.L. 1972. Geology and gold deposits of a portion of the Cuyamaca Peak quadrangle, San Diego County, California. Undergraduate Research Reports, Department of Geology, San Diego State University, California 20(1):1-21.

Cramer, S.D., and Carter, J.P. 1980. Laboratory corrosion studies in low- and high-salinity geobrines of the Imperial Valley, California. U.S. Bureau of Mines Report of Investigation 8415:1-30.

Cranswick, E., and P. Spudich 1981. What caused the large vertical arrivals observed in the 1979 Imperial Valley, California, earthquake? EOS, Transactions of the American Geophysical Union 62:972.

Creasey, S.C. 1946. Geology and nickel mineralization of the Julian-Cuyamaca area, Orange and San Diego Counties, California. California Journal of Mines and Geology 42(1):15-29.

Crockett, A.B., and G.B. Wiersma 1977. Status of baseline sampling for elements in soil and vegetations at four KGRA's in the Imperial Valley, California. Geothermal; State of the Art, Transactions of the Geothermal Resources Council 1:65-67.

Crook, C.N., R.G. Mason, and P.R. Wood 1982. Geodetic measurements of horizontal deformation on the Imperial Fault. In The Imperial Valley, California, Earthquake of October 15, 1979, U.S. Geological Survey Professional Paper 1245:183-191.

Crook, C.N., R.G. Mason, and A.D. Pullen 1982. Horizontal deformation associated with the October 15, 1979 Imperial Valley Earthquake, and in the three years following. Earthquake Notes 54:18.

Crosby, S.W. 1929. Notes on a map of the Laguna Salada basin, Baja California, Mexico. Geographical Review 19:613-620.

Crouch, R.W., and C.W. Poag 1979. *Amphistegina gibbosa* d'Orbigny from the California borderlands, the Caribbean connection. Journal of Foraminiferal Research 9(2):85-105.

Crowe, B.M. 1973a. Cenozoic volcanic geology of the southeastern Chocolate Mountains. Doctoral Dissertation, University of California, Santa Barbara 117 p.

----- 1973b. Tertiary volcanic stratigraphy of the southeastern Chocolate Mountains. Geological Society of America Abstracts 5:30-31.

----- 1975. Probable age of inception of basin and range faulting in southeasternmost California. Geological Society of America Abstracts 7:601.

----- 1978. Cenozoic volcanic geology and probable age of inception of basin-range faulting in the southeasternmost Chocolate Mountains, California. Geological Society of America Bulletin 89:251-264.

Crow, M.B., Crowell, J.C., and D. Krummenacher 1979. Regional stratigraphy, K-Ar ages, and tectonic implications of Cenozoic volcanic rocks, southeastern California. American Journal of Science 279:186-216.

Crowell, J.C. 1952. Probable large lateral displacement on the San Gabriel Fault, southern California. American Association of Petroleum Geologists Bulletin 36:2026-2035.

----- 1962. The San Andreas Fault in southern California. Twenty-first International Geological Congress 18:45-52.

----- 1966. Displacement along the San Andreas Fault, California. Geological Society of America Special Paper 71:1-61.

----- 1975. The San Andreas Fault in southern California. In San Andreas Fault in Southern California, edited by J.C. Crowell, California Division of Mines and Geology Special Report 118:7-27.

----- 1981. An outline of tectonic history of southeastern California. In The Geotectonic Development of California, edited by W.G. Ernst, Prentice-Hall, New Jersey, Ruby Volume 1:583-600.

----- 1987a. The tectonically active margin of the western U.S.A. Episodes 10(4):278-282.

----- 1987b. Late Cenozoic basins of onshore southern California; complexity is the hallmark of their tectonic history. In Cenozoic Basin Development of Coastal California, edited by R.V. Ingersoll and W.G. Ernst, Prentice-Hall, New Jersey, Ruby Volume 6:207-241.

----- 1989. The San Andreas transform belt. In Field Trip Guidebook, leaders A.G. Sylvester and J.C. Crowell, 28th International Geological Congress, American Geophysical Union, Washington, D.C. T309:1-119.

----- 1992a. Cenozoic faulting in the Little San Bernardino - Orocopia Mountains region, southern California. In Deformation Associated with the Neogene Eastern California Shear Zone, Southeastern California and Southwestern Arizona, edited by S.M. Richard, San Bernardino County Museum Special Publication 92-1:16-19.

----- 1992b. Tectonic mobility of the San Andreas Fault belt bordering the Mojave and Colorado Deserts, southern California. In Deformation Associated with the Neogene Eastern California Shear Zone, Southeastern California and Southwestern Arizona, edited by S.M. Richard, San Bernardino County Museum Special Publication 92-1:16-19.

----- 1993. The significance of the Orocopia Mountains region in the displacement history of the San Andreas Fault system. In Ashes, Faults and Basins, edited by R.E. Reynolds and J. Reynolds. San Bernardino County Museum Association Special Publication 93-1:53-58.

Crowell, J.C., and B. Baca 1979. Sedimentation history of the Salton Trough. In Tectonics of the Juncture Between the San Andreas Fault System and the Salton Trough, Southeastern California, edited by J.C. Crowell and A.G. Sylvester. Department of Geological Sciences, University of California, Santa Barbara p. 101-110.

Crowell, J.C., and T. Susuki 1959. Eocene stratigraphy and paleontology, Orocopia Mountains, southeastern California. Geological Society of America Bulletin 70:581-592.

Crowell, J.C., and A.G. Sylvester 1979a. Introduction to the San Andreas Fault-Salton Trough juncture. In Tectonics of the Juncture Between the San Andreas Fault System and the Salton Trough, Southeastern California, edited by J.C. Crowell and A.G. Sylvester. Department of Geological Sciences, University of California, Santa Barbara p. 1-13.

----- 1979b. Tectonics of the juncture between the San Andreas Fault system and the Salton Trough, Southeastern California, J.C. Crowell and A.G. Sylvester editors. Department of Geological Sciences, University of California, Santa Barbara 193 p.

Crowell, J.C., and J.W.R. Walker 1961. Displacement of anorthosite and related rocks by the San Andreas Fault, southern California. Geological Society of America, Abstracts with Programs 68:157.

Cummings, C.R. 1977. East Meas geothermal component test facility. Geothermal Energy 5(20):8-12.

Cunningham, G.D. 1984a. The Plio-Pleistocene Dipodomyinae and geology of the Palm Spring Formation, Anza-Borrego Desert, California. Master of Science Thesis, Idaho State University, Pocatello 193 p.

----- 1984b. Geology and paleoecology of the Palm Spring Formation at the Plio-Pleistocene boundary: Vallecito-Fish Creek Basin, Anza-Borrego Desert, southern California. Geological Society of America Special Paper 15:219.

Czaplewski, N.J. 1990. The Verde Local Fauna: small vertebrate fossils from the Verde Formation, Arizona. San Bernardino County Museum Association Quarterly 37(3):1-39.

Daley, T.M. 1987. Analysis of P- and S-wave VSP data from the Salton Sea geothermal field. Master's Thesis, University of California, Berkeley.

Daley, T.M., T.V. McEvilly, and E.L. Majer 1988a. Analysis of P and S wave vertical seismic profile data from the Salton Sea Scientific Drilling Project. Journal of Geophysical Research, B, Solid Earth and Planets 93:13,025-13,036.

----- 1988b. Analysis of P and S wave data from the Salton Sea Scientific Drilling Project. Transactions of the Geothermal Resources Council 12:237-243.

----- 1988c. Multicomponent VSPs at Cajon Pass and the Salton Sea, Earth Science Division 1987 Annual Report, Lawrence Berkeley Laboratory, California p. 103-107.

Dalrymple, G.B., A. Cox, and R.R. Doell 1965. Potassium-Argon age and paleomagnetism of the Bishop Tuff, California. Geological Society of America Bulletin 76:665-673.

Damiata, B.N., S.K. Parks, and L. Tien-chang 1986. Geophysical overview: Imperial Valley, southern California. In Geology of the Imperial Valley, California, edited by P.D. Guptil, E.M. Grath and R.W. Ruff. South Coast Geological Society, Annual Field Trip Guidebook, Santa Ana, California 14:116-133.

Darnell, W.I. 1959. The Imperial Valley: its physical and cultural geography. Master of Science Thesis, San Diego State University, California 127 p.

Darton, N.H. 1933. Guidebook of the western states, the Southern Pacific Lines, New Orleans to Los Angeles. U.S. Geological Survey Bulletin 845(F):1-304.

da Sie, W.J., and V.T. Hoang 1985. The well testing program at the Heber geothermal field. Society of Petroleum Engineers Proceedings 5:247-256.

Davis, A.P. 1907a. The new inland sea. National Geographic Magazine 18:36-49.

----- 1907b. The new inland sea. Nature 75:501.

Davis, B.L., C.K. Shearer, Jr., S.B. Simon, M.N. Spilde, J.J. Papike, and J.C. Laul 1990. X-ray reference-intensity and X-ray flourence analyses of Salton Sea core. American Mineralogist 75:230-236.

Davis, D.G., and S.K. Sanyal 1979. Case history report on East Mesa and Cerro Prieto geothermal fields. Los Alamos Scientific Laboratory 7889:1-182.

Davis, G.A., and J.L. Anderson 1991. Low-angle normal faulting and rapid uplift of mid-crustal rocks in the Whipple Mountains metamorphic core complex, southeastern California: discussion and field guide. In Geological Excursions in Southern California and Mexico, edited by M.J. Walawender and B.B. Hanan. Geological Society of America Annual Meeting Guidebook, San Diego p. 417-446.

Dawson, M.R., and C.E. Jacobson 1986. Trace element geochemistry of the metabasites from the Pelona-Orocopia, and Rand schists, southern California. Geological Society of America Abstracts with Programs 18:99.

----- 1989. Geochemistry and origin of mafic rocks from the Pelona, Orocopia, and Rand schists, southern

California. Earth and Planetary Science Letters 92:371-385.

Dean, C.J. 1982. Separation of geophysical anomalies by space and frequency domain filtering. Undergraduate Research Reports, Department of Geology, San Diego State University, California 40(1):1-52.

Dean, M.A. 1988. Genesis, mineralogy and stratigraphy of the Neogene Fish Creek gypsum, southwestern Salton Trough, California. Master of Science Thesis, San Diego State University, California 150 p.

----- 1990. The Neogene Fish Creek gypsum; a forerunner to the incursion of the Gulf of California into the western Salton Trough. Geological Society of America Abstracts with Programs 22(3):18.

de la Peña, I.A., and I.I. Puente 1979. The geothermal field of Cerro Prieto. In Geology and Geothermics of the Salton Trough, edited by W.A. Elders, University of California, Riverside, Campus Museum Contributions 5:20-35.

DeLattre, M. 1984. Permian miogeoclinal strata at El Volcano, Baja California, Mexico. In Geology of the Baja California Peninsula, edited by V.A. Frizzell, Jr., Society of Economic Paleontologists and Mineralogists, Pacific Section p. 23-29.

Dellinger, D.A. 1989. California's unique geologic history and its role in mineral formation, with emphasis on the mineral resources of the California desert region. U.S. Geologic Survey Circular C-1024:1-16.

Deméré, T.A. 1988. Paleontology and our local desert. Environment Southwest 522:17-19.

----- 1993. Fossil mammals from the Imperial Formation (upper Miocene-lower Pliocene), Coyote Mountains, Imperial County, California. In Ashes, Faults and Basins, edited by R.E. Reynolds and J. Reynolds. San Bernardino County Museum Association Special Publication 93-1:82-85.

DePaolo, D.J. 1981. A neodymium and strontium isotopic study of the Mesozoic calc-alkaline granitic batholiths of the Sierra Nevada and Peninsular Ranges, California. Journal of Geophysical Research 86(B11):10470-10488.

de Stanley, M. 1966. The Salton Sea yesterday and today. Triumph Press, Incorporated, Los Angeles 127 p.

Detra, D.E., and J.E. Kilburn 1985a. Analytical results and sample locality map of heavy-mineral-concentrate samples from the Jacumba/In-Ko-Pah Mountain Wilderness Study Area (CDCA 368), Imperial County, California. U.S. Geological Survey Open-File Report 85-272:1-16.

----- 1985b. Analytical results and sample locality map of heavy-mineral-concentrate samples from the Fish Creek Mountains Wilderness Study Area (CDCA 372), Imperial County, California. U.S. Geological Survey Open-File Report 85-524:1-8.

Detterman, M.E. 1984 Geology of the Metal Mountain district, In-Ko-Pah Mountains, San Diego County, California. Master of Science Thesis, San Diego State University, 216 p.

Dibblee, T.W., Jr. 1954. Geology of the Imperial Valley region, California. In Geology of Southern California, edited by R.H. Jahns, California Division of Mines and Geology Bulletin 170(2,2):21-81, plate 2.

----- 1968. Displacements of the San Andreas Fault system in the San Gabriel, San Bernardino and San Jacinto Mountains, southern California. Stanford University Publications in Geological Sciences 13:260-278.

----- 1981. Geology of the San Jacinto Mountains and adjacent areas. In Geology of the San Jacinto Mountains, edited by A.R. Brown and R.W. Ruff, South Coast Geological Society Annual Field Trip Guidebook, Santa Ana, California 9:1-47.

----- 1984. Stratigraphy and tectonics of the San Felipe Hills, Borrego Badlands, Superstition Hills, and vicinity. In The Imperial Basin, Tectonics, Sedimentation and Thermal Aspects, edited by C.A. Rigsby, Society of Economic Paleontologists and Mineralogists 40:31-44.

----- 1986. Geology of the Imperial Valley Region, California. In Geology of the Imperial Valley, California, edited by P.D. Guptil, E.M. Gath and R. W. Ruff, South Coast Geological Society Annual Field Trip Guidebook, Santa Ana, California 14:1-14.

----- 1989. Late Cenozoic tectonics of northern and western Imperial Basin. American Association of Petroleum Geologists Bulletin 73:537.

Dickerson, R.E. 1918. Mollusca of the Carrizo Creek beds and their Caribbean affinities. Geological Society of America Special Paper 29:148.

Dickinson, W.R. 1981. Plate tectonics and continental margin of California. In The Geotectonic Development of California, edited by W.G. Ernst, Prentice-Hall, Incorporated, Ruby Volume 1:1-28.

Diercks, C. 1981. The structure and petrology of the southern portion of the Jacumba roof pendant. Undergraduate Research Reports, Department of Geology, San Diego State University, California 39(2):1-36.

Dillon, J.T. 1968. Summary of the geology of the Chocolate and Cargo Muchacho Mountains, California. In Geology of the Imperial Valley, California, edited by P.D. Guptil, E.M. Gath and R. W. Ruff, South Coast Geological Society, Annual Field Trip Guidebook, Santa Ana, California 14:173-179.

----- 1975. Geology of the Chocolate and Cargo Muchacho

Mountains, southeasternmost California. Doctoral Dissertation. University of California, Santa Barbara. 405p.

----- 1988. Timing of thrusting and metamorphism along the Vincent-Chocolate Mountain thrust system, southern California. Geological Society of America Abstracts with Programs 18:101.

Dillon, J.T., and G.B. Haxel 1975. The Chocolate Mountain-Orocopia-Vincent thrust system as a tectonic element on late Mesozoic California. Geological Society of America Abstracts with Programs 7:311-312.

Dillon, J.T., G.B. Haxel, and R.M. Tosdal 1990. Structural evidence for northeastward movement on the Chocolate Mountains thrust, southeasternmost California. Journal of Geophysical Research, B, Solid Earth and Planets 95(12):19,953-19,971.

Diment, W.H., T.C. Urban, M. Nathenson, and K.E. Mathias 1977. East Mesa geothermal anomaly, Imperial County California; effects of canal leakage on shallow thermal regime. EOS, Transactions of the American Geophysical Union 58:1241.

Divis, A.F., and J. McKenzie 1973. Diagenetic mineralization in geothermal brines. Geological Society of America Abstracts with Programs 5:34.

Dockum, M.S. 1982. Greenschist-facies carbonates, eastern Coyote Mountains, western Imperial County, California. Master of Science Thesis, San Diego State University, California 89 p.

Dockum, M.S., and R.H. Miller 1982. Ordovician conodonts from the greenschist facies carbonates, western Imperial County, California. Geological Society of America Abstracts with Programs 14:160-161.

Doe, B.R., C.E. Hedge, and D.E. White 1966. Preliminary investigation of the source of lead and strontium in deep geothermal brines underlying the Salton Sea geothermal area. Economic Geology 61:462-483.

Doe, B.R., D.E. White, and C.E. Hedge 1963. Preliminary isotopic data for brine and obsidian near Niland, California. Mining Engineering 15:60.

Doering, W.P., and I. Friedman 1982. Survey of helium in natural water wells and springs in southwest Montana and vicinity and Imperial Valley, California; Part IV, January 1-December 31, 1981. U.S. Geological Survey Open-File Report 82-486:1-33.

----- 1983. Survey of helium in natural water wells and springs in southwest Montana and vicinity and Imperial Valley, California; Part V, January 1-December 31, 1982. U.S. Geological Survey Open-File Report 83-500:1-32.

----- 1984. Survey of helium in natural water wells and springs in southwest Montana and Imperial Valley, California; Part VI, Jan.-Dec. 31, 1983. U.S. Geological Survey Open-File Report 84-422:1-30.

Dokka, R.K. 1984. Fission-track geochronologic evidence for late Cretaceous mylonization and early Paleocene uplift of the northeastern Peninsular Ranges, California. Geophysical Research Letters 11:46-49.

Dokka, R.K., and R.H. Merriam 1982. Late Cenozoic extension of northeast Baja California, Mexico. Geological Society of America Bulletin 93:371-378.

Donegan, D.P. 1982. Modern and ancient marine rhythmites from the Sea of Cortez and California continental borderland: sedimentological study. Master's Thesis. Oregon State University.

Donnelly, M.G. 1934a. Geology and mineral deposits of the Julian district, San Diego County, California. California Journal of Mines and Geology 30:331-370.

----- 1934b. Economic geology of the Julian region. Pan-American Geologist 61:316.

----- 1935. Economic geology of the Julian region. Geological Society of America Proceedings p. 321.

Dorn, R.I. 1989. Cation-ratio dating of rock varnish: a geological perspective. Progress in Physical Geography 13:559-596.

Dorf, E. 1930. Pliocene floras of southern California. Carnegie Institute of Washington Publication 412:1-112.

Doser, D.I. 1990. Source characteristics of earthquakes along the southern San Jacinto and Imperial Fault zones (1937 to 1954). Seismological Society of America Bulletin 80:1099-1117.

Doser, D.I., and H. Kanamori 1984. Depth of seismicity in the Imperial Valley and southern Peninsular Ranges of California. EOS, Transactions of the American Geophysical Union 65:1118.

----- 1985a. Depth of seismicity in the Imperial Valley region (1977-1983) and its relationship to heatflow, crustal structure, and the October 15, 1979 earthquake. U.S. Geological Survey Open-File Report 85-507:30-38; and California Institute of Technology, Division of Geology and Planetary Science Contribution 4258:1-9.

----- 1985b. Seismicity of the Imperial Valley Region (1977-1983) and its relationship to the October 15,1979 earthquake. Earthquake Notes 55:1-31.

----- 1986a. Spacial and temporal variations in seismicity in the Imperial Valley (1902-1984). Seismological Society of America Bulletin 76:421-438.

----- 1986b. Depth seismicity in the Imperial Valley region, (1977-1983) and its relationship to heat flow, crustal structure, and the October 15, 1979 earthquake. Journal of Geophysical Research 91:675-688.

Douglas, A. 1971. Cenozoic marine faunas of the Imperial Formation of Whitewater and Fossil Canyon. Senior Thesis, Department of Geological Sciences, University of California, Riverside 24 p.

Douglas, B., G. Norris, L.Dodd, and J. Richardson 1984. Behavior of the Meloland overcrossing during the 1979 Imperial Valley earthquake. Earthquake Notes 55:1-26.

Douze, E.J., and G.G. Sorrells 1972. Geothermal ground-noise surveys. Geophysics 37:813-824.

Dowlen, R.J. 1969. The coral-barnacle fauna of the Imperial Formation in Imperial County, California. Undergraduate Research Reports, Department of Geology, San Diego State University, California 14:1-34.

Downs, T. 1957. Late Cenozoic vertebrates from the Imperial Valley region, California. Geological Society of America Bulletin 68(12,2):1822.

----- 1965. Pleistocene vertebrates of the Colorado Desert, California. International Association for Quaternary Research, VII International Congress General Session, Abstracts with Programs p. 107.

----- 1966. Southern California field trip, Anza-Borrego Desert and Barstow areas. Society of Vertebrate Paleontology Field Trip, Manuscript on File Anza-Borrego Desert State Park, California 12 p.

----- 1967. Airlift for fossils. Los Angeles County Museum of Natural History Alliance Quarterly 6(1):20-25.

Downs, T., and G.J. Miller 1994. Late Cenozoic equids from Anza-Borrego Desert, California. Natural History Museum of Los Angeles County Contributions in Science 440:1-90.

Downs, T., and J.A. White 1965. Late Cenozoic vertebrates of the Anza-Borrego Desert area, southern California. American Association for the Advancement of Science Meeting 1965:10-11.

----- 1966. The vertebrate faunal sequence of the Vallecito-Fish Creek area western Colorado Desert. Society of Vertebrate Paleontology Field Trip, Manuscript on File Anza-Borrego Desert State Park, California 5 p.

----- 1968. A vertebrate faunal succession in superposed sediments from late Pliocene to middle Pleistocene in California. In Tertiary/Quaternary Boundary, International Geological Congress 23, Prague 10:41-47.

Downs, T., and G.D. Woodard 1961a. Stratigraphic succession of the western Colorado Desert, San Diego and Imperial Counties, California. Geological Society of America Special Paper 68:73-74.

----- 1961b. Middle Pleistocene extension of the Gulf of California into the Imperial Valley. Geological Society of America, Abstracts with Programs 68(12):21.

Downs, W.F., J.D. Rimstidt, and H.L. Barnes 1977. Hydrothermal experiments on Salton Sea geothermal brines. Geological Society of America Abstracts with Programs 9:955.

Doyle, L.J., and D.J. Gorsline 1977. Poway Conglomerate in northwest Baja California derived from Sonora. American Association of Petroleum Geologists Bulletin 61:903-917.

Drobeck, P.A., F.L. Hillemeyer, E.G. Frost, and G.S. Leiber 1986. The Picacho Mine: a gold mineralized detachment in southeastern California. In Frontiers in Geology and Ore Deposits of Arizona and the Southwest, edited by B. Beatty and P.A.K. Wilkinson, Arizona Geological Society Digest 16:187-221,280.

Dronyk, M.P. 1977. Stratigraphy, structure and seismic refraction survey of a portion of the San Felipe Hills, Imperial Valley, California. Master of Science Thesis, Department of Geological Sciences, University of California, Riverside 141 p.

Duda, S. 1965. Regional seismicity and seismic wave propagation from records at the Tonto Forest Seismological Observatory, Payson, Arizona. Annali Geofisica 18:365-397

Dudziak, S. 1984. A gravity study of the southern portion of the Elsinore Fault zone, Ocotillo, California. Undergraduate Research Reports, Department of Geology, San Diego State University, California 44(1):27.

Dunn, F. 1963. Treasures of the badlands. Desert 26(10):12-13.

Durham, J.W. 1950. 1940 E.W. Scripps cruise to the Gulf of California. Part II, megascopic paleontology and marine stratigraphy. Geological Society of America Bulletin 43(2):1-216.

----- 1952. Early Tertiary marine faunas and continental drift. American Journal of Science 5(250):321-343.

Durham, J.W., and E.C. Allison 1961. Stratigraphic position of the Fish Creek gypsum at Split Mountain Gorge, Imperial County, California. Geological Society of America Special Paper 68:32.

Durrell, C. 1953a. Celestite deposits near Ocotillo, San Diego County, California. California Division of Mines and Geology Special Report 32:5-7.

----- 1953b. Geological investigations of strontium deposits in southern California. California Division of Mines Special Report 32:4-7.

Durrell, C., and J. Gilluly 1944. Calcite deposits in Imperial and San Diego counties, California. U.S. Geological Survey Open-File Report 77-685:1-35.

Durrenberger, R.W. 1959. The geography of California in essays and readings. Brester Publications, Los Angeles 196 p.

Earl, J. 1959. Geochemical studies in the Julian-Banner mining district, California - a field method for the determination of nickel in soils. Undergraduate Research Reports, Department of Geology, San Diego State University, California 3(2):1-23.

Eastman, B.J., and M.A. Watkins 1983. The Marcus Wash granite: its structural setting and implications for the timing of the Vincent-Chocolate Mountains thrust system. Undergraduate Research Reports, Department of Geology, San Diego State University, California 42(3):1-40.

Eaton. A.L. 1940. Flourescent minerals of the Colorado Desert. Mineralogist 8(4):1-156

Ebel, J., and D.V. Helmberger 1979. Fault roughness as inferred from the teleseismic short period P waves and strong motion recordings of the Borrego Mountain Earthquake. EOS, Transactions of the American Geophysical Union 60:896.

----- 1982. P-wave complexity and fault asperities; the Borrego Mountain, California, earthquake of 1968. Seismological Society of America Bulletin 72:413-437.

Eberly, L.D., and T.B. Stanly 1978. Cenozoic stratigraphy and geologic history of southwestern Arizona. Geological Society of America Bulletin 89:921-940.

Eckis, R. 1930. The geology of the southern part of the Indio Hills quadrangle, California. Master of Science Thesis, California Institute of Technology, Pasadena 24 p.

Edmunds, S.W. 1978. Institutional policy and planning issues in geothermal development; the case of the Imperial Valley. Geothermal Resources Council Transactions 2(1):167-170.

----- 1982. Geothermal energy development in the desert; the case for the Imperial Valley. In Alternative Strategies for Desert Development and Management; Energy and Minerals, edited by M.R. Biswas, Pergamond Press, New York p. 61-71.

Edmunds, S.W., and A.Z. Rose 1979. Geothermal energy and regional development; the case for Imperial County, California. Editors, Praeger Publications, New York 371 p.

Ehlig, P.L., K.W. Ehlert, and B.M. Crowe 1975. Offset of the upper Miocene Caliente and Mint Canyon Formations along the San Gabriel and San Andreas Faults. California Division of Mines and Geology Special Report 118:83-92.

Ehlig, P.L., and S.E. Joseph 1977. Polka dot granite and correlation of La Panza Quartz Monzonite with Cretaceous batholith rocks north of Salton Trough. In Cretaceous Geology of the California Coast Ranges, West of the San Andreas Fault, edited by D.G. Howell and J.G. Vedder, Society of Economic Paleontologists and Mineralogists, Pacific Coast Paleogeography Field Guide 2:91-96.

Einsele, G., and K. Kelts 1982. Pliocene and Quaternary mud turbidites in the Gulf of California; sedimentology, mass physical properties and significance. In Initial reports of the Deep Sea Drilling Project, edited by J.R. Curray, and D.G. Moore. U.S. Government Printing Office, Washington, D.C. 64:511-528.

Elders, W.A. 1977. Rock-water interaction and temperature distribution in the Salton Sea geothermal field, Imperial Valley, California, U.S.A. Geological Society of America Abstracts with Programs 9:965.

----- 1979a. The geological background of the geothermal fields of the Salton Trough. In Geology and Geothermics of the Salton Trough, edited by W.A. Elders, University of California, Riverside, Campus Museum Contributions 5:1-19.

----- 1979b. Historical preface; man and the nature of the Colorado River delta. Geology and Geothermics of the Salton Trough. editor, University of California, Riverside, Campus Museum Contributions 5:v-viii.

----- 1980. Magma-hydrothermal systems in a sediment-smothered crustal spreading regime; the Salton Trough. EOS, Transactions of the American Geophysical Union 61:1149.

----- 1981. Applications of petrology and geochemistry to the study of active geothermal systems in the Salton Trough of California. In Process Mineralogy; Extractive Metallurgy, Mineral Exploration, Energy Resources, edited by D.M. Hausen, et al., American Institute of Mineralogy and Metallurgy, New York p. 591-606.

----- 1982. The environment of active sulfide mineralization in the Salton Sea geothermal system. In The Genesis of Stratiform Sediment-Hosted Lead and Zinc Deposits, edited by R.J.W. Turner, Stanford University Publications, Geological Sciences 20:156-160.

----- 1983a. Models of hydrothermal systems in the Salton Trough; active ore formation in a sedimentary basin. Geological Society of America Abstracts with Programs 15:566.

----- 1983b. Salton Sea research well. EOS, Transactions of the American Geophysical Union 64:514.

----- 1984. Continental scientific drilling to 5.5 km in the Salton Sea geothermal field, California, U.S.A. International Geological Congress 27(9[1]):123-124.

----- 1985a. The Salton Trough as a delta. In Geology and Geothermal Energy of the Salton Trough, edited by L.J. Herber, National Association of Geology Teachers, Far Western Section p. 24-26.

----- 1985b. Hydrothermal alteration and ore genesis. In Geology and Geothermal Energy of the Salton Trough, edited by L.J. Herber, National Association of Geology Teachers, Far Western Section p. 140-

155.

----- 1985c. Continental scientific drilling in California; the saga of the Salton Sea Scientific Drilling Project (SSSDP). Geothermal Resources Council Transactions 9(1):107-112.

----- 1986a. The Salton Sea Scientific Drilling Project; an investigation of an active hydrothermal system in a continental rift. Jahrestagung der Deutschen Geophysikalischen Gesellschaft 46:184.

----- 1986b. The Salton Sea Scientific Drilling Project; an investigation of an active hydrothermal system in the Colorado River delta of California. International Symposium on Water-Rock Interaction 5:193-196.

----- 1987a. A natural analogue for near-field behavior in a high-level radioactive waste repository in salt; the Salton Sea geothermal field, California, U.S.A. In Natural Analogues in Radioactive Waste Disposal in the Collection Radioactive Waste Management Series, edited by B. Come and others. London p. 342-353.

----- 1987b. Research drilling in an active geothermal system; Salton Sea Scientific Drilling Project (SSSDP). American Association of Petroleum Geologists Bulletin 71:552-553.

Elders, W.A., S. Biehler, and I.A. de la Peña 1979. Geology and geothermic of the Salton Trough road log. In Geology and Geothermics of the Salton Trough, edited by W.A. Elders, University of California, Riverside, Campus Museum Contributions 5:104-108.

Elders, W.A., D.K. Bird, A.E. Williams, and P. Schiffman 1984. Hydrothermal flow regime and magmatic heat source of the Cerro Prieto geothermal system, Baja California, Mexico. Geothermics 13(1-2):27-47.

----- 1985. Hydrothermal flow regime and magmatic heat source of the Cerro Prieto geothermal system, Baja California, Mexico. In Geology and Geothermal Energy of the Salton Trough, edited by L.J. Herber, National Association of Geology Teachers, Far Western Section p. 64-84.

Elders, W.A., L.H. Cohen 1983a. The Salton Sea geothermal field, Imperial Valley, California as a site for continental scientific drilling. Geological Society of America Abstracts with Programs 15:434.

----- 1983b. The Salton Sea Scientific Drilling Project; phases 1 to 5. EOS, Transaction of the American Geophysical Union 64:864.

Elders, W.A., L.H. Cohen, and J.M. Mehegan 1986. Magmatism and Volcanism in the Salton Trough. In Geology of the Imperial Valley, California, edited by P.D. Guptil, E.M. Gath and R.W. Ruff, South Coast Geological Society Annual Field Trip Guidebook, Santa Ana, California 14:134-143.

Elders, W.A., L.H. Cohen, A.E. Williams, S. Neville, P. Collier, and C. Oakes 1984. Naturally-occurring radionuclides in the Salton Sea geothermal field, California; a near-field natural analog of a waste repository in salt. Geological Society of America Abstracts 16:500.

Elders, W.A., J. Combs, and T.B. Coplen 1974. Geophysical, geochemical and geological investigation on the Dunes geothermal system, Imperial Valley, California. Conference on Research for the Development of Geothermal Energy Resources, California Institute of Technology, Jet Propulsion Laboratory, California p. 45-72.

Elders, W.A., J.R. Hoagland, S.D. McDowell and J.M. Cobo 1979. Hydrothermal mineral zones in the geothermal reservoir of Cerro Prieto. In Geology and Geothermics of the Salton Trough, edited by W.A. Elders, University of California, Riverside, Campus Museum Contributions 5:36-43.

Elders, W.A., H.T. Meidav, R.W. Rex, and P.T. Robinson 1970. The Imperial Valley of California; the product of oceanic spreading centers acting on a continent. Geological Society of America Abstracts 2:545.

Elders, W.A., R.W. Rex, H.T. Meidav, P.T. Robinson, and S. Biehler 1972. Crustal spreading in southern California. Science 178(4056):15-23.

Elders, W.A., and J.H. Sass 1988. The Salton Sea Scientific Drilling Project. Journal of Geophysical Research, B, Solid Earth and Planets 93:12,953-12.968.

Eldridge, C.S., M.A. McKibben 1988. Sulfur isotopic systematics in the Salton Sea geothermal system; a SHRIMP ion microprobe study of micron-scale S variations. Geochemical Society, V.M. Goldschmidt Conference p. 40.

Eldridge, C.S., M.A. McKibben, A.E. Williams, W. Compston, I.S. Williams, and A.E. Walshe 1986. SHRIMP ion microprobe measurement of the sulfur isotopic compositions of sulfate and sulfide minerals from the Salton Sea geothermal system, California. Geological Society of America Abstracts with Programs 18:593.

Elliot, W.J. 1974. Seismicity of the San Diego region. In Recent Geological and Hydrological Studies, Eastern San Diego County and Adjacent Areas, San Diego Association of Geologists, California p. 61-70.

Ellis, R.S. 1981. Sedimentary petrography of heavy minerals in the Santa Ysabel/Julian district. Undergraduate Research Reports, Department of Geology, San Diego State University, California 38(1):1-31.

Ellsberg, H. 1972. Mines of Julian. La Sierra Press, Glendale, California 72 p.

Emery, K.O. 1945. Mineralogy of calcite from San Diego County, California. Southern California Academy of Sciences 44:130-135.

Engle, A.E.J., and C.G. Engle 1982. Late Mesozoic and early Cenozoic tectonics along the Salton Trough-Peninsular Range boundary. Geological Society of America Abstracts with Programs 14:162.

Engel, A.E.J., and P.A. Schultejann 1984. Late Mesozoic and Cenozoic tectonic history of south central California. Tectonics 3(6):659-675.

English, D.J. 1985. Regional structural analysis of the Santa Rosa Mountains, San Diego and Riverside Counties, California, implications for the geologic history of southern California. Master of Science Thesis, San Diego State University, California 170 p.

Erpenbeck, M.F. 1977. The sedimentation rate in a playa in Davies Valley using the lead-210 radiometric method. Undergraduate Research Reports, Department of Geology, San Diego State University, California 30(1):1-22.

Ershagi, I., and D. Abdassah 1983. Interpretation of some wireline logs in geothermal fields of the Imperial Valley, California. Society of Petroleum Engineers 53:727-736.

Erskine, B.G. 1982. A paleomagnetic, rock magnetic and magnetic mineralogic investigation of the northern Peninsular Ranges Batholith, southern California. Master of Science Thesis, San Diego State University, California 194 p.

----- 1986. Metamorphic and deformational history of the eastern Peninsular Ranges mylonite zone: Implications on tectonic reconstructions of southern California. Geological Society of America Abstracts with Programs 18:105.

Erskine, B.G., and M. Marshall 1980. A paleomagnetic and rock magnetic investigation of the northern Peninsular Ranges Batholith, southern California. EOS, Transactions of the American Geophysical Union 61:948.

Erskine, B.G., and H.R. Wenk 1985. Evidence for late Cretaceous crustal thinning in the Santa Rosa mylonite zone, southern California. Geology 13:274-277.

Espinosa, A.F. 1982. M and M determination from strong-motion accelerograms, and expected-intensity distribution. In The Imperial Valley, California, earthquake of October 15, 1979. U.S. Geological Survey Professional Paper 1254:433-438.

Everhart, D.L. 1951. Geology of the Cuyamaca Peak quadrangle. California Division of Mines Bulletin 159:51-115.

----- 1953. Geology of the Cuyamaca Peak quadrangle, San Diego county, California. Doctoral Dissertation, Harvard University, Cambridge, Massachusetts.

Eyman, J.L. 1953. A study of sand dunes in the Colorado and Mojave Deserts. Master's Thesis, University of Southern California, Los Angeles 91 p.

Faught, M., A. Jeton, and K. Gregory 1988. Kiwi conveyor belt hill site soil chronosequence Anza-Borrego Desert State Park. Geosciences 650, University of Arizona, Tucson, Manuscript on File Anza-Borrego Desert State Park, California 21 p.

Fairbanks, H.W. 1893. Geology of San Diego County, also portions of Orange and San Bernardino Counties. California State Mineral Bureau, 11th Annual Report of the State Mineralogist p. 88-90.

----- 1910. Some topographical features of the western side of the Colorado Desert. Science, New Science, 32(31):31, and Geological Society of America Bulletin 21:793.

Farrel, B.H. 1965. The use of geothermal steam in California. Geographical Review 55:109-110.

Farrel, W.E. 1969. A gyroscopic seismometer: measurements during the Borrego Earthquake. Seismological Society of America Bulletin 59:1239-1245.

Fasano, G.A. 1982. Analysis and proposed origin of a peridotitie-talc body within the Julian Schist, Julian, California. Undergraduate Research Reports, Department of Geology, San Diego State University, California 40(2):1-26.

Fenneman, N.M. 1931. Salton Basin. In Physiography of the Western United States, McGraw-Hill Book Company, New York p. 377-379.

Feragen, E.S. 1986. Geology of the southeastern San Felipe Hills, Imperial Valley, California. Master of Science Thesis, San Diego State University, California 144 p.

Feragen, E.S., and D.L. Wells 1986. Structural fabrics developed in the Borrego Formation, eastern San Felipe Hills, Imperial Valley, California. Geological Society of America Abstracts with Programs 18:600.

Fergusson, G.J., and W.F. Libby 1963. UCLA radiocarbon dates II. Radiocarbon 5:1-22.

Fernelius, W.A. 1975. Production of fresh water by desalting geothermal brines; a pilot desalting program at the Mesa geothermal field, Imperial Valley, California. United Nations Symposium on the Development and Use of Geothermal Resources, Abstracts 2.

Fernelius, W.A., M.K. Fulcher 1979. East Mesa geothermal test site. American Society of Engineers, Journal of the Environmental Engineering Division 105(10):13-32.

Fernelius, W.A., and K.E. Mathias 1974. Geothermal; the

resource from within. U.S. Department of the Interior, Conservation Yearbook Series 10:42-45.

Fife, D.L., J.A. Minch, and P.J. Crampton 1967. Late Jurassic age of the Santiago Peak volcanics, California. Geological Society of America Bulletin 78:299-304.

Filson, J.R., R.L. Porcella, and R.B. Matthiesen 1984. U.S. Geological Survey strong-motion records from the Imperial Valley Earthquake, October 15, 1979; preliminary summary. In Wind and Seismic Effects, edited by E.V. Leyendecker, et al., U.S. National Bureau of Standards Special Publication 655:29-39.

Finch, M.O. 1988. Damage to irrigation facilities in Imperial Valley; Superstition Hills Earthquakes of November 1987, Imperial County, California. California Geology 41:85-90.

Fleming, R. F. 1993a. Palynological data from the Imperial and Palm Spring Formations, Anza-Borrego Desert State Park, California. U.S. Geological Survey Open-File Report 93-678:1-29.

----- 1993b. Cretaceous pollen and Pliocene climate. The American Association of Stratigraphic Palynologists, Program and Abstracts, Annual Meeting 26: 24.

----- 1994. Cretaceous pollen in Pliocene rocks: implications for Pliocene climate in the southwestern United States. Geology 22:787-790.

----- (in press). Palynological records from Pliocene sediments in the California region: Centerville Beach, DSDP site 32, and the Anza-Borrego Desert. In Pliocene Terrestrial Environments and Data/Model Comparisons, edited by R.S. Thompson. U.S. Geological Survey Open-File Report 94.

Fleming, R. F., and P. Remeika 1994. Pliocene climate of the Colorado Plateau and age of the Grand Canyon: evidence from Anza-Borrego Desert State Park, California. 4th NPS Conference on Fossil Resources, Colorado Springs, Co. (unpaginated).

Flessa, K.W., and A.A. Ekdale 1987. Paleoecology and taphonomy of Recent to Pleistocene intertidal deposits, Gulf of California. In Geologic Diversity of Arizona and Its Margins: Excursions to Choice Areas, edited by G.H. Davis and E.M. VandenDolder, Arizona Bureau of Geology and Mineral Technology Geological Survey Branch Special Paper 5:295-307.

Fletcher, J.B., R.L. Zepeda, and D.M. Boore 1981. Digital seismograms of aftershocks of the Imperial Valley, California, earthquake of October 15, 1979. U.S. Geological Survey Open-File Report 81-655:1-130.

Flynn, C. 1965. Geology of a portion of the Elsinore Fault Zone between Temecula and Pauma Valley, San Diego County, California. Undergraduate Research Reports, Department of Geology, San Diego State University, California 9(1):1-26.

Flynn, M.R., and R.J. Dowlen 1973. A bibliography of San Diego County geology, 1963-1973. In Studies on the Geology and Geologic Hazards of the Greater San Diego area, California, edited by A. Ross and R.J. Dowlen, San Diego Association of Geologists, California p. 133-140.

Ford, E.E., D. London, A.R. Kampf, J.E. Shingley, and L.W. Snee 1991. Gem-bearing pegmatites of San Diego County, California. In Geological Excursions in Southern California and Mexico, edited by M.J. Walawender and B.B. Hanan, Geological Society of America, San Diego Annual Meeting Guidebook p. 128-146.

Forester, R.M. 1991. Pliocene-climate history of the western United States derived from lacustrine ostracodes. Quaternary Science Reviews 10(2-3):133-146.

Foster, A.B. 1979. Environmental variation in a fossil scleractinian coral. Lethaia 12:245-264.

----- 1987. Pliocene reef-corals north of the Gulf of California and their biogeography. Geological Society of America Abstracts with Programs 19(7):666-667.

Fournier, R.B. 1976. A study of the mineralogy and lithology of cuttings from the U.S. Bureau of Reclamation Mesa 6-2 drillhole, Imperial County, California. U.S. Geological Survey Open-File Report 76-88:1-57.

Fourt, R. 1979. Post-batholithic geology of the Volcanic Hills and vicinity, San Diego County, California. Master of Science Thesis, San Diego State University, California 66 p.

Fox, C.K. 1936. The Colorado delta: a discussion of the Spanish explorations and maps, the Colorado River silt load, and its seismic effects on the southwest. (private printing) Los Angeles, California 75 p.

Fradkin, P.L. 1964. A river no more, the Colorado River and the west. University of Arizona Press, Tucson 360 p.

Frankel, A., J. Fletcher, F. Vernon, L. Haar, J. Berger, T. Hanks, and J. Brune. 1968. Rupture characteristics and tomographic source of $M_L=3$ earthquakes near Anza, southern California. Journal of Geophysical Research 91:12633-12650.

Frankel, A., and L. Wennerberg 1989. Rupture process of the Ms 6.6 Superstition Hills, California, earthquake determined from strong-motion recordings: application of tomographic source inversion. Seismological Society of America Bulletin 79:515-541.

Frazier, D.M. 1931. Geology of the San Jacinto quadrangle south of San Gorgonio Pass, California. California Division of Mines and Geology Report 27:494-540.

Freckman, J.T. 1978. Fluid inclusion and oxygen isotope geothermometry of rock samples from Sinclair 4 and Elmore 1 boreholes, Salton Sea geothermal field, Imperial Valley, California. U.S.A. Master's Thesis, University of California, Riverside.

Free, E.E. 1914. Sketch of geology and soils of the Cahuilla basin. In The Salton Sea, Carnegie Institute of Washington Publication 193:21-33.

Freeman, W.B., and R.H. Bolster 1910a. Surface water supply of the United States, 1907-07, part IX, Colorado River basin. U.S. Geological Water-Supply Paper 249:46-54.

----- 1910b. Surface water supply of the United States, 1907-07, part IX, Colorado River basin. U.S. Geological Water-Supply Paper 251:38-46.

Frez, J., and J.J. Gonzalez-Garcia 1991. Seismo-tectonics of the border region between southern and northern Baja California. Geological Society of America Abstracts with Programs p. A24.

Frick, C. 1921. Extinct vertebrate faunas of the badlands of Bautista Creek and San Timoteo Canyon, southern California. University of California Publications, Department of Geological Sciences Bulletin 12:277-409.

----- 1933. New remains of Trilophodont-Tetrabelodont Mastodons. American Museum of Natural History Bulletin 59:505-652.

----- 1937. Horned ruminants of North America. Bulletin of the American Museum of Natural History 69:1-669.

Frith, R.B. 1978. A seismic refraction investigation of the Salton Sea geothermal area, Imperial Valley, California. Master's Thesis, University of California, Riverside.

Frizzell, V.A., Jr. 1984. Geology of the Baja California Peninsula, an introduction. In Geology of the Baja California Peninsula, edited by V.A. Frizzell, Jr., Society of Economic Paleontologists and Mineralogists Pacific Section p. 19-27.

Frost, E.G., D.M. Frost, and C.E. Houser 1989. Mid-Tertiary ductile formation below Trigo Mountains detachment fault, southwestern Arizona. Geological Society of America Abstracts with Programs 21(5):81.

Frost, E.G., L.A. Heizer, R.G. Blom, and R.L. Crippen 1993. Western Salton Trough detachment system: its geometric role in localizing the San Andreas system. In The San Andreas Fault System, edited by F.V. Corona. Ninth Thematic Conference on Geologic Remote Sensing, Pasadena, California, p. 186-197.

Frost, E.G., and D.L. Martin 1982a. Mesozoic- Cenozoic tectonic evolution of the Colorado River region, California, Arizona, and Nevada. editors, Cordilleran Publishers, San Diego, California 608 p.

----- 1982b. Comparison of Mesozoic compressional tectonics with mid-Tertiary detachment faulting in the Colorado River area, California, Arizona, and Nevada. In Geological Excursions in the California Desert, Geological Society of America, Cordilleran Section Guidebook p. 113-159.

----- 1983. Overprint of Tertiary detachment on the Mesozoic Orocopia Schist and Chocolate Mountains thrust. Geological Society of America Abstracts with Programs 14:164.

Frost, E.G., D.L. Martin, and D. Krummenacher 1982. Mid-Tertiary detachment faulting in southwestern Arizona and southeastern California and its overprint on the Vincent-Orocopia thrust system. Geological Society of America Abstracts with Programs 14:164.

Frost, E.G., D.A. Okaya, T.V. McEvilly, E.C. Hauser, G.S. Galvan, J. McCarthy, G.S. Fuis, C.M. Conway, R.G. Blom, and T.L. Heidrick 1987. Crustal transect: Colorado Plateau -- detachment terrain -- Salton Trough. In Geologic Diversity of Arizona and Its Margins: Excursions to Choice Areas, edited by G.H. Davis and E.M. VandenDolder, Arizona Bureau of Geology and Mines Technical Special Paper 5:398-422.

Frost, E.G., and M. Shafiqullah 1989. Pre-San Andreas opening of the Salton Trough as an extensional basin; K-Ar ages on regional detachment faults along the western margin of the Salton Trough. Geological Society of America Abstracts with Programs 21(5):81.

Fuis, G.S. 1982. Displacement on the Superstition Hills Fault triggered by the earthquake. In The Imperial Valley, California, Earthquake of October 15, 1979. U.S. Geological Survey Professional Paper 1254:145-154.

Fuis, G.S., C.E. Johnson, and D.J. Jenkins 1978a. Preliminary catalog of earthquakes in northern Imperial Valley, California, October 1977-December 1977. U.S. Geological Survey Open-File Report 78-673:1-43.

----- 1978b. Preliminary catalog of earthquakes in northern Imperial Valley, California, October 1978-March 1978. U.S. Geological Survey Open-File Report 78-671:1-23.

Fuis, G.S., C.E. Johnson, and K.J. Richter 1980. Preliminary catalog of earthquakes in northern Imperial Valley, California, April 1978-June 1978. U.S. Geological Survey Open-File Report 80-1167:1-21.

Fuis, G.S., and M.W. Kohler 1984. Crustal structure and tectonics of the Imperial Valley region, California. In The Imperial Basin, Tectonics, Sedimentation and Thermal Aspects, edited by C.A. Rigsby,

Society of Economic Paleontologists and Mineralogists 40:1-14.

----- 1985. Crustal structure and tectonics of the Imperial Valley region, California. In Geology and Geothermal Energy of the Salton Trough, edited by L.J. Herber, National Association of Geology Teachers, Far Western Section p. 95-107.

Fuis, G.S., W.D. Mooney, J.H. Healey, W.J. Lutter, and G.A. McMechan 1981. Seismic-refraction results in the Imperial Valley region, California, and implications for plate tectonics. EOS, Transactions of the American Geophysical Union 61:1025.

Fuis, G.S., W.D. Mooney, J.H. Healy, G.A. McMechan, and W.J. Lutter 1981. Seismic refraction studies of the Imperial Valley region, profile model, a traveltime contour map, and a gravity model. U.S. Geological Survey Open-File Report 80-270:1-83.

----- 1982. Crustal structure of the Imperial Valley region. In The Imperial Valley, California, Earthquake of October 15, 1979. U.S. Geological Survey Professional Paper 1254:25-49.

----- 1984. A seismic refraction survey of the Imperial Valley region, California. Journal of Geophysical Research 89:1165-1189.

Fuis, G.S., and M. Schnapp 1977. The November-December 1976 earthquake swarms in northern Imperial Valley, California; seismicity on the Brawley Fault and related structures. EOS, Transactions of the American Geophysical Union 58:1181.

Fuis, G.S., A.W. Walter, W.D. Mooney, and J. McCarthy 1986. Crustal velocity structure of the Salton Trough, western Mojave Desert and Colorado Desert, from seismic refraction. Geological Society of America Abstracts with Programs 18:107.

Fujino, Y., T. Yokota, Y. Hamazaki, and R. Inque 1984. Multiple event analysis of 1979 Imperial Valley, California earthquake using district phases in near-field accelerograms. Proceedings of the World Conference on Earthquake Engineering 8:385-392.

Fulcher, M.K. 1974. Overview of Reclamation's geothermal program in Imperial Valley, California. Conference on Research for the Development of Geothermal Energy Resources, California Institute of Technology, Jet Propulsion Laboratory, California p. 33-40.

Full, W.E. 1980. Processes of lithosphere thinning and crustal rifting in the Salton Trough. Master's Thesis, University of Illinois, Chicago.

Full, W.E., and L.W. Younker 1981. Processes of lithosphere thinning and crustal rifting in the Salton Trough, California. Geological Society of America Abstracts with Programs 13:456.

Gale, H.S. 1912. Nitrate deposits. U.S. Geological Survey Bulletin 523:1-36.

Gallup, D.L. 1989. The solubility of amorphous silica in geothermal brines. Transactions of the Geothermal Resources Council 13:241-245.

Ganus, W.J. 1973. Groundwater resources in crystalline rock, Mount Laguna area, San Diego County, California. Association of Engineering Geologists, Abstracts p. 27.

----- 1974. Groundwater occurrence in the Mount Laguna area. In Recent Geological and Hydrological studies, Eastern San Diego County and Adjacent Areas, San Diego Association of Geologists, California p. 86-92.

Garner, K.B. 1936. Concretions near Mount Signal, lower California. American Journal of Science 184:301-311.

Garth, J.S. 1955. The case for a warm-temperate marine fauna on the west coast of North America. In Essays in the Natural Sciences in Honor of Captain Allan Hancock, University of Southern California Press, Los Angeles p. 19-27.

Gastil, R.G. 1961. The elevated erosion surfaces. In Geological Society of America, Cordilleran Section, Guidebook for Field Trips p. 1-14.

----- 1975. Plutonic zones in the Peninsular Ranges of southern California and northern Baja California. Geology 3:361-363.

----- 1983. Mesozoic and Cenozoic granitic rocks in southern California and western Mexico. In Circum-Pacific Plutonic Terrains, edited by J.A. Roddick, Geological Society of America Memoir 159:265-275.

----- 1985. Terranes of peninsular California and adjacent Sonora. In Tectonostratigraphic terranes of the circum-Pacific region, edited by D.G. Howell. Circum-Pacific Council for Energy and Mineral Resources Earth Science Series 1:273-283.

----- 1991. Is there a Oaxaca-California megashear? Conflict between paleomagnetic data and other elements of geology. Geology 19:502-505.

Gastil, R.G., K.K. Bertine, J. Criscione, and T. Davis 1981. Sr87/Sr86 ratio from hot and carbonated springs in the Peninsular Ranges of southern and Baja California, U.S.A. and Mexico. Geological Society of America Abstracts with Programs 13:57.

Gastil, R.G., and J. Bushee 1961. Geology and geomorphology of eastern San Diego County, field trip No. 1, road log.. In Guidebook for Field Trips, edited by B.E. Thomas. Geological Society of America, Cordilleran Section, p. 8-22.

Gastil, G., and T. Davis (in press). $^{87}Sr/^{86}Sr$ ratios in prebatholithic strata of peninsular California.

Gastil, G., J. Diamond, C. Knaack, M. Walawender, M. Marshall, C. Boyles-Reaber, B. Chadwick, and B.

Erskine 1990. The problem of the magnetite/ilmenite boundary in southern and Baja California. Geological Society of America Memoir 174:19-32.

Gastil, R.G., R. Kies, and D.J. Melius 1979. Active, and potentially active faults; San Diego County and northernmost Baja California. In Earthquakes and Other Perils -- San Diego Region, edited by P.L. Abbott and W.J. Elliott, Geological Society of America Field Trip, San Diego Association of Geologists p. 47-60.

Gastil, R.G., D. Krummenacher, and J. Bushee 1974. The batholith belt of southern California and western Mexico. Pacific Geology 8:73-78.

Gastil, R.G., D. Krummenacher, and J. Minch 1981. The record of Cenozoic volcanism around the Gulf of California. Geological Society of America Bulletin 90(1):839-857.

Gastil, R.G., and R.H. Miller 1981. Lower Paleozoic strata on the Pacific Plate of North America. Nature 292:828-830.

----- 1983a. Prebatholithic terrains of peninsular California north of the 28th parallel U.S.A. and Mexico, August 1983 status. In Proceedings of the Circum-Pacific Terrains Conference, edited by D.G. Howell, D.L. Jones, A. Cox, and A.M. Nur, Stanford University Publications, Geological Sciences 18:93-97.

----- 1983b. Prebatholithic terrains of southern and peninsular California, U.S.A. and Mexico; status report. In Pre-Jurassic Rocks in Western North American Suspect Terrains, edited by C.H. Stevens, Society of Economic Paleontologists and Mineralogists, Pacific Section p. 49-61.

----- 1984. Prebatholithic paleogeography of peninsular California and adjacent Mexico. In Geology of the Baja California Peninsula, edited by V.A. Frizzell, Jr., Society of Economic Paleontologists and Mineralogists, Pacific Section 39:9-16.

Gastil, R.G., G.J. Morgan, and D. Krummenacher 1978. Mesozoic history of peninsular California and related areas east of the Gulf of California. In Mesozoic paleogeography of the western United States, edited by D.G. Howell and K.A. McDougall. Society of Economic Paleontologists and Mineralogists, Pacific section, Pacific Coast Paleogeography Symposium 2:107-115.

----- 1981. The tectonic history of peninsular California and adjacent Mexico. In The Geotectonic Development of California, edited by W.G. Ernst, Prentice-Hall, Incorporated, New Jersey, Ruby Volume 1:284-305.

Gastil, G., M. Wardlaw, and G. Girty (in press). U-Pb systematics in detrital zircon populations from the Triassic/Jurassic flysch-type sequences of peninsular California.

Gastil, G., M. Wracher, G. Strand, L.L. Kear, D. Eley, D. Chapman, and C. Anderson 1992. The tectonic history of the southwestern United States and Sonora, Mexico, during the past 100 m.y. Tectonics 11:990-997.

Gath, E., R.W. Ruff, R. McElwain, and M.L. Raub 1986. Geology of the Salton Trough Field Trip Roadlog. In Geology of the Imperial Valley, California, edited by P.D. Guptil, E.M. Gath and R.W. Ruff, South Coast Geological Society, Annual Field Trip Guidebook, Santa Ana, California 14:181-225.

Gazin, C.L. 1942. Late Cenozoic faunas from San Pedro Valley, Arizona. Proceedings of the U.S. National Museum 92:475-518.

Geis, P. 1982. Geochemical analysis of manganese ore deposits associated with detachment faulting, Midway Mountains, Imperial County, California. Undergraduate Research Reports, Department of Geology, San Diego State University, California 40(2):1-39.

George, P.G., R.K. Dokka, and D.J. Henry 1991. Major late Cretaceous thrusting and subsequent extension in the eastern Peninsular Ranges of southern California. Abstract with programs. Geological Society of America annual meeting. San Diego p. A480.

Gergen, J.G. 1978. Horizontal displacement in the earth's crust in the vicinity of El Centro, California. EOS, Transactions of the American Geophysical Union 58:242.

Germinario, M.P. 1982a. The depositional and tectonic environments of the Julian Schist, Julian, California. Master of Science Thesis, San Diego State University, California 95 p.

----- 1982b. The depositional and tectonic environments of the Julian Schist, Julian, California. Geological Society of America Abstracts with Programs 14:165.

Gester, K., S. Lindval, J. Brake, and K. Rexrode 1986. The Yaqui Ridge detachment fault, south Borrego Valley, California. Unpublished Research Report, San Diego State University, California.

Geyer, B.L. 1962. Geology of the Palm Desert region, Riverside County, California. Undergraduate Research Report, Department of Geology, San Diego State University, California, 20p.

Gibson, L.M. 1983. The configuration of the Vallecito-Fish Creek Basin, western Imperial Valley, California, as interpreted from gravity data. Master of Arts Thesis, Dartmouth College, New Hampshire 87 p.

Gibson, L.M., L.L. Malinconico, T. Downs, and N.M. Johnson 1984. Structural implications of gravity data from the Vallecito-Fish Creek Basin, western Imperial Valley, California. In The Imperial Basin, Tectonics, Sedimentation and Thermal Aspects, edited by C.A. Rigsby, Society of Economic Paleontologists and Mineralogists 40:15-29.

Gilbert, J.D. 1966. Geology of the Ocotillo Badlands. Undergraduate Research Reports, Department of Geology, San Diego State University, California 10(2):1-37.

Gillies, W.D. 1958. The geology of a portion of the Cottonwood Springs quadrangle, Riverside County, California. Master's Thesis, University of California, Los Angeles 70 p.

Gilluly, J. 1965. Volcanism, tectonism and plutonism in the western United States. Geological Society of America Special Paper 80:68.

Gilman, R.C., N.R. Goulty, C.R. Allen, R.P. Keller, and R.O. Burford 1977. Fault creep in the Imperial Valley, California. EOS, Transactions of the American Geophysical Union 58:1226.

Gilmore, T.D. 1985. Episodic regional uplift in southeastern California preceding major historic earthquakes in the Imperial Valley. EOS, Transactions of the American Geophysical Union 66:856.

Gilmore, T.D., and R.O. Castle 1983a. A contemporary tectonic boundary coincident with the Arizona-California border. Geological Society of America Abstracts with Programs 15:315.

----- 1983b. A proposed tectonic contribution to the preservation of the divide between the Salton Basin and the Gulf of California. Geological Society of America Abstracts with Programs 15:439.

----- 1983c. Tectonic preservation of the divide between the Salton Basin and the Gulf of California. Geology 11:474-477.

Gilpin, B.E., IV 1977. A microearthquake study in the Salton Sea geothermal area, California. Master's Thesis, University of California, Riverside.

Gilpin, B.E., IV, and T.C. Lee 1978. A microearthquake study in the Salton Sea geothermal area, California. Seismological Society of America Bulletin 68:441-450.

Girty, G.H., and A. Armitage 1989. Composition of Holocene Colorado River sand: an example of mixed-provenance sand derived from multiple tectonic elements of the Cordilleran continental margin. Journal of Sedimentary Petrology 59:597-604.

Given, D.D. 1981. Seismicity of the San Jacinto Fault Zone. In Geology of the San Jacinto Mountains, edited by A.R. Brown and R.W. Ruff, South Coast Geological Society Annual Field Trip Guidebook, Santa Ana, California 9:55-60.

----- 1983. Seismicity and structure of the trifurcation in the San Jacinto Fault Zone, southern California. Master of Science Thesis, Los Angeles, California State University. 73p.

Given, D., and W. Stuart 1988. A fault interaction model for triggering of the Superstition Hills Earthquake of November 24, 1987. Seismology Research Letters 59:48.

Gjerde, M.W. 1982. Petrology and geochemistry of the Alverson Formation, Imperial County, California. Master of Science Thesis, San Diego State University, California 85 p.

Gobalet, K.W. 1992. Colorado River fishes of Lake Cahuilla, Salton Basin, southern California: a cautionary tale for zooarchaeologists. Bulletin of the Southern California Academy of Science 91(2):70-83.

Gochenaur, D. 1993. Pegmatite minerals near Aguanga, Riverside County, California. In Ashes, Faults and Basins, edited by R.E. Reynolds and J. Reynolds. San Bernardino County Museum Association Special Publication 93-1:97.

Goldman, H.B. 1957. Carbon dioxide in mineral commodities of California. California Division of Mines and Geology Bulletin 17:1-736.

----- 1968. Sand and gravel in California, part C -- southern California. California Division of Mines and Geology Bulletin 180:1-56.

Goldsmith, M. 1982. Geothermal development and the Salton Sea. California Institute of Technology, Environmental Quality Laboratory Memorandum 17:1-15.

Golz, D.J., G.T. Jefferson, and M.P. Kennedy 1977. Late Pleistocene vertebrate fossils from the Elsinore Fault Zone. Journal of Paleontology 51(4):864-867.

Goodmacher, J.C. 1990. Soil development along the Elsinore Fault, Coyote Mountains, California. Master of Science Thesis, San Diego State University, California 148 p.

Goodmacher, J.C., and T.K. Rockwell 1990. Changes in geomorphic surface roughness over time and its inferred affects on the chemistry and development of a soil chronosequence, Coyote Mountains, southern California. Geological Society of America Abstracts with Programs 22(3):26.

----- 1990. Properties and inferred ages of soils developed in alluvial deposits in the southwestern Coyote Mountains, Imperial Valley, California. In Western Salton Trough Soils and Neotectonics: Friends of the Pleistocene Field Trip Guidebook, p. 43-104.

Goodreau, D.D., and J.C. Savage 1972. Strain accumulation in Imperial Valley. EOS, Transactions of the American Geophysical Union 53:1116.

Goodwin, L.B., and P.R. Renne 1991. Effects of progressive mylonitization on Ar retention in biotites from the

Santa Rosa mylonite zone, California, and thermochronologic implications. Contributions to Mineralogy and Petrology 108:283-297.

Gookin, E. 1967. Geology of the Indian Pass area, Imperial County, California. Undergraduate Research Reports, Department of Geology, San Diego State University, California 11(3):1-45.

Gonzalez-Garcia, J.J., F. Suarez-Vidal, A.P. Cicese, J.G. Quiñones-Rodriguez, T. Rockwell and A. Thomas 1991. The Imperial Valley Fault in Mexicali and its recent activity near Ejido Saltillo, Baja California, Mexico. Geological Society of America Abstracts with Programs p. A84.

Goss, R., and J. Combs 1974. Thermal conductivity of sediments from drill chips, cores, and geophysical log parameters. EOS, Transactions of the American Geophysical Union 55:423.

Goulty, N.R., R.O. Burford, C.R. Allen, R. Gilman, C.E. Johnson, and R.P. Keller 1978. Large creep events on the Imperial Fault, California. Seismological Society of America Bulletin 68:517-521.

Grabau, A.W. 1920. Formation of the Salton Sink by the Colorado delta. Principles of salt deposition, McGraw-Hill Book Company, New York p. 146-151.

Graber, D. 1970. Aerial geology and deformation of the Yaqui Ridge area, east San Diego County, California. Undergraduate Research Reports, Department of Geology, San Diego State University, California 16(1):1-15.

Grannell, R.B. 1985. Land deformation in the Salton-Mexicali Trough. In Geology and Geothermal Energy of the Salton Trough, edited by L.J. Herber. National Association of Geology Teachers, Far Western Section p. 136-139.

Grannell, R.B., D.W. Tarman, R.C. Clover, R.M. Leggewie, N.E. Goldstein, D.S. Chase, and J. Eppink 1980. Precision gravity studies at Cerro Prieto. Geothermics 9:89-99.

Grant, U.S., IV, and H.R. Gale 1931. Pliocene and Pleistocene mollusca of California and adjacent regions. San Diego Society of Natural History Memoirs 1:1-1036.

Grant, U.S., IV, and L.G. Hertlein 1956. *Schizaster morlini*, a new species of echinoid from the Pliocene Imperial County, California. Southern California Academy of Sciences Bulletin 55:107-110.

Gray, C.H., Jr. 1954. Geology of the Corona-Elsinore-Murrieta area, Riverside County, California. California Division of Mines Bulletin 170, Map Sheet 21.

Gray, J. 1961. Early Pleistocene paleoclimatic record from Sonoran Desert, Arizona. Science 133:38-39.

Grayzer, V.M., and S.G. Molotkov 1983. Analiz akselerogram sil'nykh dvizheniy zemletryaseniya 15 oktyabrya 1979 g. v Impirial Velli. Voprosy Inzhenernoy Seismology 24:62-73.

Green, K.E. 1988a. Salton Basin regional setting. In Geologic Field Guide to the Western Salton Basin Area, edited by S.M. Testa and K.E. Green, American Institute of Professional Geologists p. 1-6.

----- 1988b. Field trip log. In Geologic Field Guide to the Western Salton Basin Area, edited by S.M. Testa and K.E. Green, American Institute of Professional Geologists p. 31-36.

Greenwood, R.B., D.M. Morton, R.C. Jachens, and R.H. Chapman 1991. Geology and geophysics of the Santa Ana 1:100,000 quadrangle, southern California: a progress report. Abstract with programs Geological Society of America annual meeting. San Diego p. A478.

Gregory, K., A. Jeton, and M. Faught 1988. Kiwi conveyor belt lithology study, Anza-Borrego State Park. Geosciences 650, University of Arizona, Tucson, Manuscript on File Anza-Borrego Desert State Park, California 26 p.

Griscom, A., and L.J.P. Muffler 1971a. Aeromagnetic map and interpretation of the Salton Sea geothermal area, California. U.S. Geological Survey Geophysical Investigations Maps GP-0754:1-4.

----- 1971b. Aeromagnetic survey of the Salton Sea geothermal area, southeastern California. Geological Society of America Abstracts 3:129.

Groh, J.M., A. J. Ciccateri, and R.E. Casey 1991. Foraminifera and other organisms associated with the hydrothermal vent systems of the Salton Sea. Geological Society of America Abstracts with Programs p. A36.

Gromet, L.P. 1979a. Profile of rare earth element characteristics across the Peninsular Ranges Batholith near the international border, southern California, U.S.A. and Baja California, Mexico. In Mesozoic Crystalline Rocks: Peninsular Ranges Batholith and Pegmatites, Point Sal Ophiolite, edited by P.L. Abbott and V.R. Todd, Department of Geological Sciences, San Diego State University, California p. 133-142.

----- 1979b. Rare earth's abundances and fractionations and their implications for batholithic petrogenesis in the Peninsular Ranges Batholith, California, U.S.A., and Baja California, Mexico. Doctoral Dissertation, California Institute of Technology, Pasadena 348 p.

Gromet, L.P., and L. T. Silver 1983. Rare earth element distributions among minerals in a granodiorite and their petrogenetic implications. Geochimica et Cosmochimica Acta 47:925-939.

----- 1987. REE variations across the Peninsular Ranges

Batholith: implications for batholithic petrogenesis and crustal growth in magmatic arcs. Journal of Petrology 28(1):75-125.

Gross, W.K., M. Lisowski, and J.C. Savage 1988. Marginally significant temporal variations in the rate of shear-strain accumulation at two sites along the San Andreas Fault in southern California. EOS, Transactions of the American Geophysical Union 69:1432.

Gross, W.W. 1984. Geochemistry and origins of dark inclusions and dikes in the granitic rocks of a portion of the Peninsular Ranges Batholith, San Diego County, California. Master of Science Thesis, Department of Geology, San Diego State University, California 103 p.

Grubbs, D.K. 1963. Ore-bearing magmatic and metamorphic brine from the Salton Sea volcanic domes geothermal area, Imperial County, California. Master's Thesis, University of Virginia, Charlottesville.

----- 1964. Ore-bearing magmatic and metamorphic brine from the Salton Sea volcanic domes geothermal area, Imperal County, California. Virginia Journal of Science 15:333.

Guffanti, M. 1988. Geothermal Energy sources. U.S. Geological Survey Bulletin 1850:71-72.

Gupta, H.K., and D. Nyman 1977. Short period surface wave dispersion studies in the East Mesa geothermal field, California. Geothermal Resources Council Transactions 1:123-125.

Haak, B. and S. Jampoler 1987. Landsat TM and MSS digital data comparison; Imperial Valley. Proceedings of the International Symposium on Remote Sensing of Environment 21:853-862.

Haar, S.V., and I.P. Cruz 1979. Hybrid transform faults and fault intersections in the southern Salton Trough geothermal area, Baja California, Mexico. In Geology and Geothermics of the Salton Trough, edited by W.A. Elders, University of California, Riverside Campus Museum Contributions 5:95-100.

Habermann, R.E., and M. Wyss 1982. Background seismicity rates and precursory seismic quiescence; Imperial Valley, California. Seismological Society of America Bulletin 74:1743-1755.

Hadjian, A.H., R.B. Fallgren, and L. Lau 1990. Imperial County Services Building revisited; a reevaluation with pile-soil-structure interaction. Proceedings of the U.S. National Conference on Earthquake Engineering 4(3):835-844.

Hadley, J.B. 1942. Manganese deposits in the Paymaster Mining District, Imperial Valley, California. U.S. Geological Survey Bulletin 931-S:459-473.

Hadley, O.J., F.K. Wyatt, and D.C. Agnew 1988. The steadiness of fault zone strain; evidence from Pinon Flat Observatory. EOS, Transactions of the American Geophysical Union 69:1432.

Hail, W.R., W. Randall, and R.W. Rex 1970. Possible sea-floor spreading in the Imperial Valley of California; II, configuration of the basement. EOS, Transactions of the American Geophysical Union 51:421-422.

Hagerty, S., H. Maciolek, and S. Sun 1986. Federal geothermal activities in the Imperial Valley, California. In Geology of the Imperial Valley, California, edited by P.D. Guptil, E.M. Gath and R.W. Ruff, South Coast Geological Society, Annual Field Trip Guidebook, Santa Ana, California 14:104-115.

Hagstrum, J.T., M. McWilliams, D.G. Howell, and S. Gromme 1985. Mesozoic paleomagnetism and northward translation of the Baja California peninsula. Geological Society of America Bulletin 96:1077-1090.

Hagstrum, J.T., M.G. Sawlan, B.P. Hausback, J.G. Smith, and C.S. Gromme 1987. Miocene paleomagnetism and tectonic setting of the Baja California peninsula, Mexico. Journal of Geophysical Research 92(B3):2627-2639.

Hall, D.A. 1978. Analysis of geothermal scale deposits, Imperial Valley, California. Undergraduate Research Reports, Department of Geology, San Diego State University, California 33:1-39.

Halliburton, R.A. 1980. Magnetic survey of fault splinters in the Elsinore Fault zone. Undergraduate Research Reports, Department of Geology, San Diego State University, California 36(2):1-27.

Hamblin, W.K. 1976. Late Cenozoic volcanism in the western Grand Canyon. In Geology of the Grand Canyon, edited by W.J. Breed and E. Roat, Museum of Northern Arizona and Grand Canyon Natural History Association, Flagstaff p. 142-169.

Hamelehle, D.M. 1982. Geology of The Narrows, Vallecito Mountains and Yaqui Ridge, eastern San Diego County, California. Undergraduate Research Reports, Department of Geology, San Diego State University, California 41(2):1-40.

Hamilton, R.M. 1968. Time-term analysis of explosion data from the vicinity of Borrego Mountain, California, earthquake of 9 April 1968. EOS, Transactions of the American Geophysical Union 49:711.

----- 1969. Aftershocks of the Borrego Mountain, California, earthquake of 9 April 1968. Earthquake Notes 40:26.

----- 1970. Time-term analysis of explosion data from the vicinity of the Borrego Mountain, California, earthquake of April 9, 1968. Seismological Society of America Bulletin 60:367-381.

----- 1972. Aftershocks of the Borrego Mountain Earthquake for April 12 to June 12, 1968. U.S. Geological Survey Professional Paper 787:31-54.

Hamilton, W.B. 1961. Origin of the Gulf of California. Geological Society of America Bulletin 72:1307-1318.

----- 1966. Geology of the Salton Trough. California Division of Mines and Geology Bulletin 191:73-76.

----- 1969. Geology of the Colorado Desert (Salton Through). California Division of Mines Mineral Information Service 22:96-98.

----- 1978. Mesozoic tectonics of the western United States. In Mesozoic Paleogeography of the Western United States, edited by D.G. Howell and K.A. McDougall, Pacific Coast Paleogeography Symposium 2, Society of Economic Paleontologists and Mineralogists, Pacific Section p. 33-70.

Hamlin, H. 1917. Miscellaneous earthquakes in southern and eastern California. Seismological Society of America Bulletin 7:113-118.

Hammond, D.E., J.G. Zukin, and T.L. Ku 1988. The kinetics of radioisotope exchange between brine and rock in a geothermal system. Journal of Geophysical Research, B, Solid Earth and Planets 93:13,175-13,186.

Hanks, H.G. 1882. On the occurrence of salt in California and its manufacture. California Mining Bureau Report 2:217-226.

Hanks, T.C., and C.R. Allen 1989. The Elmore Ranch and Superstition Hills Earthquakes of 24 November 1987. Seismological Society of America Bulletin 79:231-238.

Hanna, G.D. 1926. Paleontology of Coyote Mountain, Imperial County, California. Proceeding of the California Academy of Sciences, 4th Series, 14:427-503.

Hannibal, H. 1912. A synopsis of Recent and Tertiary freshwater Mollusca of the California province, based on ontogenetic classification. Proceedings of the Malacological Society 10:112-211.

Hansen, C.L., and A.O. Clarke 1987. Origin of gravel-capped pediments in the Anza-Borrego Desert, California. Association of Pacific Coast Geographers, Annual Meeting, Davis, California.

Hanson, J.M. and R.L. Newmark 1989. Three-dimensional geophysical modeling at the Salton Sea geothermal field. EOS, Transactions of the American Geophysical Union 70:603.

Harding, M.B. 1981. Erosional history of the Painted Gorge area, Coyote Mountains, Imperial County, California. Undergraduate Research Reports, Geology Department, San Diego State University, California 39(2):1-28.

Harrar, J.E., and others, 1979. Studies of brine chemistry, precipitation of solids, and scale formation at the Salton Sea geothermal field. Lawrence Livermore Laboratory, Berkeley, California, Report 52640:1-16.

Harris, A.H. 1985. Late Pleistocene paleoecology of the southwest. University of Texas Press, Austin 293 p.

Harris, M.E. 1987. Geology of San Diego County, California, a bibliography with subject index. California Division of Mines and Geology Special Publication 096:1-57.

----- 1991. Geology of Imperial County: a bibliography. Friends of the Library, San Diego State University Library p. (not numbered).

Harris, M., and S.D. Van Nort 1975. Geology and mineralization of the Picacho gold prospect. Mew Mexico Geological Society Annual Field Conference Guidebook 26:339-340.

Harsh, P.W. 1982. Distribution of afterslip along the Imperial Fault. In The Imperial Valley, California, Earthquake of October 15, 1979, U.S. Geological Survey Professional Paper 1254:193-203.

Hart, E.W. 1981. Preliminary map of October 1979 fault ruptures, Imperial County, California. California Division of Mines and Geology Open-File Report DMG OFR-81-05.

Hart, E.W., W.A. Bryant, and J.A. Treiman 1993. Surface faulting associated with the June 1992 Landers Earthquake, California. California Geology 46(1):10-16.

Hart, M.W. 1964. The geology of the Elsinore Fault between Banner Grade and Vallecito Valley, San Diego County, California. Undergraduate Research Reports, Department of Geology, San Diego State University, California 8(3):1-14.

----- 1991a. Landslides in the Peninsular Ranges, southern California. In Geological Excursions in Southern California and Mexico, edited by M.J. Walawender and B.B. Hanan, Geological Society of America, San Diego Annual Meeting Guidebook p. 349-371.

----- 1991b. Failure mechanisms of mega-landslides in the Peninsular Ranges of southern California. Geological Society of America Abstracts with Programs p. A126.

Hart, M.W., and A.E. Farcas 1983. Reconnaissance of landslides and debris flows in the Cuyamaca Peak gabbro, San Diego County, California. Association of Engineering Geologists 26:71.

Harthill, N. 1978. Aquadripole resistivity survey of the Imperial Valley, California, earthquake of October 15, 1979. Geophysics 43:1485-1500.

Hartzell, S.H., and J.N. Brune 1977. Source parameters for the January 1975 Brawley-Imperial Valley Earthquake swarm. Pure Applied Geophysics 115:333-355.

Hartzell, S.H., and T.H. Heaton 1983. Inversion of strong ground motion and teleseismic waveform data for the fault rupture history of the 1979 Imperial Valley, California, earthquake. Seismological Society of America Bulletin 73:1553-1583.

Hartzell, S., and D.V. Helmberger 1981. Strong ground motion modeling of the October 15, 1979 Imperial Valley Earthquake. EOS, Transactions of the American Geophysical Union 61:1036.

----- 1982. Strong-motion modeling of the Imperial Valley Earthquake of 1979. Seismological Society of America Bulletin 72:571-596.

----- 1983. Strong-motion modeling of the Imperial Valley Earthquake of 1979. In The Dynamic Characteristics of Faulting Inferred from Strong Ground Motion, edited by J. Boatwright, U.S. Geological Survey Open-File Report 82-591:839-876.

Havholm, K.G., and G. Kocurek 1986. Draa dynamics; a study of a modern draa, Algodones Dune field, California. International Sedimentological Congress, Canberra, Australia, Bureau of Mineral Resources, Geology, and Geophysics p. 134.

----- 1988. A preliminary study of the dynamics of a modern draa, Algodones, southeastern California, U.S.A. Sedimentology 35:649-669.

Hawkins, J.W. 1970. Petrology and possible tectonic significance of late Cenozoic volcanic rocks, southern California and Baja California. Geological Society of America Bulletin 81:3323-3338.

----- 1970. Petrology and possible tectonic significance of late Cenozoic volcanic rocks, southern California and Baja California. Scripps Institution of Oceanography Contributions 2862:1-15.

Hawkins, J.W., and T.M. Atwater 1971. Mesozoic-Cenozoic vulcanism and plutonism in southern California; implications for zones of subduction and crustal dilation. Geological Society of America Abstracts with Programs 3:134.

Haxel, G.B. 1977. The Orocopia Schist and the Chocolate Mountain thrust, Picacho-Peter Kane Mountain area, southeasternmost California. Doctoral Dissertation, University of California, Santa Barbara 277 p.

Haxel, G.B., J.R. Budahn, T.L. Fries, B.W. King, L.D. White, and P.J. Aruscavage 1987. Geochemistry of the Orocopia Schist, southern California. In Mesozoic Rocks of Southern Arizona and Adjacent Areas, edited by W.R. Dickinson and M.A. Klute, Arizona Geological Society Digest 18:49-64.

Haxel, G.B., and J. Dillon 1978. The Pelona-Orocopia Schist and Vincent-Chocolate Mountain thrust system, southern California. In Mesozoic Paleogeography of the Western United States, edited by D.G. Howell and K.A. McDougall, Pacific Coast Paleogeography Symposium 2, Society of Economic Paleontologists and Mineralogists, Pacific Section p. 453-469.

Haxel, G.B., and R.M. Tosdal 1986. Significance of the Orocopia Schist and Chocolate Mountains thrust in the late Mesozoic tectonic evolution of the southeastern California-southwestern Arizona region. In Frontiers in Geology and Ore Deposits of Arizona and the Southwest, edited by B. Beatty and P.A.K. Wilkinson, Arizona Geological Society Digest 16:52-61.

Haxel, G.B., R.M. Tosdal, and J.T. Dillon 1985. Tectonic setting and lithology of the Winterhaven Formation, a new Mesozoic stratigraphic unit in southeasternmost California and southwestern Arizona. U.S. Geological Survey Bulletin 1599:1-19.

----- 1986. Field guide to the Chocolate Mountain thrust and Orocopia Schist, Gavilan Wash area, southeasternmost California. In Frontiers in Geology and Ore Deposits of Arizona and the Southwest, edited by B. Beatty and P.A.K. Wilkinson, Arizona Geological Society Digest 16:52-61.

Hayba, D.O. 1979. Characterization of scales formed from Salton Sea geothermal brines. Master's Thesis, Pennsylvania State University, University Park.

Hayes, E.M. 1989. Mid-crustal Mesozoic plutonism in the Cargo Muchacho Mountains, southeasternmost California. Geological Society of America Abstracts with Programs 21(5):92

Hays, W.H. 1957. Geology of the central Mecca Hills, Riverside County, California. Doctoral Dissertation, Yale University, New Haven, Connecticut.

----- 1958. Structure of the central Mecca Hills, Riverside County, California. Geological Society of America Special Paper 12(2):1687.

Hazen, G.E. 1954. Sandstone concretions of the Colorado Desert. Earth Science Digest 7(9):13-16.

Heath, E.G. 1980. Evidence of faulting along a projection of the San Andreas Fault, south of the Salton Sea. In Geology and Mineral Wealth of the California Desert, edited by D.L. Fife and A.R. Brown, South Coast Geological Society, Santa Ana, California p. 467-474.

Heaton, K.M. 1978. Geology of Grapevine Canyon to San Felipe Creek narrows, Anza-Borrego State Park, California. Undergraduate Research Reports, Department of Geology, San Diego State University, California 32(1):1-19.

----- 1981. K-Ar age patterns in the Cibbets Flat pluton, Mount Laguna, California. Master of Science Thesis. San Diego State University. 73 p.

Heaton, T.H., J.G. Anderson, and P.T. German 1983. Ground failure along the New River caused by the October 1979 Imperial Valley Earthquake sequence. Seismological Society of America Bulletin 73:1161-1171.

Heaton, T.H., and D.V. Helmberger (no date). A study of the ground strong motion of the Borrego Mountain, California, earthquake. California Institute of Technology 2769:1-18.

----- 1977. Predictability of strong ground motion in the Imperial Valley; modeling the M 4.9, November 4, 1976 Brawley Earthquake. EOS, Transactions of the American Geophysical Union 58:445.

----- 1978a. Predictability of strong ground motion in the Imperial Valley; modeling the M 4.9, November 4, 1976 Brawley Earthquake. California Institute of Technology, Division of Geological and Planetary Sciences Contribution 2929:1-18.

----- 1978b. Predictability of strong ground motion in the Imperial Valley; modeling the M 4.9, November 4, 1976 Brawley Earthquake. Seismological Society of America Bulletin 68:31-48.

Heck, N.H. 1940. The Imperial Valley Earthquake. Scientific Monthly 51:91-94.

Hedel, C.W. 1979. Index maps for large-scale vertical black and white aerial photographs of the Elsinore, San Jacinto, and Imperial Faults, California. U.S. Geological Survey Open-File Report 79-285:3 sheets.

Heffner, J.D. 1980a. Salton Sea 1° x 2° NTMS area, California and Arizona; hydrogeochemical and stream sediment reconnaissance. U.S. Department of Energy Report GJBX-217-80, Report DPST-80-164-2.

----- 1980b. Salton Sea 1° x 2° NTMS quadrangle area; supplemental data report; California and Arizona; hydrogeochemical and stream sediment reconnaissance. U.S. Department of Energy Report GJBX-113-80, DPST-80-164-25:1-100.

Heizler, M.T., Harrison, T.M., W.A. Elders, and C.T. Herzig 1988. Hydrothermal system provenance ages given by $^{40}Ar/^{39}Ar$ results from drill core sandstones from the Salton Sea system southern California. Geological Society of America Abstracts with Programs 20(7):97-98.

----- 1989. $^{40}Ar/^{39}Ar$ microcline thermochronology for the Salton Sea geothermal field, southern CA; aspects on heating durations and provenance ages. EOS, Transactions of the American Geophysical Union 70:721.

Helferty, M.G., and C.F. Erdman 1985. Principal facts for 113 gravity stations in the Salton Sea 1° by 2° quadrangle, southern California. U.S. Geological Survey Open-File Report 85-254:1-13.

Helgeson, H.C. 1967. Solution chemistry and metamorphism. In Researches in Geochemistry, edited by P.H. Ableson, John Wiley and Sons, New York p. 362-404.

Hely, A.G. 1969. The lower Colorado River water supply - its magnitude and distribution. U.S. Geological Survey Professional Paper 486D:1-54.

Hely, A.G., and G.H. Hughes, and B. Irelan 1966. Hydrologic regime of the Salton Sea, California. U.S. Geological Survey Professional Paper 486C:1-32.

Hely, A.G., and E.L. Peck 1964. Precipitation runoff and water loss in the lower Colorado River-Salton Sea area. U.S. Geological Survey Professional Paper 486-B:1-16.

Hempy, D.W. 1981. Fish Creek fluvial system and associated tectonically influenced morphology, Salton Trough, California. Master of Science Thesis, San Diego State University, California 86 p.

Henderson, W.H. 1919. Possibilities of oil in Imperial Valley. Oil Age 15:10-14, and 1921 Oil Age 17:18-22.

Henry, A.J. 1907. The possibilities of Salton Sea. Popular Science Monthly 70(1):5-18.

Henry, D.J. 1957. California gem trails. J.D. Simpson and Company, Spokane, Washington 101 p.

Henshaw, P.C. 1942. Geology and mineral deposits of the Cargo Muchacho Mountains, Imperial County, California. California Journal of Mines and Geology 38:147-196.

Henyey, T.L., and J.L. Bischoff 1973. Tectonic elements of the northern part of the Gulf of California. Geological Society of America Bulletin 84:315-330.

Herber, L. 1985. Road log and geology and geothermal energy of the Salton Trough. National Association of Geology Teachers Guidebook, Far Western Section 171 p.

Hertlein, L.G., and E.K. Jordan 1927. Paleontology of the Miocene of lower California. Proceedings of the California Academy of Sciences, 4th Series 16(19):607-611.

Herzig, C.T., and W.A. Elders 1988a. Nature and significance of igneous rocks cored in the State 2-14 research borehole; Salton Sea Scientific Drilling Project, California. Journal of Geophysical Research, B, Solid Earth and Planets 93:13,069-13,080.

----- 1988b. Probable occurrence of the Bishop Tuff in the Salton Sea Scientific Drilling Project borehole, Salton Sea geothermal system, California. Transactions of the Geothermal Resources Council 12:115-120.

Herzig, C.T., and J.M. Mehegan 1987. Quaternary depositional history of an active pull-apart basin,

Salton Trough, California. Geological Society of America Abstracts with Programs 19:701.

Herzig, C.T., J.M. Mehegan, and Sterling 1988. Lithostratigraphy of the State 2-14 borehole; Salton Sea Scientific Drilling Project. Journal of Geophysical Research, B, Solid Earth and Planets 93:12,969-12,980.

Herzig, C.T., and D.C. Jacobs 1991. Late Cenozoic basaltic magmatism in the Salton Trough. In The Diversity of Mineral and Energy Resources of Southern California, edited by M.A. McKibben. Society of Economic Geologists 12:104-116.

----- 1994. Cenozoic volcanism and two-stage extension in the Salton Trough, southern California and northern Baja California. Geology 22(11):991-994.

Higgins, C.T. 1990. Mesquite Mine a modern example of the quest for gold, Imperial County, California. California Geology 43(3):51-56, 67.

Higley, R.T. 1975. A groundwater hydrology study of the Cuyamaca Peak area, San Diego County, California. Undergraduate Research Reports, Department of Geology, San Diego State University, California 27:1-14.

Hileman, J.A. 1979. Seismicity of the San Diego region. In Earthquakes and Other Perils -- San Diego Region, edited by P.L. Abbott and W.J. Elliott, Geological Society of America Field Trip, San Diego Association of Geologists, California p. 11-20.

Hileman, J.A., C.R. Allen, J.N. Brune, and M. Wyss 1971. Measurement of active slip on faults in the Imperial Valley region, California. Geological Society of America Abstracts 3:136.

Hileman, J.A., and G.C. Laing 1981. Near-field particle motions, 1979 Imperial Valley Earthquake. EOS, Transactions of the American Geophysical Union 62:972.

Hilgard, W.E., F.J. Snow, and G.W. Shaw 1902. Lands of the Colorado delta in the Salton Basin. University of California, Berkeley, Agricultural Experiment Station Bulletin 140:1-51.

Hill, D.P. 1977. A model for earthquake swarms. Journal of Geophysical Research, B, Solid Earth and Planets 82:11,347-11,352.

----- 1986. Landsliding and faulting in Mason Valley eastern San Diego County, California. Undergraduate Research Reports, Department of Geology, San Diego State University, California 47:1-17.

Hill, D.P., F.G. Fischer, K.M. Lahr, and J.M. Coakley 1975. Earthquake sounds generated by body waves from local earthquakes. EOS, Transactions of the American Geophysical Union 56:1023.

Hill, D.P., P. Mowinckel, and K.M. Lahr 1975. Catalog of earthquakes in the Imperial Valley, California, June 1973-May 1974. U.S. Geological Survey Open-File Report 75-401:1-30.

Hill, D.P., P. Mowinckle, and L.G. Peake 1974. Earthquakes, active faults, and geothermal areas in the Imperial Valley, California. Science 188:1306-1308.

Hill, D.P., L.G. Peake, P. Mowinckle, and J.A. Hileman 1974. Seismicity of the Imperial Valley, California, 1973. EOS, Transactions of the American Geophysical Union 55:346.

Hill, J.S. 1912. The mining districts of the western United States, with a geologic introduction by Waldemar Lindgren. U.S. Geological Survey Bulletin 507:130-131.

Hill, R.I. 1988. San Jacinto intrusive complex 1. Geology and mineral chemistry, and a model for intermittent recharge of tonalitic magma chambers. Journal of Geophysical Research 93(B9):10325-10348.

Hill, R.I., B.W. Chappell, L.T. Silver 1988. San Jacinto intrusive complex 2. Geochemistry. Journal of Geophysical Research 93(B9):10349-10372.

Hill, R.I., and L.T. Silver 1988. San Jacinto intrusive complex 3. Constraints on crustal magma chamber processes from strontium isotope heterogeneity. Journal of Geophysical Research 93(B9):10373-10388.

Hill, R.I., L.T. Silver, and H.P. Taylor 1986. Coupled Sr-O isotope variations as an indicator of source heterogeneity for the northern Peninsular Ranges Batholith. Contributions to Mineralogy and Petrology 92:351-361.

Hillemeyer, F.L. 1984. Geometric and kinematic analysis of high- and low-angle normal faults with the Mojave-Sonoran detachment terrain. Master's Thesis, San Diego State University, California 209 p.

Hillemeyer, F.L., M.D. Johnson, R.R. Kern 1991. Geology, alteration and mineralization of the Modoc Hot Springs gold prospect, Imperial County, California. In The Diversity of Mineral and Energy Resources of Southern California, edited by M.A. McKibben. Society of Economic Geologists 12:139-155.

Hinds, N.E.A. 1952. Evolution of the California landscape. California Division of Mines Bulletin 158:1-240.

Hinricks, T. and H.W. Falk, Jr. 1977. The east mesa "Magamax Process" power generation plant. Transactions of the Geothermal Resources Council 1:141-142.

Hoagland, J.R. 1976a. Petrology and geochemistry of hydrothermal alteration in borehole 6-2, East Mesa geothermal area, Imperial Valley, California. Master's Thesis, University of California, Riverside.

----- 1976b. Petrology and geochemistry of hydrothermal alteration in borehole 6-2, East Mesa geothermal area, Imperial Valley, California. Institute of Geophysics and Planetary Physics Report 76(12):1-90.

----- 1977. Chemical change in thermal fluids produced from East Mesa geothermal system. Geological Society of America Abstracts with Programs 9:1016-1017.

Hodgson, S.F. 1989. The Algodones Dune field, east of East Mesa. Geothermal Hotline 1-2:49-51.

Hoggatt, W.C. 1979. Geologic map of Sweeney Pass quadrangle, San Diego County, California. U.S. Geological Survey Open-File Report 79-754:1-34.

Hoggatt, W.C., and V.R. Todd 1976. Potassium-argon dating of metamorphic and plutonic rocks from Cuyamaca Peak and Mt. Laguna 15-minute quadrangles, San Diego County, California. Geological Society of America Abstracts with Programs 8:382-382.

Holder, C.F. 1901a. A remarkable salt deposit, Salton, California. Scientific American 84:217.

----- 1901b. Salton, California. National Geographic 12:391.

Ho-Lui, P.H.-Y. 1988. I. Attenuation tomography; II. Modeling regional Love waves; Imperial Valley to Pasadena. Doctoral Dissertation, California Institute of Technology, Pasadena 166 p.

Ho-Liu, P.H.-Y., and D.V. Helmberger 1989. Modeling regional Love waves; Imperial Valley to Pasadena. Seismological Society of America Bulletin 79:1194-1209.

Ho-Liu, P.H.-Y., H. Kanamori, and R.W. Clayton 1988. Applications of attenuation tomography to Imperial Valley and Coso-Indian Wells region, southern California. Journal of Geophysical Research, B, Solid Earth and Planets 93:10,521-10,540.

Holmes, J.G., and party 1903. Soil survey of the Imperial area, California. U.S. Department of Agriculture, Bureau of Soils Report 5:1219-1248.

----- 1904. Soil survey of the Indio area, California. U.S. Department of Agriculture, Bureau of Soils Report 5:1249-1262, maps 75-76.

Holzer, T.L., T.L. Youd, and T.C. Hanks 1989. Dynamics of liquefaction during the 1987 Superstition Hills, California, earthquake. Science 244:56-59.

Hoover, R.A. 1965. Areal geology and physical stratigraphy of a portion of the southern Santa Rosa Mountains, San Diego County, California. Master of Science Thesis, Department of Geological Sciences, University of California, Riverside 81 p.

Hopkins, V. 1985. The Colorado River. Chartwell Books, Incorporated, New Jersey 176 p.

Housner, G.W. 1966. Intensity of earthquake ground shaking near the causative fault. New Zealand Institute of Engineers Proceedings 1:III-94-III-115.

Howard, A.D., and J. Mercado 1970. Low-sun-angle vertical photography versus thermal infrared scanning imagery. Geological Society of America Bulletin 81:521-523.

Howard, H.H. 1950. Wonder bird of the Ice Age. Los Angeles County Museum Leaflet Series, Science 3:1-3.

----- 1963. Fossil birds from the Anza-Borrego Desert. Los Angeles County Museum Contributions in Science 73:1-33.

----- 1972. The incredible teratorn again. The Condor 74(3):341-344.

Howard, J.H., J.A. Apps, S.M. Benson, N.E. Goldstein, N.A. Grand, J.P. Hamey, D.D. Jackson, S. Juprasert, E.L. Majer, D.G. McEdwards, T.V. McEvilly, T.N. Narasimhan, B. Schechter, R.C. Schroeder, R.W. Taylor, F.C. van de Kamp, and T.J. Wolery 1978. Geothermal resources and reservoir investigations of the U.S. Bureau of Reclamation leaseholds at East Mesa, Imperial Valley, California. Lawrence Laboratory, Berkeley, California, Report 7094:1-305.

Howell, N. 1964. Geology of Quartz Peak quadrangle SE, Imperial County, California. Undergraduate Research Reports, Department of Geology, San Diego State University, California 2:8.

Huang, M.J., T.Q. Cao, C.E. Ventura, D.L. Parke, and A.F. Shakal 1987. CSMIP strong motion records from the Superstition Hills, Imperial County, California earthquakes of 23 and 24 November 1987. California Division of Mines and Geology Earthquake Data Reports SMS 87-06:1-42.

Huang, M.J., M.A. Haroun, and G.W. Housner 1982. A study on strong motion data of the 1979 Imperial Valley Earthquake. In Third International Earthquake Microzonation Conference Proceedings, edited by M. Sherif, National Science Foundation, Washington, D.C. 1:401.

Hubbs, C.L., and G.S. Bien 1967. La Jolla natural radiocarbon measurements V. Radiocarbon 9:261-294.

Hubbs, C.L., G.S. Bien, and H.E. Suess 1960. La Jolla natural radiocarbon measurements. American Journal of Science, Radiocarbon Supplement 4:197-223.

----- 1963. La Jolla natural radiocarbon measurements III. Radiocarbon 5:254-272.

----- 1965. La Jolla natural radiocarbon measurements IV. Radiocarbon 7:99-117.

Hubbs, C.L., and R.R. Miller 1948. The Great Basin. Part II, the zoological evidence. University of Utah Bulletin 38:103-113.

----- 1965. La Jolla radiocarbon measurements IV. Radiocarbon 7:66-117.

Huberty, M.R., A.F. Pilsbury, and V.P. Sokoloff 1948. Hydrologic studies in Coachella Valley, California. University of California, College of Agriculture, Berkeley 31 p.

Hudnut, K.W. 1989. Active tectonics of the Salton Trough, southern California. Doctoral Dissertation, Columbia University, New York.

Hudnut, K.W., and J. Beavan 1989. Vertical deformation (1952-1987) in the Salton Trough, California from water level recordings. Journal of Geophysical Research, B, Solid Earth and Planets 94(7):9463-9467.

Hudnut, K.W., J. Beavan, and R. Bilham 1985a. Salton Sea level data; active transpression on the southern San Andreas Fault. EOS, Transactions of the American Geophysical Union 66:383.

----- 1985b. Water level recording for the study of crustal deformation; Salton Sea, California. EOS, Transactions of the American Geophysical Union 66:856.

Hudnut, K.W., and M.M. Clark 1989. New slip along parts of the 1968 Coyote Creek Fault rupture, California. Seismological Society of America Bulletin 79(2):451-465.

Hudnut, K.W., and L. Seeber 1986. Astroazimuth geodetic measurements of block roatation in the southern San Jacinto Fault Zone, California. EOS, Transactions of the American Geophysical Union 68:287.

Hudnut, K.W., L. Seeber, and T. Rockwell 1989. Slip on the Elmore Ranch Fault during the past 300 years and its relation to slip on the Superstition Hills Fault. Seismological Society of America Bulletin 79(2):330-341.

Hudnut, K.W., L. Seeber, T. Rockwell, J. Goodmacher, R. Klinger, S. Lindvall, and R. McElwain 1989. Surface ruptures on cross-faults in the 24 November 1987 Superstition Hills, California earthquake sequence. Seismological Society of America Bulletin 79(2):282-296.

Hudnut, K.W., L. Seeber, and J. Pacheo 1989. Cross-fault triggering in the November 1987 Superstition Hills Earthquake sequence, southern California. Geophysical Research Letters 16(2):199-202.

Hudnut, K.W., L. Seeber, J. Pacheo, J. Armbruster, L. Sykes, G. Bond, and M. Kominz 1989. Cross faults and block rotation in southern California, earthquake triggering and strain distribution. Lamont-Doherty Geological Observatory Yearbook p. 44-48.

Hudnut, K.W., K.E. Sieh 1989. Behavior of the Superstition Hills Fault during the past 330 years. Seismological Society of America Bulletin 79:304-329.

Hudnut, K.W., and P.L. Williams 1987. Evidence for large pre-historic earthquakes on the Superstition Hills Fault, southern California. Geological Society of America Abstracts with Programs 19(7):710.

Hudnut, K.W., P.L. Williams, K. Sieh, and L. Seeber 1987. Geomorphic evidence for a pre-historic earthquake on the Superstition Mountain Fault, Southern California. EOS, Transactions of the American Geophysical Union 68:1369.

Hudson, F.S. 1920. Geology of the Cuyamaca region of California. Doctoral Dissertation, University of California, Berkeley 78 p.

----- 1922. Geology of the Cuyamaca region of California with special reference to the origin of the nickeliferous pyrrhotite. University of California Publications in Geological Science 13:175-252.

Hughes, G.H. 1967. Analysis of techniques used to measure evaporation from Salton Sea, California. U.S. Geological Survey Professional Paper 272-H:151-176.

Hughes, K.M. 1989. The Bear Canyon Conglomerate as a record of tectonics and sedimentation during initiation of the Salton Trough. Master's Thesis, San Jose State University, California.

Hughes, K.M., and E.G. Frost 1989. The Bear Canyon Fanglomerate as a record of the sedimentary and tectonic history of the southeastern margin of the Salton Trough and the southern Chocolate Mountains of southeasternmost California. Geological Society of America Abstracts with Programs 21(5):96.

Hughes, R.E. 1986. Trace element composition of Obsidian Butte, Imperial County, California. Southern California Academy of Sciences Bulletin 85:35-45.

Hull, A.G. 1991. Pull-apart basin evolution along the northern Elsinore Fault Zone, southern California. Geological Society of America Abstracts with Programs p. A257.

Hull, C.D., and W.A. Elders 1982. Clinopyroxene compositions as indicators of magma types in the Salton Sea magma-hydrothermal system of southern California. EOS, Transactions of the American Geophysical Union 63:1150

----- 1983. Possible affinities of basaltic xenoliths in the Salton Buttes sodic rhyolites, southern California. Geological Society of America Abstracts with Programs 15:421.

Humphreys, G. 1978a. Telluric mapping in the vicinity of the Salton geothermal area. EOS, Transactions of the American Geophysical Union 59:1201.

Humphreys, G. 1978b. Telluric sounding and mapping in the vicinity of the Salton Sea geothermal area, Imperial Valley, California. Master's Thesis, University of California, Riverside.

Hutchings, L.J., S. Jarpe, P. Kasameyer, and T. Hauk 1988. Microseismicity of the Salton Sea geothermal field.

EOS, Transactions of the American Geophysical Union 69:1312.

Hutton, L.K., C.E. Johnson, J.C. Pechman, J.E. Ebel, D.W. Given, P.T. German 1980. Epicentral locations for the Homestead Valley Earthquake sequence, March 15, 1979. California Geology 33(5):110-114.

Hutton, L.K., and C.E. Johnson 1981. Preliminary study of the Westmorland, California, earthquake swarm. EOS, Transactions of the American Geophysical Union 62:957.

Hutton, L.K., Allen, C.R., and C.E. Johnson 1985. Seismicity of southern California; earthquakes of ML 3.0 and greater, 1975 through 1983. Division of Geological and Planetary Sciences, Seismological Laboratory, California Institute of Technology, Pasadena 142p.

Hutton, L.K., L.M. Jones, E. Hauksson, and D.D. Given 1991. Seismotectonics of southern California. In Neotectonics of North America, edited by D.B. Slemmons, E.R. Engdahl, M.D. Zoback and D.D. Blackwell, Geological Society of America, the Geology of North America Decade Map 1:133-152.

Hwang, L.J., H.W. Magistrale, and H. Kanamori 1990. Teleseismic source parameters and rupture characteristics of the 24 November 1987, Superstition Hills Earthquake. Seismological Society of America Bulletin 80:43-56.

Iacopi, R. 1964. Earthquake country. Lane Book Company (Sunset Books), Menlo Park, California 192 p.

Ingle, J.C. 1974. Paleobathymetric history of Neogene marine sediments, northern Gulf of California. In Geology of Peninsular California. American Association of Petroleum Geologists, Pacific Section, Guidebook for Field Trips p. 121-138.

----- 1984. Neogene marine stratigraphy and history of the Gulf of California. Society of Economic Paleontologists and Mineralogists, Pacific Section 59:76-77.

Ingram, S.H. 1974. I was there. California Geology 27(3):57-63.

Irelan, B. 1971. Salinity of surface water in the Lower Colorado River - Salton Sea Area. U.S. Geological Survey Professional Paper 486-E:1-40.

Isaac, S. 1982. Pearluminous granitoid rocks in the Tule Canyon area of the In-Ko-Pah Mountains, southern California. Undergraduate Research Reports, Department of Geology, San Diego State University, California 41(2):1-25.

----- 1987. Geology and structure of the Yuha Desert between Ocotillo, California, U.S.A., and Laguna Salada, Baja California, Mexico. Master of Science Thesis, San Diego State University, California 165 p.

Isaac, S., T.K. Rockwell, and G. Gastil 1986. Plio-Pleistocene detachment faulting, Yuha Desert region, western Salton Trough, northern Baja California. Geological Society of America Abstracts with Programs 18(2):120.

Ives, R.L. 1951. Mud volcanoes of the Salton depression [California and Mexico]. Rocks and Minerals 26:227-235.

Iyer, H.M. 1972. Seismic noise in geothermal areas. Society of Exploration Geophysicists, Annual International Meeting 42:38.

----- 1973. Seismic noise in geothermal areas. Geophysics 38:177-178.

----- 1975. Search for geothermal seismic noise in the East Mesa area, Imperial Valley, California. Geophysics 40:1066-1072.

Izett, G.A. 1981. Volcanic ash beds: recorders of upper Cenozoic silicic pyroclastics volcanism in the western United States. Journal of Geophysical Research 86(B11):10200-10222.

Izett, G.A., and C.W. Naeser 1976. Age of the Bishop Tuff of eastern California as determined by the fission-track method. Geology 4(10):587-590.

Izett, G.A., J.D. Obradovich, and H.H. Mehnert 1988. The Bishop ash bed (middle Pleistocene) and some older (Pliocene and Pleistocene) chemically and mineralogically similar ash beds in California, Nevada, and Utah. U.S. Geological Survey Bulletin 1675:1-37.

Jachens, R.C. 1991. Geophysical studies in the southern California areal mapping project (SCAMP). Geological Society of America Abstracts with Programs, San Diego p. A476.

Jachens, R.C., R.W. Simpson, A. Griscom, and J. Mariano 1986. Plutonic belts in southern California defined by gravity and magnetic anomalies. Geological Society of America Abstracts with Programs 18:120.

Jackson, D., A.J. Piwinskii, and D. Miller 1976. Computer modelling of geothermal brines from the Salton Sea geothermal field, California. EOS, Transactions of the American Geophysical Union 57:1016.

Jackson, G., and G. Calderon 1988. Late Quaternary slip rates on the Clark Fault: inferences from offset fluvial landforms, SW 1/4 Font's Point quadrangle, California. Field Studies in Geomorphology, University of Arizona, Tucson, Manuscript on File Anza-Borrego Desert State Park, California 24 p.

Jackson, R.T. 1917. Fossil echini of the Panama Canal Zone and Costa Rica. Proceedings of the U.S. National Museum 53:489-501.

Jacobson, G.R. 1980. Geology adjacent to a portion of the Superstition Mountain Fault, southwestern Imperial Valley, California. Undergraduate Research Reports, Geology Department, San Diego State

University, California 37:18.

Jacobson, C.E., M.R. Dawson, and C.E. Postlethwaite 1987. Evidence for late-stage normal slip on the Orocopia Thrust and implications for the Vincent Chocolate Mountains thrust problem. Geological Society of America Abstracts with Programs 19:714.

----- 1988. Structure, metamorphism, and tectonic significance of the Pelona, Orocopia, and Rand Schists, southern California. In Metamorphism and Crustal Evolution of the Western United States, edited by W.G. Ernst, Prentice Hall, New Jersey, Ruby Volume 7:976-997.

Jacobson, C.E. 1990. The $^{40}Ar/^{39}Ar$ geochronology of the Pelona Schist and related rocks, Southern California. Journal of Geophysical Research, B, Solid Earth and Planets 95:509-528.

Jaeger, E.C. 1955. The California deserts. Stanford University Press, California 308 p.

Jahns, R.H. 1954a. Geology of the Peninsular Range province. In Geology of Southern California, edited by R.H. Jahns, California Division of Mines Bulletin 170(2):29-52.

----- 1954b. Pegmatites of southern California. In Geology of Southern California, edited by R.H. Jahns, California Division of Mines Bulletin 170(7):37-50.

----- 1969. Long-term behavior of some faults in southern California. EOS, Transactions of the American Geophysical Union 50:382-384.

----- 1979. Gem-bearing pegmatites in San Diego County, California: the Stewart Mine, Pala district, and the Himalaya Mine, Mesa Grande district. In Mesozoic Crystalline Rocks: Peninsular Ranges Batholith and Pegmatites, Point Sal Ophiolite, edited by P.L. Abbott and V.R. Todd, Department of Geological Sciences, San Diego State University, California p. 3-38.

James, G.W. 1906. The wonders of the Colorado Desert: Its rivers and its mountains, its canyons and its springs, its life and its history. Little, Brown and Company, Boston, Massachusetts 547 p.

Jansen, L. 1983. Ground water conditions near the Elsinore Fault Zone in the Ocotillo-Coyote Wells basin, Imperial County, California. Master of Science Thesis, San Diego State University, California 143 p.

Jansen, E.J. 1978. A paleomagnetic investigation of the Jacumba basalts. Undergraduate Research Reports, Department of Geology, San Diego State University, California 32(1):1-43.

Jarpe, S.P., P.W. Kasameyer, C. Johnston 1989. Passive seismic monitoring of a flow test in the Salton Sea geothermal field. Transactions of the Geothermal Resources Council 13:265-270.

Jarzabek, D.P. 1980. A geochemical reconnaissance of thermal waters along portions of the San Jacinto and San Andreas Fault Zones, Southern California. Master of Science Thesis, San Diego State University, California 148 p.

Jarzabek, D.P., and J. Combs 1976. Microearthquake survey of the Dunes KGRA, Imperial Valley, southern California. Geological Society of America Abstracts with Programs 8:939.

Jefferson, G.T. 1962. Geology of the Painted Hill area Whitewater, California. Geology 118, Department of Geological Sciences, University of California, Riverside 15 p.

----- 1966. The paleoecology of a "borrow horizon" in the base of the Imperial Formation at Whitewater, California. Geology 260, Department of Geological Sciences, University of California, Riverside 13 p.

----- 1989. Late Cenozoic tapirs (Mammalia: Perissodactyla) of western North America. Natural History Museum of Los Angeles County Contributions in Science 406:1-21.

----- 1991a. A catalogue of late Quaternary vertebrates from California: part one, nonmarine lower vertebrate and avian taxa. Natural History Museum of Los Angeles County Technical Reports 5:1-59.

----- 1991b. A catalogue of late Quaternary vertebrates from California: part two, mammals. Natural History Museum of Los Angeles County Technical Reports 7:1-129.

Jefferson, G.T., and P. Remeika 1994. The Mid-Pleistocene stratigraphic co-occurrence of *Mammuthus columbi* and *M. imperator* in the Ocotillo Formation, Borrego Badlands, Anza-Borrego Desert State Park, California. Current Research in the Pleistocene 11:89-92.

Jefferson, G.T., and A. Tejada-Flores 1995. Late Blancan *Acinonyx* (Carnivora, Felidae) from the Vallecito Creek Local Flora of Anza-Borrego Desert State Park, California. Abstracts of Proceedings. 9th Annual Mojave Desert Quaternary Research Symposium, San Bernardino County Museum Quarterly 42(2):33.

Jennings, C.W. 1967. Geologic map of California, Salton Sea sheet. California Division of Mines and Geology, Scale 1:250,000.

----- 1982. Salton Sea sheet, Bouguer gravity map of California. California Division of Mines and Geology, Gravity Survey Map 1.

Jennings, S. 1983. Burial metamorphism in Plio-Pleistocene sediments of the Colorado River delta. Master's Thesis, University of Montana, Missoula 40 p.

Jensen, G. 1988. Geoelectric modeling near the Salton Sea Geothermal Field, California. Master of Science Thesis, San Diego State University, California 187 p.

Jensen, G., G.R. Jiracek, K.M. Johnson, M. Martinez, and J.M. Romo 1990. Magnetotelluric modeling in the vicinity of the Salton Sea Scientific Drilling Project. Society of Economic Geologists 60:521-515.

John, B.E., and K.A. Howard 1985. Brittle extension of the continental crust along a rooted system of low-angle normal faults; Colorado River extensional corridor. Conference on heat and detachment in crustal extension on continents and planets, LPI Contributions 575:58.

John, B.E., and J. McCarthy 1987. Geologic and geophysical evidence for the nature of upper and mid-crustal continental extension in the Colorado River area, California and Arizona. Geological Society of America Abstracts with Programs 19:717.

Johnson, C., and D.G. Miller 1980. Late Cenozoic alluvial history of the lower Colorado River. In Geology and Mineral Wealth of the California Desert, edited by D.L. Fife and A.R. Brown, South Coast Geological Society, Santa Ana, California p. 441-446.

Johnson, C.E. 1977. Swarm tectonics of the Imperial and Brawley Faults of southern California. EOS, Transactions of the American Geophysical Union 58:1188.

----- 1978. A deterministic model for earthquake swarm sequences in the Imperial Valley, California. EOS, Transactions of the American Geophysical Union 59:1205.

----- 1979. CEDAR - An approach to the computer automation of short period local seismic networks; seismotectonics of the Imperial Valley of southern California. Doctoral Dissertation, California Institute of Technology, Pasadena 343 p.

Johnson, C.E., and D.M. Hadley 1975. Tectonic implications of the Brawley Earthquake swarm, Imperial Valley, California, January 1975. EOS, Transactions of the American Geophysical Union 56:1022.

----- 1976. Tectonic implications of the Brawley Earthquake swarm, Imperial Valley, California, January 1975. Seismological Society of America Bulletin 66:1133-1144.

Johnson, C.E., and D.P. Hill 1982. Seismicity of the Imperial Valley. In The Imperial Valley, California, Earthquake of October 15, 1979. U.S. Geological Survey Professional Paper 1254:15-24.

Johnson, C.E., and L.K. Hutton 1981. Imperial Valley seismicity and the San Andreas Fault. EOS, Transactions of the American Geophysical Union 62:957.

----- 1982. Aftershocks and pre-earthquake seismicity. In The Imperial Valley, California, earthquake of October 15, 1979. U.S. Geological Survey Professional Paper 1254:59-76.

----- 1986. A tectonic model for the Imperial Valley and its relation to seismic risk on the southern San Andreas Fault. EOS, Transactions of the American Geophysical union 67:1200.

----- 1988. Tectonic implications of the November 24, 1987, Superstition Hills Earthquakes, Imperial Valley, California. Seismological Research Letters 59:48.

Johnson, C.E., C. Rojahn, and R.V. Sharp 1982. The Imperial Valley, California, earthquake of October 15, 1979; introduction. In The Imperial Valley, California, Earthquake of October 15, 1979. U.S. Geological Survey Professional Paper 1254:1-3.

----- 1985. The Imperial Valley, California, earthquake of October 15, 1979. In Strong-motion Program Report, January-December, 1982, edited by R.L. Porcella, U.S. Geological Survey Circular 0965:1-2.

Johnson, D.A., and T.C. Hanks 1976. Strong-motion earthquake accelerograms at Brawley, California; January 25, 1975. Seismological Society of America Bulletin 66:1155-1158.

Johnson, N.M., C.B. Officer, N.D. Opdyke, G.D. Woodard, P.K. Zeitler, and E.H. Lindsay 1983a. Rates of late Cenozoic tectonism in the Vallecito-Fish Creek Basin, western Imperial County, California. Geology 11:664-667.

----- 1983b. Rates of tectonic displacement and rotation of the western Imperial Valley, California. Geological Society of America Abstracts with Programs 15:605.

Johnson, N.M., N.D. Opdyke, and E. Lindsay 1975. Magnetic polarity stratigraphy of Plio-Pleistocene terrestrial deposits and vertebrate fauna, San Pedro Valley, Arizona. Geological Society of America Bulletin 86:5-11.

Johnston, I.M. 1954. Geology special studies report age determination by the use of foraminifera. Undergraduate Research Reports, Geology Department, San Diego State University, California 2:1-12.

Jones, E., Jr. 1919. Deposits of manganese in southeastern California. U.S. Geological Survey Bulletin 710-E:185-189, 200-201.

Jones, J.C. 1913. Origin of travertine tufa deposits of Salton Sea Sink. Carnegie Institute of Washington Yearbook 12:60-61.

----- 1914. The tufa deposits of the Salton Sink. Carnegie Institute of Washington Publication 193:79-83.

Jones, L.M. 1988. Focal mechanisms and the state of stress on the San Andreas Fault in southern California. Journal of Geophysical Research, B, Solid Earth and Planets 93:8869-8891.

Jones, L.M., L.K. Hutton, D.D. Given, and C.R. Allen 1986. The North Palm Springs, California, earthquake sequence of July 1986. Seismological Society of America Bulletin 76:1830-1837.

Jones, M.L. 1972. The Panamic biota: some observations prior to a sea-level canal. Bulletin of the Biological Society of Washington 2:1-270.

Jorgensen, M., C.J. Natenstedt, and P.N. Trumbly 1982. Possible relationship between Miocene crustal extension/detachment faulting and the deposition of the Tolbard Fanglomerate in the Mid Way and western Palo Verde Mountains, Imperial County, California. In Mesozoic-Cenozoic Tectonic Evolution of the Colorado River Region, California, Arizona, and Nevada, edited by E.G. Frost and D.L. Martin, Cordilleran Publishers, San Diego, California p. 313-315.

Jorgensen, M.C. 1991. Paleontology resource management plan; draft. California Department of Parks and Recreation, Colorado Desert District, Anza-Borrego Desert State Park, California p. 1-21.

Jowett, E.C., R.L. Sherlock, and S. Losh 1990. Gold precipitation by fluid oxidation during extension; the detachment-related Picacho deposit, California. Geological Association of Canada, Mineralogical Association of Canada, Canadian Geophysical Union Program with Abstracts 15:67.

Joyner, W.B., and D.M. Boore 1981. Peak horizontal acceleration and velocity from strong-motion records including records from the 1979 Imperial Valley, California, earthquake. U.S. Geological Survey Open-File Report 81-0365:-147.

----- 1982. Peak horizontal acceleration and velocity from strong-motion records including records from the 1979 Imperial Valley, California, earthquake. Seismological Society of America Bulletin 71:2011-2038.

Joyner, W.B., D.M. Moore, and R.L. Purcella 1981. Peak Horizontal acceleration and velocity from strong motion records including records from the 1979 Imperial Valley, California, earthquake. U.S. Geological Survey Open-File Report 81-0365:1-47.

Julian, B.R., M. Zirbes, and R. Needham 1982. The focal mechanism from the Global Digital Seismograph Network. In The Imperial Valley, California, earthquake of October 15, 1979. U.S. Geological Survey Professional Paper 1254:77-82.

Kahle, J.E., et al. 1984. Preliminary geologic map of the California-Baja border region, Imperial and San Diego Counties, California. California Division of Mines and Geology Open-File Report DMG OFR-84-59.

Kahle, J.E., C.J. Wills, E.W. Hart, J.A. Treiman, R.B. Greenwood, and R.S. Kaumeyer 1988. Surface rupture, Superstition Hills Earthquake of November 23 and 24, 1987, Imperial County, California. California Geology 42:75-84.

Kam, M.N.S. 1980. Determination of Curie isotherm from aeromagnetic data in the Imperial Valley. Master's Thesis, University of California, Riverside.

Kamp, R.L., Earley, P.J., Sturz, A. 1991. Clay mineralogy of sediments and chemical composition of interstitial waters from mud volcanoes, Salton Sea geothermal area, Imperial County, California. Geological Society of America Abstracts with Programs p. A155.

Karfunkel, B.S. 1982. Geology and mineral resources of the Los Angeles, Needles, Salton Sea, San Bernardino, and Trona 1° x 2° NTMS quadrangles. compiler, U.S. Department of Energy, Grand Junction, Colorado 186 p.

Karig, D.E., and W. Jensky 1972. The proto-Gulf of California. Earth and Planetary Science Letters 17:169-174.

Karpinski, J.C. 1971. Geology of the southwest corner of the Seventeen Palms quadrangle, San Diego County, California. Undergraduate Research Reports, Department of Geology, San Diego State University, California 18:1-22.

Kasameyer, P.W., and J.R. Hearst 1988. Borehole gravity measurements in the Salton Sea Scientific Drilling Project. Journal of Geophysical Research, B, Solid Earth and Planets 93:13,037-13,045.

Kasameyer, P.W., and L.W. Younker 1978. Natural fluid flow patterns in the Salton Sea geothermal field. Transactions of the Geothermal Resources Council 2(1):359-361.

Kasameyer, P.W., L.W. Kounker, and M.M. Hanson 1984. Development and application of a hydrothermal model for the Salton Sea geothermal field, California. Geological Society of America Bulletin 95:1242-1252.

Kassoy, D.R., and K.P. Goyal 1979. Modeling heat and mass transfer at the Mesa geothermal anomaly, Imperial Valley, California. Lawrence Berkeley Laboratory, California, Report 8784:1-171.

Katz, L.J., W.D. Wagner, and H.M. Iyer 1976. Search for geothermal seismic noise in the East Mesa area, Imperial Valley, California. Geophysics 41:45-60.

Kawane, Y. 1984. Geoscientific outline of Imperial Valley. Chinetsu, Journal of the Japan Geothermal Energy Association 21(5[85]):87-90.

Kazmerski, K. 1974. Geology of the Tule Mountain area, Jacumba, California. Undergraduate Research Reports, Department of Geology, San Diego State University, California 24(1):1-12.

Keith, T.E.C., L.J.P. Muffler, and M. Cremer 1968. Hydrothermal epidote formed in the Salton Sea geothermal system, California. American Mineralogist 53:1635-1644.

Keller, B. 1979a. Imperial Valley Earthquake Swarms. In Tectonics of the Juncture Between the San Andreas Fault System and the Salton Trough, Southeastern California, edited by J.C. Crowell and A.G. Sylvester, Department of Geological Sciences, University of California Santa Barbara p. 53-56.

----- 1979b. Structure of the Salton Trough from Gravity and Seismic Refraction Data. In Tectonics of the Juncture Between the San Andreas Fault system and the Salton Trough, Southeastern California, edited by J.C. Crowell and A.G. Sylvester, Department of Geological Sciences, University of California, Santa Barbara p. 57-64.

Kelley, R.E. 1976. Atmospheric dispersion and noise propagation at Imperial Valley geothermal fields. University of California, Riverside 52053:1-29.

----- 1977. Predicting the effect of geothermal emissions on the temperature and relative humidity in the Imperial Valley of California. Geothermal Energy 5:18-22.

Kelley, V.C. 1936. Occurrence of claudetite in Imperial County, California. American Mineralogist 21:137-138.

Kelley, V.C., and J.L. Soske 1936. Origin of the Salton volcanic domes, Salton Sea, California. Journal of Geology 44:496-509.

Kelm, D.L. 1971. A gravity and magnetic study of the Laguna Salada area, Baja California, Mexico. Master of Science Thesis, San Diego State University, California, 103 p.

Kendall, C. 1976a. Isotope exchange in well cuttings from Magmamax number 2, Magmamax number 3, and Woolsey number 1, Salton Sea geothermal field, California. Geological Society of America Abstracts with Programs 8:952.

----- 1976b. Petrology and stable isotope geochemistry of three wells in the Buttes area of the Salton Sea geothermal field, Imperial Valley, California, U.S.A. Master's Thesis, University of California, Riverside.

Kendrick, K.L., L.D. McFadden, and D.M. Morton 1993. Soil development in the northern part of the San Timoteo Badlands, San Bernardino and Riverside counties, California. In Ashes, Faults and Basins, edited by R.E. Reynolds and J. Reynolds, San Bernardino County Museum Association Special Publication 93-1:106-107.

Kennedy, J.S. 1981. Restivity and seismic investigations in Chihuahua Valley, San Diego County, California. Undergraduate Research Reports, Department of Geology, San Diego State University, California 39(3):1-57.

Kennedy, M.P. 1973. Bedrock lithologies San Diego area, California. In Studies on the Geology and Geologic Hazards of the Greater San Diego Area, California, edited by A. Ross and R.J. Dowlen, San Diego Association of Geologists, Guidebook p. 1-15.

----- 1977. Recency and character of faulting along the Elsinore Fault Zone in southern Riverside County, California. California Division of Mines and Geology, Special Report 131:1-12.

Kennedy, M.P., and G.W. Moore 1971. Stratigraphic relations of upper Cretaceous and Eocene formations, San Diego coastal area, California. American Association of Petroleum Geologists Bulletin 55:709-722.

Kerr, D.R. 1982a. Early Neogene continental sedimentation, western Salton Trough, California. Master of Science Thesis, San Diego State University, California 138 p.

----- 1982b. Early Neogene continental sedimentation, western Salton Trough, California. Geological Society of America Abstracts with Programs 14:177.

----- 1984. Early Neogene continental sedimentation in the Vallecito and Fish Creek Mountains, western Salton Trough, California. Sedimentary Geology 38:217-246.

Kerr, D.R., and S.M. Kidwell 1991. Late Cenozoic sedimentation and tectonics, western Salton Trough, California. In Geological Excursions in Southern California and Mexico, edited by M.J. Walawender and B.B. Hanan, Geological Society of America Annual Meeting Guidebook, San Diego p. 379-416.

Kerr, D.R., and S. Pappajohn 1982. Sedimentologic, stratigraphic, and tectonic significance of Neogene sedimentary megabreccias, western Salton Trough, California. American Association of American Petroleum Geologists Bulletin 66:1692.

Kerr, D.R., S. Pappajohn, and P.J. Bell 1979. Neogene continental rifting and sedimentation in the western Salton Trough, California. Geological Society of America Abstracts with Programs 11:457.

Kerr, D.R., S. Pappajohn, and G.L. Peterson 1979. Neogene stratigraphic section at Split Mountain, eastern San Diego County, California. In Tectonics of the Juncture Between the San Andreas Fault System and the Salton Trough, Southeastern California, edited by J.C. Crowell and A.G. Sylvester, Geological Society of America Annual Meeting Guidebook p. 111-124.

Kerstitch, A. 1989. Sea of Cortez marine invertebrates. Sea Challenges, Monterey, California 114 p.

Kew, W.S.W. 1914. Tertiary echnoids of the Carrizo Creek region in the Colorado Desert. California University Publications, Bulletin of the Department of Geology 8:39-60.

----- 1920. Cretaceous and Cenozoic echinoidea of the

Pacific coast of North America. University of California Publication of the Geology Department 12(2):32-137.

Kidwell, S.M. 1988. Taphonomic comparison of passive and active continental margins: Neogene shell beds of the Atlantic coastal plain and northern Gulf of California. Palaeogeography, Palaeoclimatology, Palaeoecology 63:201-223.

Kidwell, S.M., C.D. Winker, and E. D. Gyllenhaal 1988. Transgressive stratigraphy of a marine rift basin: Neogene Imperial Formation, northern Gulf of California. Geological Society of America Abstracts with Programs 20:A380.

Kim, J. 1973. Ecosystem of the Salton Sea. American Geophysical Union, Geophysical Monograph 17:601-605.

Kimsey, J.A. 1982. Petrography and geochemistry of the La Posta granodiorite. Master of Science Thesis, San Diego State University, California, California 81 p.

King, J.L., and B.E. Tucker 1982. Analysis of differential array data from El Centro, U.S.A., and Garm, U.S.S.R. In Third International Earthquake Microzonation Conference Proceedings, edited by M. Sherif, National Science Foundation, Washington, D.C. 2:611-622.

King, N.E., and J.C. Savage 1983. Strain-rate profile across the Elsinore, San Jacinto, and San Andreas Faults near Palm Springs, California, 1973-81. Geophysical Research Letters 10(1):55-57.

King, V.L. 1932. Some observations of Imperial Valley, California. Oil Age 20:12.

Kirkpatrick, J.C. 1958. A study of some Eocene formations in southern California. Master's Thesis, University of California, Los Angeles 75 p.

Kissel, R. 1977. A geophysical study of an alpine meadow in the Laguna Mountains. Undergraduate Research Reports, Department of Geology, San Diego State University, California 30(2):1-23.

Klinger, R. 1986. Recency of faulting within the Anza seismic gap, San Jacinto Fault Zone, southern California. Undergraduate Research Report, San Diego State University, California, 18p.

Klinger, R.E., and T.K. Rockwell 1989. Flexural-slip folding along the eastern Elmore Ranch Fault in the Superstition Hills earthquake sequence of November 1987. Seismological Society of America Bulletin 79:297-303.

Kniffen, F.B. 1932. Lower California studies IV, the natural landscape of the Colorado delta. University of California Publications in Geology 5(4):149-244.

Koch, F.W. 1907. California's inland sea. California Physical Geography Club Bulletin 1:4-7.

Kocher, A.E., and W.B. Harper 1927. Soil survey of the Coachella Valley area, California. U.S. Department of Agriculture, Bureau of Chemistry and Soils 1923:485-535.

Koenig, J.B. 1967. The Salton-Mexicali geothermal province. Mineral Information Service 20(7):75-81.

----- 1969. Salton Sea; a new approach to environmental geologic problems. Association of Engineering Geologists, Program National Meeting p. 26.

----- 1971. History of geothermal exploration in the Salton Trough. Geological Society of America Abstracts 3:143-144.

Kofron, R.J. 1981. The geochemistry of the west contact of the La Posta pluton. Undergraduate Research Reports, Department of Geology, San Diego State University, California 39(3):1-22.

----- 1984. Age and origin of gold mineralization in the southern portion of the Julian mining district, southern California. Master of Science Thesis, Geology Department, San Diego State University, California 75 p.

Kofron, R.J., and M.J. Walawender 1984. Age and style of mineralization, Julian gold district, southern California. Geological Society of America Abstracts with Programs 16:563.

Kohler, W.M., and G.S. Fuis 1986. Travel-time, time-term, and basement depth maps for the Imperial Valley region, California, from explosions. Seismological Society of America Bulletin 76:1289-1303.

Korenic, J.M. 1978. Gravity survey of the southern portion of the Elsinore Fault zone, Coyote Mountains-Palm Canyon Wash, Imperial County, southern California. Undergraduate Research Reports, Department of Geology, San Diego State University, California 32(2):1-22.

Korsch, R.J. 1979. Cenozoic volcanic activity in the Salton Trough region. In Tectonics of the Juncture Between the San Andreas Fault system and the Salton Trough, Southeastern California, edited by J.C. Crowell and A.G. Sylvester, Department of Geological Sciences, University of California Santa Barbara p. 87-100.

Kovach, R.L. 1961. Some preliminary gravity results in Imperial Valley. American Association of Petroleum Geologists 45:127.

----- 1962. Geophysical invstigations in the Colorado Delta region. Doctoral Dissertation, California Institute of Technology, Pasadena.

Kovach, R.L., C.R. Allen, and F. Press 1962. Geophysical investigations in the Colorado Delta region. Journal of Geophysical Research 67(7):2845-2871.

Koxhwe, A.E., and party 1923. Soil survey of the Brawley area, California. U.S. Department of Agriculture map.

Krinsley, D.H., and R.I. Dorn 1991. New eyes on eastern California rock varnish. California Geology

May:107-114.

Krumbein, W.C., and M. Potts 1979. Girvanelle-like structures formed by *Plectonema gloeophilium* Borzi (Cyanophyta) from the Borrego Desert in southern California. Geomicrobiology 1:211-217.

Krummenacher, D., R.G. Gastil, J. Bushee, and J. Doupont 1975. K-Ar apparent ages, Peninsular Ranges Batholith, southern California and Baja California. Geological Society of America Bulletin 86:760-768.

Kundert, C.L. 1955. Geologic map of California, Santa Ana sheet. California Division of Mines and Geology, Scale 1:250,000.

Kupferman, S.A. 1982. Gypsum Deposits of the Fish Creek Mountains Imperial and San Diego Counties, California. American Association of Petroleum Geologists Bulletin 66:1693.

----- 1986. Gypsum Deposits of the Fish Creek Mountains Imperial and San Diego Counties, California. In Geology of the Imperial Valley, California, edited by P.D. Guptil, E.M Gath and R. W. Ruff, South Coast Geological Society Annual Field Trip Guidebook, Santa Ana, California 14:49-55.

----- 1988. Gypsum deposits of the Fish Creek Mountains Imperial and San Diego Counties, California. In Geologic Field Guide to the Western Salton Basin Area, edited by S.M. Testa and K.E. Green, American Institute of Professional Geologists p. 15-22.

Kurtén, B., and E. Anderson 1980. Pleistocene mammals of North America. Columbia University Press, New York 442 p.

Lachenbruch, A.H., J.H. Sass, and S.P. Galanis 1985. Heat flow in southernmost California and the origin of the Salton Trough. Journal of Geophysical Research 90(B8): 6709-6736.

Lachenbruch, A.H., J.H. Sass, S.P. Galanis, and B.V. Marshall 1983. Heat flow, crustal extension, and sedimentation in the Salton Trough. EOS, Transactions on the American Geophysical Union 64:836.

Lamanuzzi, V., and C.E. Johnson 1979. Preliminary catalog of earthquakes in northern Imperial Valley, California, July 1978-September 1978. U.S. Geological Survey Open-File Report 79-931:1-14.

Lamanuzzi, V., C.E. Johnson, and P.T. German 1979. Preliminary catalog of earthquakes in northern Imperial Valley, California, October 1978-December 1978. U.S. Geological Survey Open-File Report 79-930:1-13.

Lamar, D.L., and P.M. Merifield 1974. Investigation of lineaments on Skylab and ERTS images of Peninsular Ranges, southwestern California. California Earth Science Corporation Technical Report 74-5:1-16.

Lamar, D.L., and T.K. Rockwell 1986. An overview of the tectonics of the Elsinore Fault Zone. In Neotectonics and Faulting in Southern California, edited by P.L. Ehlig, Geological Society of America, Cordilleran Section, Annual Meeting Guidebook and Volume 82:149-158.

Lampe, C.M. 1988. Geology of the Granite Mountain area: implications of the extent and style of deformation along the southeast portion of the Elsinore Fault. Master of Science Thesis, San Diego State University, California, 150p.

Lampe, C.m., M.J. Walawender, and T.K. Rockwell 1988. Contacts between La Posta-type plutonic rocks and stromatic magmatites offset across the Elsinore Fault, southern California. Geological Society of America, Abstracts with Programs 20(3):174-175.

Lande, D. 1979. A history of geothermal drilling in the Imperial Valley. In Geology and Geothermics of the Salton Trough, edited by W.A. Elders, University of California, Riverside, Campus Museum Contributions 5:45-46.

Langbein, J.O., M.J.S. Johnston, and A. McGarr 1982. Geodetic observations of postseismic deformation around the north end of surface rupture. In The Imperial Valley, California, earthquake of October 15, 1979. U.S. Geological Survey Professional Paper 1254:205-212.

Langbein, J.O., A. McGarr, M.J.S. Johnston, and P.W. Harsh 1983. Geodetic measurements of postseismic crustal deformation following the 1979 Imperial Valley Earthquake, California. Seismological Society of America Bulletin 73:1203-1224.

Langenkamp, D.F. 1973. Microearthquake study of the Elsinore Fault Zone, southern California. Master of Science Thesis, Department of Geological Sciences, University of California, Riverside 102 p.

Langenkamp, D.F., and J. Combs 1974. Microearthquake study of the Elsinore Fault Zone, southern California. Seismological Society of America Bulletin 64:187-202.

Lasby, R.L. 1969. The stratigraphy and paleontology of the Imperial Formation at Cabazon, California. Senior Thesis, Department of Geological Sciences, University of California, Riverside 15 p.

Larson, E.S., Jr. 1941. The batholith of southern California. Science 93:442-443.

----- 1948. Batholith and associated rocks of Corona, Elsinore, and San Luis Rey quadrangles, southern California. Geological Society of America Memoirs 29:1-182.

----- 1951. Crystalline rocks of the Corona, Elsinore and San Lewis Rey quadrangles, southern California. California Division of Mines Bulletin 159:7-50.

----- 1954. The batholith of southern California. California

Division of Mines Bulletin 170(7):25-30.

Larson, E.S., Jr., and W.M. Draisin 1948. Composition of the minerals of the rocks of the southern California batholith. International Geological Congress, Report of the 18th Session, 1(2):66-79.

Larson, K., D.C. Agnew, and F. Webb 1987. Lake Cahuilla shorelines; constraints on crustal models. EOS, Transactions of the American Geophysical Union 68:1465.

Larson, R.L., H.W. Menard, and S.M. Smith 1968. Gulf of California: a result of ocean floor spreading and transform faulting. Science 161:781-784.

Layton, D.W. 1978. Water for long-term geothermal energy production in the Imperial Valley. Lawerence Livermore Laboratory, Berkeley, California, Report 52576:48.

Le Bras, R. 1981. A preliminary study for the inversion of strong ground motion data from the 1979 Imperial Valley earthquake. EOS, Transactions of the American Geophysical Union 62:972.

LeConte, J.L. 1851. (Observations on the geology of California and adjacent regions). Proceedings of the Academy Natural Sciences, Philadelphia 5:264-265.

----- 1855. Account of some volcanic springs in the desert of the Colorado in southern California. American Journal of Science and Arts, 2nd Series 19(55):1-6.

Lee, T.C. 1976. Earthquake risk in geothermal energy extraction from the Salton Sea Field, California. EOS, Transactions of the American Geophysical Union 52:963.

----- 1977. On shallow-hole temperature measure-ments; a test study in the Salton Sea geothermal field. Geophysics 42:572-583.

Lee, T.C., and L.H. Cohen 1979. Onshore and offshore measurements of temperature gradients in the Salton Sea geothermal area, California. Geophysics 44:206-215.

Lee, T.C., and P.A. Witherspoon 1983. Suggested research program for the Salton Sea deep hole. EOS, Transactions of the American Geophysical Union 64:864.

Lee, W.B., and D.D. Jackson 1979. Crustal structure and vertical crustal movements in Imperial Valley, California. EOS, Transactions of the American Geophysical Union 60:810.

Leeds, D.J., editor 1980. Imperial County, California, earthquake, October 15, 1979. Earthquake Engineering Research Institute, Berkeley, California 194 p.

Lei, H.H. 1987. Porosity and hydrothermal alteration determined from wireline logs from the Salton Sea geothermal field, California, U.S.A. Master's Thesis, University of California, Riverside 168 p.

Leivas, E., E.W. Hart, R.D. McJunkin, and C.R. Real 1980. Geological setting, historical seismicity and surface effects of the Imperial Valley Earthquake October 15, 1979, Imperial County, California. In Imperial County, California, Earthquake, October 15, 1979, edited by D.J. Leeds, Earthquake Engineering Research Institute, Berkeley, California p. 5-19.

Leveille, G. 1982. Mid-Tertiary detachment faulting and manganese mineralization in the Midway Mountains, southeastern California. Undergraduate Research Reports, Department of Geology, San Diego State University, California 40:(unpaginated).

Lillegraven, J.A. 1973. Terrestrial Eocene vertebrates from San Diego County, California. In Studies on the Geology and Geologic Hazards of the Greater San Diego Area, California, edited by A. Ross and R.J. Dowlen, San Diego Association of Geologists, Guidebook p. 27-32.

Lin, W., and W. Daily 1988. Laboratory-determined transport properties of core from the Salton Sea Scientific Drilling Project. Journal of Geophysical Research, B, Solid Earth and Planets 93:13.

Lindblad, L., and S. Lindblad 1987. Baja California. Rizzoli International Publications, Incorporated, New York 184 p.

Lindsay, E.H. 1978. Late Cenozoic vertebrate faunas of southeastern Arizona. New Mexico Geological Society Guidebook, 29th Field Conference, Land of Cochise p. 269-275.

Lindsay, E.H., and N.T. Tessman 1974. Cenozoic vertebrate localities and faunas in Arizona. Journal of the Arizona Academy of Science 9(1):3-24.

Lindsay, E.H., N.M. Johnson, and N.D. Opdyke 1975. Preliminary correlation of Northern American Land Mammal Ages and geomagnetic chronology. In Studies on Cenozoic Paleontology and Stratigraphy, Claude W. Hibbard Memorial Volume 3, edited by G.R. Smith and N.E. Friedland, University of Michigan Papers in Paleontology 12:111-119.

Lindsay, E.H., N.D. Opdyke, and N.M. Johnson 1980. Pliocene dispersal of the horse *Equus* and late Cenozoic mammalian dispersal events. Nature 287:135-138.

Lindsay, E.H., and J.S. White 1993. Biostratigraphy and magnetostratigraphy in the southern Anza-Borrego area. In Ashes, Faults and Basins, edited by R. E. Reynolds and J. Reynolds, San Bernardino County Museum Association Special Publication 93-1:86-87.

Lindvall, S.C. 1988. Evidence of prehistoric earthquakes on the Superstition Hills Fault, southern California and a Holocene slip rate of the San Andreas Fault at Gorman Creek, southern California. Master of Science Thesis, San Diego State University, California 102 p.

Lindvall, S.C., T.K. Rockwell, and K.W. Hudnut 1989a. Evidence for prehistoric earthquakes on the Superstition Hills Fault from offset geomorphic features. Seismological Society of America Bulletin 79:342-361.

----- 1989b. Slip distribution of prehistorical earthquakes on the Superstition Hills Fault, San Jacinto Fault zone, southern California, based on offset geomorphic features. Geological Society of America Abstracts with Programs 21(5):107.

Link, M.H., G.L. Peterson, and P.L. Abbott 1979. Eocene depositional systems, San Diego, California. In Studies on the Geology and Geologic Hazards of the Greater San Diego Area, California, edited by A. Ross and R.J. Dowlen, San Diego Association of Geologists, Guidebook p. 1-8.

Lintz, J, Jr. 1969. Distribution of concretions at Sabin's anticline, Imperial County, California. Geological Society of America Abstracts with Programs p. 35.

Lippmann, M.J., and K.P. Goyal 1980. Numerical modeling studies of the Cerro Preito Reservoir. Lawerence Berkeley Laboratory, California, Energy and Environment Division, Report p. 44-52.

Lippmann, M.J., and G.S. Bodvarsson 1985. The Heber geothermal field, California; natural state and exploitation modeling studies. Journal of Geophysical Research 90:745-758.

Lippmann, M.J., N.E. Goldstein, S.E. Halfman, and P.A. Witherspoon 1985. Exploration and development of the Cerro Prieto geothermal field. In Geology and Geothermal Energy of the Salton Trough, edited by L.J. Herber, National Association of Geology Teachers, Far Western Section p. 27-39.

Lippmann, M.J., and A. Manon M. 1985. Status of the Cerro Prieto Project. In Geology and Geothermal Energy of the Salton Trough, edited by L.J. Herber, National Association of Geology Teachers, Far Western Section p. 40-63.

Lipton, D.A. 1985. Mineral resources of the Carrizo Gorge wilderness study area (BLM No. CA-060-025), San Diego County, California. U.S. Bureau of Mines Report MLA 23-85:1-14.

Lisouski, M., and J.C. Savage 1988. Deformation associated with the Superstition Hills, California, earthquakes of November 1987. Seismological Research Letters 59:35.

Littlefield, W.M. 1966. Hydrology and physiography of the Salton Sea, California. U.S. Geological Survey Hydrological Investigations Atlas HA-222.

Littleton, R.T., and E. Burnett 1976. The salinity profile of the East Mesa field as determined from dual induction resistivity and SP logs. Geothermal Energy Development Report EPA-600/9-76-0111:114-125.

Liu, H., and D.V. Helmberger 1982. Modeling the near-source ground motion for the 23.19 aftershock of the October 15, 1979 Imperial Valley Earthquake. EOS, Transactions of the American Geophysical Union 63:1038.

----- 1985. The 23.19 aftershock of the October 15, 1979 Imperial Valley earthquake; more evidence for an asperity. Seismological Society of America Bulletin 75:689-708.

LKB Resources 1979. NURE aerial gamma-ray and magnetic reconnaissance survey; Colorado-Arizona area; Salton Sea NI 11-9, Phoenix NI 12-7, El Centro NI 11-12, AJO NI 12-10, Lukeville NH 12-1, quadrangles. U.S. Department of Energy, Report GJBX 12-80.

Locke, F., L.B. Owen, and R. Quong 1979. Lawrence Livermore Laboratory test facilities at the Salton Sea geothermal field, southern California. In Geology and Geothermics of the Salton Trough, edited by W.A. Elders, University of California, Riverside, Campus Museum Contributions 5:62-69.

Lockwood, J.P. 1961. Geology of an area in the western Santa Rosa Mountains of southern California, with special reference on a previously unreported mylonite belt. Senior Thesis, Department of Geological Sciences, University of California, Riverside.

Loeltz, O.J., B. Irelan, J.H. Robinson, and F.H. Olmsted 1975. Geohydrologic reconnaissance of the Imperial Valley, California. U.S. Geological Survey Professional Paper 486K:1-54.

Loeltz, O.J., and S.A. Leake 1979. Relation between proposed developments of water resources and seepage from the All-American Canal, eastern Imperial Valley, California. U.S. Geological Survey Open-File Report 79-744:1-60.

----- 1983a. A method for estimating ground-water return flow to the lower Colorado River in the Yuma area, Arizona and California. U.S. Water-Resources Investigations 83-4220:1-94.

----- 1983b. A method for estimating ground-water return flow to the lower Colorado River in the Yuma area, Arizona and California; executive summary. U.S. Water-Resources Investigations 83-4221:1-13.

Loew, O. 1876a. Report on the geological and mineralogical character of southeastern California and adjacent country. In Annual Report of the Geographical Surveys West of the 100th Meridian, Lieutenant G.W. Wheeler 1876:173-188.

----- 1876b. Report on the alkaline lakes, thermal springs, and brackish waters of southern California. In Annual Report of the Geographical Surveys West of the 100th Meridian, Lieutenant G.W. Wheeler 1876:188-199.

Lofgren, B.E. 1973. Monitoring ground movement in geothermal areas. American Society of Civil Engineers, Hydraulics Division Annual Specialty Conference Proceedings 21:437-447.

----- 1974. Measuring ground movement in geothermal areas of Imperial Valley, California. Conference on Research for the Development of Geothermal Energy Resources, California Institute of Technology, Jet Propulsion Laboratory-National Science Foundation, Pasadena, California p. 128-138.

----- 1978a. Measured crustal deformation in Imperial Valley, California. U.S. Geological Survey Open-File Report 78-910:1-7.

----- 1978b. Salton Trough continues to deepen in Imperial Valley, California. EOS, Transactions of the American Geophysical Union 59:1051.

----- 1979a. Measured crustal deformation in Imperial Valley, California. Geothermics 8:267-272.

----- 1979b. Monitoring crustal deformation in the Imperial Valley-Mexicali Valley structural trough. In Geology and Geothermics of the Salton Trough, edited by W.A. Elders, University of California, Riverside, Campus Museum Contributions 5:47.

----- 1979c. Measured crustal deformation in Imperial Valley, California. In Geology and Geothermics of the Salton Trough, edited by W.A. Elders, University of California, Riverside, Campus Museum Contributions 5:48-52.

Lofgren, B.E., and B.L. Massey 1979. Monitoring crustal strain, Cerro Prieto geothermal field, Baja California, Mexico. In Geology and Geothermics of the Salton Trough, edited by W.A. Elders, University of California, Riverside, Campus Museum Contributions 5:53-57.

Logan, C.A. 1947. Limestone in California. California Division of Mines Report 43:300-303.

Lomnitz, C., F. Mooser, C.R. Allen, J.N. Brune, and W. Thatcher 1970. Seismicity and tectonics of northern Gulf of California region, Mexico--preliminary results. Geofisica International 10:37-48.

Long, J.T., and R.P. Sharp 1964. Barchan-dune movement in Imperial Valley, California. Geological Society of America Bulletin 75:149-156.

Longwell, C.R. 1954. History of the lower Colorado River and the Imperial depression. In Geology of Southern California, edited by R.H. Jahns, California Division of Mines Bulletin 170(5):53-56.

Lonsdale, P. 1989. Geology and tectonic history of the Gulf of California. In The Eastern Pacific Ocean and Hawaii, edited by E.L. Winterer, D.M. Hussong, and R.W. Decker. Geological Society of America, the Geology of North America N:499-521.

Lorwoski, R.M. 1981. Petrologic description and possible source area for the Table Mountain Formation of southeastern San Diego County, California. Master of Science Thesis, San Diego State University, California 88 p.

Losh, S., and E.C. Jowett 1989. The detachment related Picacho gold deposit; structural setting and fluid-rock interaction. Geological Association of Canada; Canadian Geophysical Union, Joint Annual Meeting 14:3-4.

Lothringer, C.J. 1984. Geology of a lower Ordovician allochton, Rancho San Marcos, Baja California, Mexico. In Geology of the Baja California Peninsula, edited by V.A. Frizzell, Jr., Society of Economic Paleontologists and Mineralogists, Pacific Section p. 17-22.

Lougee, R.J. 1938. [Review of] The Colorado Delta by Godfrey Glenton Sykes, 1937. Geomorphology 1:79-80.

Lough, C.F. 1974. Water budget for Borrego Valley. In Recent Geologic and Hydrologic Studies, Eastern San Diego County and Adjacent Areas, edited by M.W. Hart and R.J. Dowlen, San Diego Association of Geologists Field Trip Guidebook p. 93-101.

Lough, C.F., and A.L. Stinson 1991. Structural evolution of the Vallecito Mountains, southwest California. Geological Society of America Abstracts with Programs p. A246.

Loughman, C.C. 1987. Neotectonics and geo-morphology of a portion of the San Jacinto Fault Zone in the Anza seismic gap of southern California. Master of Science Thesis, San Diego State University, California 152 p.

Lower, S.R. 1977. The use of springs in the analysis of the groundwater system at Mount Laguna, San Diego County, California. Master of Science Thesis, San Diego State University, California 235 p.

Lowman, P.D. 1977. The Elsinore Fault in southern California: a re-evaluation based on orbital photography. Geological Society of America Abstracts with Programs 9:1076-1077.

----- 1980. Vertical displacement on the Elsinore Fault of southern California: Evidence from orbital photographs. Journal of Geology 88:415-432.

Lowman, P.D., Jr., N.M. Johnson, C.B. Officer, N.D. Opdyke, G.D. Woodward, P.K. Zeitler, and E.H. Lindsay 1983. Rates of late Cenozoic tectonism in the Vallecito-Fish Creek Basin, western Imperial Valley, California; discussion and reply. Geology 12:319-320.

Lowry, W.D., and R.M. Givetti 1979. The Salt-Gila River, Arizona source of the late Eocene Poway Conglomerate of the San Diego area and the limited offset of the San Andreas Fault. Geological Society of America Abstracts with Programs 11:469.

----- 1981. Specific Arizona sources of the late Eocene Poway Conglomerate of the San Diego area and the great competence of the Salt-Gila river system. Geological Society of America Abstracts with Programs 13:68.

Lucchitta, I. 1972. Early history of the Colorado River in the Basin and Range Province. Geological Society of America Bulletin 83:1933-1948.

----- 1979. Late Cenozoic uplift of the southwestern Colorado Plateau and adjacent lower Colorado River region. Tectonophysics 61:63-95.

Lundberg, E.A. 1975. Utilization of the earth's natural heating system to desalt geothermal brines for augmentation of the Colorado River system. National Water Supply Improvement Association Journal 1:39-51.

Lundelius, E.L., Jr., T. Downs, E.H. Lindsay, H.A. Semken, R.J. Zakrzewski, C.S. Churcher, C.R. Harrington, G.E. Schultz, and S.D. Webb 1987. The North American Quaternary sequence. In Cenozoic Mammals of North America, Geochronology and Biostratigraphy, edited by M.O. Woodburne, University of California Press, Berkeley p. 211-235.

Luyendyk, B.P., and J.S. Hornafius 1982. Paleolatitude of the Point Sal ophiolite. Geological Society of America Abstracts with Programs 14:182.

Luyendyk, B.P., M.J. Kamerling, R.R. Terres, and J.S. Hornafius 1985. Simple shear of southern California during the Neogene time suggested by paleomagnetic declinations. Journal of Geophysical Research 90:12454-12466.

Maas, J.P. 1973. A shallow seismic velocity structure in the lower Borrego Valley, California. Master of Science Thesis, Department of Geological Sciences, University of California Riverside 78 p.

----- 1975. Telluric mapping over the Mesa geothermal anomaly, Imperial Valley, California. In Paleogene Symposium and Selected Technical Papers; Conference on Future Energy Horizons of the Pacific Coast, edited by D.W. Weaver, et al., American Association of Petroleum Geologists, Pacific Section p. 325-342.

----- 1976. Telluric mapping over the Mesa geothermal anomaly, Imperial Valley, California. Doctoral Dissertation, University of California, Riverside 152 p.

Maas, J.P., and J. Combs 1975. Field results of the telluric method over the Mesa geothermal field, Imperial Valley, California. United Nations Symposium on the Development and Use of Geothermal Resources, Abstracts 2.

MacDonald, R.B. 1971. The application of automatic recognition techniques in the Apollo IX SO-65 experiment. Earth Resources Program Review 1-3:39-1-39-17.

MacDougal, D.T. 1907. The desert basins of the Colorado delta. American Geographical Society Bulletin 39:705-729.

----- 1914. The Salton Sea; a study of the geography, the geology, the floristics, and ecology of a desert basin. Carnegie Institute of Washington Publication 193:1-182.

----- 1915. The Salton Sea. American Journal of Science, 4th series 39:231-250.

----- 1916. The recession of the Salton Sea. Carnegie Institution of Washington Yearbook 14:90.

----- 1917. A decade of the Salton Sea. Geographical Review 3:457-473.

Macdougal, D.T., and G. Sykes 1915. The travertine record of Blake Sea. Science, New Series, 42:113-134.

----- 1922. The travertine record of Blake Sea. Geologisches Zentralblatt, Anzieger fur Geologie 27:303.

MacCellan, D.D. 1936. The geology of the East Coachella Tunnel of the Metropolitan Water District of soutern California. Doctoral Dissertation, California Institute of Technology, Pasadena.

Mace, N.W. 1981. A paleomagnetic study of the Miocene Alverson Volcanics of the Coyote Mountains, western Salton Trough, California. Master of Science Thesis, San Diego State University, California 142 p.

Magistrale, H.L., and H. Kanamori 1988. Superstition Hills Earthquakes and basement structure of the western Imperial Valley. Seismological Research Letters 59:48.

Magistrale, H.L., L. Jones, and H. Kanamori 1989. The Superstition Hills, California, earthquake of 24 November, 1987. Seismological Society of America Bulletin 79:239-251.

Maimoni, A. 1982. Minerals recovery from the Salton Sea geothermal brines; a literature review and proposed cementation process. Geothermics 11:239-258.

Majer, E.L., and T.V. McEvilly 1980. Induced seismicity studies at the Cerro Prieto, Mexico, and East Mesa, California, geothermal fields. Lawerence Berkeley Laboratory, Energy and Environment Division, Berkeley 10686:123-125.

Malevskaya, O.Y. 1983. Velichiny srednekvadratichnykh uskorenly grunta pri zemletryasenii v Impirial Velli (SShA). Voprosy Inzhenernoy Seismology 24:73-80.

Maley, R.P. 1970a. Results from the strong-motion seismograph records of the Borrego Mountain, California, earthquake of 9 April 1968. Geological Society of America Abstracts with Programs 13:93.

----- 1970b. Strong-motion seismograph records of the Borrego Mountain Earthquake from multistory buildings in the Los Angeles area. Geological Society of America Abstracts with Programs 2:114.

Maloney, N.J. 1981. Holocene deformation of Lake Cahuilla shoreline, northeastern Salton Basin, southern California. EOS, Transactions of the American Geophysical Union 62:840.

----- 1986a. Late Holocene uplift of Bat Cave Buttes, Salton Trough, California. Geological Society of America Abstracts with Programs 18:153.

----- 1986b. Coastal landforms of Holocene Lake Cahuilla, northeastern Salton Basin, California. In Geology of the Imperial Valley, California, edited by P.D. Guptil, E.M Gath and R. W. Ruff, South Coast Geological Society, Annual Field Trip Guidebook, Santa Ana, California 14:151-158.

Mann, J.F. Jr. 1951. Cenozoic geology of the Temecula region, Riverside County, California. Manuscript on File, University of Southern California, Los Angeles 136 p.

----- 1955. Geology of a portion of the Elsinore Fault Zone, California. California Division of Mines and Geology Special Report 43:1-22.

Manske, S.L. 1990. The relative timing and phase assemblages of vein-controlled hypogene mineralization and alteration in the Mesquite deposit, Imperial County, California. Geological Society of America Abstracts with Programs 22(3):63.

Manske, S.L., W.F. Matlack, M.W. Springlet, A.E. Strakele, S.N. Watowich, B. Yeomans, and E. Yeomans 1988. Geology of the mesquite deposit, Imperial County, California. Mining Engineering 40:439-444.

Mariano, J., and V.J.S. Grauch 1988. Aeromagnetic maps of the Colorado River region including the Kingman, Needles, Salton Sea, and El Centro 1° by 2° quadrangles, California, Arizona, and Nevada. U.S. Geological Survey Miscellaneous Field Studies Map, Report MF-2023.

Mark, D.L. 1987. An evaluation of potential lateral saltwater intrusion in the Ocotillo-Coyote Wells groundwater basin, Imperial County, California. Master of Science Thesis, San Diego State University, California 169 p.

Marshall, S.A., C.D. Nuffer, B.A. Sentianin, and G.H. Girty 1991. Compositional characteristics of Holocene sand, Salton Basin. Geological Society of America Abstracts with Programs p. A109.

Martin-Barajas, A., G. Rendon-Marquez, and M. Lopez-Martinez 1991. A Neogene marine sequence in the Puertecitos volcanic province, Baja California: new evidence of early basin stage in the northern Gulf of California. Geological Society of America Abstracts with Programs p. A196.

Martin, R.A. 1993. Late Pliocene and Pleistocene cotton rats in the southwestern United States. In Ashes, Faults and Basins, edited by R.E. Reynolds and J. Reynolds, San Bernardino County Museum Association Special Publication 93-1:88-89.

Martin, R.A., and R.H. Prince 1989. A new species of early Pleistocene cotton rat from the Anza-Borrego Desert of southern California. Southern California Academy of Sciences Bulletin 88(2):80-87.

Mason, R.G., J.L Brander, and M.G. Bill 1979. Mekometer measurements in the Imperial Valley, California. In Recent Crustal Movements, edited by C.A. Whitten, R. Green, and B.K. Meade, Tectonophysics 52:497-503.

Mason, R.G., and C.N. Crook 1987. Deformation on and around the Imperial Fault 1975-1987 as measured with a mekometer. EOS, Transactions of the American Geophysical Union 69:254.

Mason, R.G., C.N. Crook, and P.R. Wood 1980. Mekometer measurements in the Imperial Valley. U.S. Geological Survey Open-File Report 80-2008:1-26.

Mathias, K., and E.A. Lundberg 1975. East Mesa test wells. Geothermal Energy 3:23-29.

Mathias, K., 1974a. Five geothermal wells, East Mesa KGRA, Imperial Valley, California. EOS, Transactions of the American Geophysical Union 56:1198.

----- 1974b. Preliminary results of geothermal wells Mesa 6-1 and Mesa 6-2, East Mesa KGRA, Imperial Valley, California. Geothermal Energy Magazine 2(6):8-17.

----- 1974c. Preliminary results of geothermal wells Mesa 6-1 and Mesa 6-2, East Mesa KGRA, Imperial Valley, California. EOS, Transactions of the American Geophysical Union 55:489.

Matti, J.C., D.M. Morton, and M.J. Rymer 1991. Geologic map of the Palm Springs and Big Bear Lake 30' X 60' quadrangles, California: a progress report. Abstract with programs Geological Society of America annual meeting. San Diego p. A478.

Matti, J.C., and D.M. Morton 1975. Geologic history of the San Timoteo Badlands, southern California. Geological Society of America Abstracts with Programs 7(3):344.

----- 1993. Paleogeographic evolution of the San Andreas Fault in southern California: a reconstruction based on a new cross-fault correlation. In The San Andreas Fault System; Displacement, Palinspastic Reconstruction and Geologic Evolution, edited by R.E. Powell, R.J. Weldon, and J.C. Matti, Geological Society of America Memoir 178:107-160.

Mattick, R.E. 1961. Isostasy determination in the Jacumba Mountains and Imperial Valley regions of southern

California. Undergraduate Research Reports, Department of Geology, San Diego State University, California 5(2):1-14.

Mattick, R.E., F.H Olmstead, and A.A.R. Zohdy 1973. Geophysical studies in the Yuma Area, Arizona and California. U.S. Geological Survey Professional Paper 726-D:1-36.

May, D.J. 1989. Late Cretaceous intra-arc thrusting in southern California. Tectonics 8:1159-1173.

May, S.R., and C.A. Repenning 1982. New evidence for the age of the Mount Eden fauna, southern California. Journal of Vertebrate Paleontology 2(1):109-113.

Mayo, A.L. 1975. Julian basic data report. San Diego County Environmental Development Agency Report, Groundwater (3):19-64.

McAnn, M.W., Jr., R.J. Geller, and H.C. Shah 1981. Normal mode modeling of strong motion accelerogram for the 1968 Borrego Mountain Earthquake. EOS, Transactions on the American Geophysical Union 62:1036.

McArthur, D.S. 1984. Geomorphology of San Diego County. In San Diego, an Introduction to the Region, edited by P.R. Pryde, Kendall/Hunt Publishing Company, Dubuque, Iowa p. 13-29.

McCamley, M.J. 1982. Geology of the Mollusk Wash area, Vallecito Mountains, San Diego County, California. Undergraduate Research Reports, Department of Geology, San Diego State University, California 40(4):1-22.

McCormik, W.V., III 1983. Aftershock analysis and comparison of the Imperial Valley Earthquake of October 15, 1979 and the Borrego Mountain Earthquake of April 9, 1968. Undergraduate Research Reports, Department of Geology, San Diego State University, California 42(5):1-9.

McCown, B.E. 1955. The Lake LeConte beach line survey. The Masterkey 29:89-92.

McCoy, F.W., Jr., W.J. Nokleberg, and R.M. Norris 1967. Speculations on the origin of the Algodones dunes, California. Geological Society of America Bulletin 78:1039-1044.

McCullogh, J. 1984. Geology of Lost Valley. Undergraduate Research Reports, Department of Geology, San Diego State University, California 44(2).

McDonald, C.C., and O.J. Loeltz 1976. Water resources of lower Colorado River-Salton Sea area of 1971. U.S. Geological Survey Professional Paper 486-A:1-34.

McDonald, H.G. 1985. The Shasta ground sloth *Nothrotheriops shastense* (Xenarthra, Megatheriidae) in the middle Pleistocene of Florida. In The Evolution of Armadillos, Sloths and Vermilinguas, edited by G. Montgomery, Smithsonian Institution Press, Washington D.C. p. 95-104.

----- 1993. Harlan's Ground Sloth, *Glossotherium harlani*, from Pauba Valley, Riverside County, California. In Ashes, Faults and Basins, edited by R.E. Reynolds and J. Reynolds, San Bernardino County Museum Association Special Publication 93-1:101-103.

McDonald, J.N. 1984. The recorded North American selection regime and late Quaternary megafaunal extinctions. In Quaternary Extinctions, a Prehistoric Revolution, edited by P.S. Martin and R.G. Klein, University of Arizona Press, Tucson p. 404-439.

McDougall, K., C.L. Powell, J.C. Matti, and R.Z. Poore 1994. The Imperial Formation and the opening of the ancestral Gulf of California. Geological Society of America Abstracts with Programs 26(2):71.

McDowell, S.D., and M. McCurry 1977. Active metamorphism in the Salton Sea geothermal field, California; mineralogical and mineral chemical changes with depth and temperature in sandstone. Geological Society of America Abstracts with Programs 9:1088.

McDowell, S.D. 1981. Alteration of smectite to illite in the Salton Sea geothermal area. Clay Minerals Society, Annual Clay Minerals Conference 30:6.

----- 1984. Composition and structural state of coexisting feldspars, Salton Sea geothermal field. EOS, Transactions of the American Geophysical Union 65:1134.

----- 1986. Composition and structural state of coexisting feldspars, Salton Sea geothermal field. Mineralogical Magazine 50(1):75-84.

McDowell, S.D., and W.A. Elders 1978. Distribution and chemistry of layer-silicate minerals in the Salton geothermal field, California. Geological Society of America Abstracts with Programs 10:452-453.

----- 1979a. Geothermal metamorphism of sandstone in the Salton Sea geothermal system, edited by W.A. Elders. In Geology and Geothermics of the Salton Trough, edited by W.A. Elders, University of California, Riverside, Museum Contribution 5:70-76.

----- 1979b. Geothermal metamorphism of sandstone in the Salton Sea geothermal system. In Geology and Geothermics of the Salton Trough, edited by W.A. Elders, University of California, Riverside, Campus Museum Contributions 5:70-76.

----- 1980a. Zonation of active greenschist facies metamorphism and physical properties of sandstone in Salton Sea geothermal field. Geological Society of America Abstracts with Programs 12:480.

----- 1980b. Authigenic layer silicate minerals in borehole Elmore 1, Salton Sea geothermal field, California, U.S.A. Beitraege zur Mineralolgie und Petrologie 74:293-310.

----- 1983. Allogen layer silicate minerals in borehole Elmore 1, Salton Sea geothermal field, California. American Mineralogist 68:1146-1159.

McDowell, S.D., and J.B. Paces 1985. Carbonate alteration minerals in the Salton Sea geothermal system, California, U.S.A. Mineralogical Magazine 49:469-479.

McEldowney, R.C. 1970. An occurrence of Paleozoic fossils in Baja California, Mexico. Geological Society of America Special Paper 2:117.

McEuen, R.B., C.W. Mase, and W.E. Loomis 1977. Geothermal tectonics of the Imperial Valley as deduced from earthquake occurrence date. Geothermal Resources Council 1:211-213.

McEuen, R.B., and C.J. Pickney 1972. Seismic risk in San Diego. San Diego Society of Natural History Transactions 17(4):1-62.

----- 1973. Seismic risk in San Diego. In Studies on the Geology and Geologic Hazards of the Greater San Diego Area, California, edited by A. Ross, and R.J. Dowlen, San Diego Association of Geologists, California p. 89-103.

McGill, S.F., C.R. Allen, D. Johnson, and K.E. Sieh 1988. Slip on the Superstition Hills Fault and on near by faults associated with the November 23 and 24, 1987, earthquakes, southern California. Seismological Research Letters 59:49.

McGill, S.F., C.R. Allen, K.W. Hudnut, D.C Johnson, W.F. Miller, and K.E. Sieh 1989. Slip on the Superstition Hills Fault and on near by faults associated with the 24 November 1987 Elmore Ranch and Superstition Hills earthquakes, southern California. Seismological Society of America Bulletin 79:362-375.

McGranahan, M.D. 1979. A magnetic survey of the southwest flank of the Elsinore Fault. Undergraduate Research Reports, Department of Geology, San Diego State University, California 34(2):1-31.

McGuire, M.D. 1980. Thermal springs of the Elsinore Fault Zone; relation of groundwater recharge and structural geology. Master of Science Thesis, Department of Geology, San Diego State University, California 167 p.

McHugh, E.L. 1985. Mineral resources of the Jacumba Study Area, Imperial County, California. U.S. Bureau of Mines MLA 55-85:1-27.

McJunkin, R.D., and J.T. Ragsdale 1980. Compilation of strong-motion records and preliminary data from the Imperial Valley Earthquake of 15 October 1979. California Division of Mines and Geology Preliminary Report 28:1-91.

McKee, E.D. 1939. Some types of bedding in the Colorado River delta. Journal of Geology 47:64-81.

McKee, E.D., R.F. Wilson, W.J. Breed, and C.S. Breed 1967. Evolution of the Colorado River in Arizona: an hypothesis developed at the symposium on Cenozoic geology on the Colorado Plateau in Arizona, August, 1964. Museum of Northern Arizona Bulletin 44:1-67.

McKibben, M.A. 1979a. Ore minerals in the Salton Sea geothermal system, Imperial Valley, California, U.S.A. Master of Science Thesis, Department of Geological Sciences, University of California, Riverside 79/17:1-99.

----- 1979b. Sulfide minerals in the Salton Sea geothermal system, Imperial Valley, California. Geological Society of America Abstracts with Programs 11:91.

----- 1983. Active ore mineralization in the Salton Sea geothermal system, Imperial Valley, California. Geological Society of America Abstracts with Programs 15:640.

----- 1986. Hydrothermal sulfide minerals in the Salton Sea geothermal system, Imperial Valley, California. In The Genesis of Stratiform Sedimente-hosted Lead and Zinc Deposits, edited by R.J.W. Turner, and others, Stanford University Publications, Geological Sciences 20:161-164.

----- 1991a. Quaternary lakes and geothermal activity in the Salton Trough. Mojave Desert Quaternary Research Symposium, Abstracts of Proceedings, San Bernardino County Museum Association Quarterly 39(2):50-51.

----- 1991b. Introduction to the Salton Trough rift. In The Diversity of Mineral and Energy Resources of Southern California, edited by M.A. McKibben. Society of Economics Geologists 12:77-86.

----- 1991c. The Salton Sea geothermal brines. In The Diversity of Mineral and Energy Resources of Southern California, edited by M.A. McKibben. Socitey of Economic Geologists 12:127-138.

----- 1993. The Salton Trough rift. In Ashes, Faults and Basins, edited by R.E. Reynolds and J. Reynolds, San Bernardino County Museum Association Special Publication 93-1:76-80.

McKibben, M.A., J.P. Andes, Jr., and A.E. Williams 1987. Ore-forming processes in the Salton Sea geothermal system, California; new insights from the SSSDP cores. Geological Society of America Abstracts with Programs 19:766-767.

----- 1988. Active ore formation at a brine interface in metamorphosed deltaic lacustrine sediments; the Salton Sea Geothermal System, California. Economic Geology and the Bulletin of the Society of Economic Geologists 83:511-523.

McKibben, M.A., and W.A. Elders 1985. Fe-Zn-Cu-Pb mineralization in the Salton Sea geothermal system, Imperial Valley, California. Economic Geology and the Bulletin of the Society of Economic Geologists 80:539-559.

McKibben, M.A., W.A. Elders, and A.E. Williams 1986. Saline brines and ore-forming processes in the geothermal system of the Salton Trough. In Geology of the Imperial Valley, California, edited by P.D. Guptil, E.M Gath and R. W. Ruff, South Coast Geological Society, Annual Field Trip Guidebook, Santa Ana, California 14:144-150.

McKibben, M.A., and C.S. Eldridge 1989. Sulfur isotopic variations among minerals and aqueous species in the Salton Sea geothermal system; a SHRIMP ion microprobe and conventional study of active ore genesis in a sediment-hosted environment. American Journal of Science 289:661-707.

McKibben, M.A., C.S. Eldridge, and A.E. Williams 1988. Sulphur and base metal transport in the Salton Sea geothermal system. Transactions, Geothermal Resources Council 12:121-125.

McKibben, M.A., and S. Okubo 1988. Metamorphosed Plio-Pleistocene evaporites and the origins of hypersaline brines in the Salton Sea geothermal system, California fluid inclusion evidence. Geochimica et Cosmochimica Acta 52:1047-1056.

McKibben, M.A., A.E. Williams, J.P. Andes, Jr., C.S Oakes, S. Okubo, and W.A. Elders 1986. Metamorphosed Plio-Pleistocene evaporites in the Salton Sea geothermal, Salton Trough rift, California. Geological Society of America Abstracts with Programs 18:690.

McKibben, M.A., A.E. Williams, and G.E.M. Hall 1989. Precious metals in the Salton Sea geothermal brines. Transactions, Geothermal Resources Council 13:45-48.

----- 1990. Precious metals in the Salton Sea geothermal brines. Geological Society of America Abstracts with Programs 22(3):67.

McLaughlin, P.V. 1988. Stratigraphy and depositional environments of late Holocene deposits in the north-central Imperial Valley with implications for liquefaction, Imperial County, California. Master's Thesis, San Jose State University, California.

McLoughlin, P.R. 1977. A chemical analysis of thermal waters at Agua Caliente County Park and Warner's resort with emphasis on the relationship of chemistry to rock weathering, geothermometry, and location of springs associated with faults. Undergraduate Research Reports, Department of Geology, San Diego State University, California 30(2):1-53.

McMahan, A.B., and C.M. Rumsey 1984. Mineral resources of the Picacho Peak Wilderness Study Area, Imperial County, California. U.S. Bureau of Mines Report MLA 38-84:1-31.

McMecham, G.A., and W.D. Mooney 1980. Asymptotic ray theory and synthetic seismograms for laterally varying structures; theory and application to the Imperial Valley, California. Seismological Society of America Bulletin 70:2021-2035.

McNitt, J.R. 1963a. Exploration and development of geothermal power in California. California Division of Mines and Geology Special Report 75:1-45.

----- 1963b. Structural environment of geothermal areas under development in California. Mining Engineering 15:1-55.

Meade, B.K. 1948a. Earthquake investigation in the vicinity of El Centro, California; horizontal movement. EOS, Transactions of the American Geophysical Union 29:27-31.

----- 1948b. Earthquake investigation in the vicinity of El Centro, California; horizontal movement. In Reports on geodetic measurements of crustal movement, 1906-71, National Oceanic and Atmospheric Administration, National Ocean Survey-National Geodetic Survey 4 p.

----- 1965. Horizontal movement along the San Andreas Fault system. Royal society of New Zealand Bulletin 9:175-179.

----- 1971a. Crustal movement investigations. EOS, Transactions of the American Geophysical Union 52:7-9.

----- 1971b. Horizontal movement along the San Andreas Fault system. Royal society of New Zealand Bulletin 9:175-179.

----- 1973a. Crustal movement investigations. In Reports on Geodetic Movements, 1906-71, National Oceanic and Atmospheric Administration, National Ocean Survey-National Geodetic Survey 3 p.

----- 1973b. Report on special purpose survey, Anza-Borrego Desert area, southern California. In Reports on Geodetic measurements of crustal movement, 1906-71, National Oceanic and Atmospheric Administration, National Ocean Survey-National Geodetic Survey.

----- 1973c. Horizontal crustal movements in the United States. In Reports on Geodetic Measurements of Crustal Movement, 1906-71, National Oceanic and Atmospheric Administration National Ocean Survey-National Geodetic Survey 23 p.

Meade, B.K., and J. Small 1966. Current and recent movement on the San Andreas Fault. California Division of Mines and Geology Bulletin 190: 385-391.

----- 1973. Current and recent movement on the San Andreas Fault. In Report on geodetic measurements of crustal movement, 1906-71, National Oceanic and Atmospheric Administration-National Ocean Survey National Geodetic Society 7 p.

Means, T.H., and J.G. Holmes 1902. Soil survey around Imperial, California. U.S. Department of Agriculture, Bureau of Soils Report 3 p.

Meckel, L.D. 1975. Holocene sand bodies in the Colorado delta area, northern Gulf of California. In Deltas, Models for Exploration, edited by M.L.S. Broussard, Houston Geological Society, Texas p. 239-265.

Medcalf, M. 1982. Chemical composition of coexisting hornblende and biotite in the La Posta granodiorite. Undergraduate Research Reports, Department of Geology, San Diego State University, California 41(2):1-10.

Meehan, J., E. Hart, and A. Schiff 1988. Superstition Hills earthquakes, November 23 and 24, 1987, Imperial County, California. Newsletter - Earthquake Engineering Research Institute 22(2):1-8.

Mehegan, J.M., C.T. Herzig, W.A. Elders, L.H. Cohen, and A.L. Quintanilla-Montoya 1987. REE geochemistry of continental rift lavas of the Salton Trough, California and Mexico. Geological Society of America Abstracts with Programs 19:769.

Meidav, H.T. 1968. Structural characteristics of the Salton Sea, California. EOS, Transactions of the American Geophysical Union 49:758.

----- 1969. Geoelectrical exploration of a geothermal area in southern California. EOS, Transactions of the American Geophysical Union 50:348.

----- 1970. Possible sea-floor spreading in the Imperial Valley, I, structural setting. EOS, Transactions of the American Geophysical Union 51:421.

----- 1971. Application of a electrical resistivity and gravimetry in deep geothermal exploration. United Nations Symposium on the Development and Utilization of Geothermal Resources Proceedings 2(1):303-310.

----- 1973. Utilization of gravimetric data for estimation of convective heat flow - Utilizacion de la informacion gravimetrica para la estimacion del flujo de calor por conveccion. Society of Exploration geophysicists Annual International Meeting 43:1-43.

Meidav, H.T., and R.W. Rex 1970a. Geophysical investigations for geothermal energy sources, Imperial Valley, California; Phase 1, 1968 Field Project. University of California Riverside, Institute of Geophysics and Planetary Physics Technical Report 3:1-54.

----- 1970b. Geothermal exploration in the Salton Trough, California. Society for Exploration Geophysicists, Abstracts 119-120.

----- 1970c. Geothermal exploration in the Salton Trough, California. Geophysics 35:1169.

Meidav, H.T., and R. Furgerson 1972. Resistivity studies of the Imperial Valley geothermal area, California. Geothermics 1(2):47-62.

Meidav, H.T., and J.H. Howard 1979. An update of tectonics and geothermal resource magnitude of the Salton Sea geothermal resource. In Geology and Geothermics of the Salton Trough, edited by W.A. Elders, University of California, Riverside, Campus Museum Contributions 5:58-61.

Meidav, H.T., and N. Harthill 1979. A quadripole resistivity survey of the Imperial Valley. Geophysics 44:2012-2020.

Meidav, H.T., R. West, A. Katzenstein 1976. Resistivity survey of the Salton Sea Geothermal area. Society of Exploration Geophysicists, Abstracts 46:106.

Meldahl, K.H., and A.H. Cutler 1991. Neotectonics and taphonomy: Pleistocene molluscan shell accumulations in the northern Gulf of California, Sonora, Mexico. Geological Society of America abstracts with programs, San Diego p. A346.

Mendenhall, W.C. 1906. The Colorado Desert. National Geographic Magazine 20:861-701.

----- 1909. Ground waters of the Indio region, California, with a sketch of the Colorado Desert. U.S. Geological Survey Water-Supply Paper 225:1-56.

----- 1909. The Colorado Desert. National Geographic Magazine 20:681-701.

----- 1910. Notes on the geology of Carrizo Mountain, San Diego County, California. Journal of Geology 18:337-355.

----- 1916. Notes on the geology of Carrizo Mountain and vicinity, San Diego County, California. Journal of Geology 18(4):336-355.

Mendez, A.J., and J.E. Lugo 1990. Steady state near source models of the Parkfield, Imperial Valley and Mexicali Valley earthquakes. Journal of Geophysical Research, B, Solid Earth and Planets 95:327-340.

Merifield, P.M., and D.L. Lamar 1986. Faults and lineaments on Skylab photographs of the Salton Trough area, southern California. In Geology of the Imperial Valley, California, edited by P.D. Guptil, E.M. Gath and R.W. Ruff, South Coast Geological Society Annual Field Trip Guidebook, Santa Ana, California 14:15-36.

Merriam, R.H. 1954. A typical portion of the southern California batholith, San Diego County, California. In Geology of Southern California, edited by R.H. Jahns, California Division of Mines Bulletin 170, Map Sheet 22.

----- 1958. Geology of the Santa Ysabel quadrangle, San Diego County, California. California Division of Mines Bulletin 159:117-128.

----- 1965. San Jacinto Fault in northwestern Mexico. Geological Society of America Bulletin 76:1051-

1054.

----- 1969. Source of sand dunes of southwestern California and northwestern Sonora, Mexico. Geological Society of America Bulletin 80:531-534.

----- 1979. Petrographic evidence of a Sonoran source for upper Paleocene conglomerates in southern California. In Eocene Depositional Systems, San Diego, California, edited by P.L. Abbott, Society of Economic Paleontologists and Mineralogists, Pacific Section p. 119-120.

Merriam, R.H., and O.L. Bandy 1965. Source of upper Cenozoic sediments in the Colorado delta region. Journal of Sedimentary Petrology 35:911-916.

Merriam, R.H., and J.L. Bischoff 1975. Bishop Ash: a widespread volcanic ash extended to southern California. Journal of Sedimentary Petrology 45:207-211.

Merrill, F.J.H. 1914. Geology and mineral resources of San Diego and Imperial Counties. California State Mining Bureau 113 p.

----- 1916. The counties of San Diego, Imperial, California. California Mining Bureau Report 14:723-743.

----- 1919. Los Angeles, Orange and Riverside Counties. California Mining Bureau Report 15:461-581.

Metropolitan Water District of Southern California 1930-1939. Reports and geologic maps by consulting geologists on proposed routes of the Colorado River Aqueduct. Manuscripts on File Metropolitan Water District of Southern California, Los Angeles.

Metzger, D.G. 1968. The Bouse Formation (Pliocene) of the Parker-Blythe-Cibola area, Arizona and California. U.S. Geological Survey Professional Paper 500-D:126-136.

Metzger, D.G., O.J. Loeltz, and B. Irelan 1969. Geohydrology of the Parker-Blythe-Cibola area, Arizona and California. U.S. Geological Survey Professional Paper 486-G:1-130.

Meyer, G.L. 1979. Aeolian features of northern Coachella Valley and landforms and tectonic features of the San Andreas Fault Zone in the Indio Hills. National Association of Geology Teachers, Far Western Section, Field Guide 25:1-16.

Meyer, M.J. 1976. Spring discharge characteristics of the East Mesa area, Rancho Cuyamaca State Park, San Diego County, California. Undergraduate Research Reports, Department of Geology, San Diego State University, California 28(2):1-28.

Michel, R.L., R.A. Schroeder, J.G. Setmire, and S.S. Hall 1988. Soluble salts and tritium concentrations in irrigation drainwaters from the Imperial Valley, California. EOS, Transactions of the American Geophysical Union 79:1181.

Michels, D.E. 1985. Fluid drift speed and permeability asymmetry; injection research at East Mesa. Transactions, Geothermal Resources Council 9:341-345.

----- 1988. Salinity stabilization for non-advecting brine in the temperature gradient with application to the Salton Sea Geothermal System. Transactions, Geothermal Resources Council 12:127-130.

Michniuk, D.S. 1980. Observations of the Imperial Valley earthquakes. EOS, Transactions of the American Geophysical Union 61:108.

Midorikawa, S., H. Kobayashi, and J. -Y. Le Doare 1984. Characteristics of near-field ground motions during the Imperial Valley Earthquake of 1979, Earthquake Engineering Research Institute, Proceedings of the World Conference on Earthquake Engineering 8(2):265-272.

Miele, M.J. 1986. A magnetotelluric profiling and geophysical investigation of the Laguna Salada basin, Baja California. Master of Science Thesis, San Diego State University, California 154 p.

Miller C.D. 1989. Potential hazards from future volcanic eruptions in California. U.S. Geological Survey Bulletin 1847:1-17.

Miller, D.E. 1972. Geology of a portion of the Painted Gorge quadrangle, Imperial County, California. Undergraduate Research Reports, Department of Geology, San Diego State University, California 20:1-32.

Miller, D.E., and T. Kato 1990. Field evidence for late Cenozoic kinematic transition in the Coyote Mountains, Imperial County, California. Geological Society of America Abstracts with Programs 22(3):68.

Miller, F.K., and D.M. Morton 1974. Comparison of granitic intrusion in the Orocopia and Pelona schists, southern California. Geological Society of America Abstracts 6:220-221.

----- 1977. Comparison of granitic intrusions in the Pelona and Orocopia schists, southern California. U.S. Geological Survey, Journal of Research 5:643-649.

Miller, F.S. 1937. Petrology of the San Marcos gabbro, southern California. Geological Society of America Bulletin 48:1397-1425.

----- 1962. Rockslide in Earthquake Valley quadrangle, San Diego County, California. Undergraduate Research Reports, Department of Geology, San Diego State University, California 6(2):1-23.

Miller, G.J. 1977a. Preliminary report on a new cervid from Anza-Borrego Desert, Imperial Valley, California. Society of Vertebrate Paleontology, Abstracts with Programs 37th Annual Meeting, Los Angeles.

----- 1977b. A new cervid (Mammalia; Cervidae) from Anza-Borrego Desert, Imperial Valley, California. Manuscript on File, Anza-Borrego Desert State Park, California 15 p.

----- 1979. A new subgenus and two new species of *Odocoileus* (Mammalia; Cervidae) with a study of variation in antlers. Society of Vertebrate Paleontology, Abstracts with Programs 39th Annual Meeting.

----- 1985a. A look into the past of the Anza-Borrego Desert. San Diego Natural History Museum Environment Southwest 510:12-17.

----- 1985b. A look into the past of the Anza-Borrego Desert. Educational Foundation of the Desert Protective Council, Incorporated, Educational Bulletin 85-3:1-7 (reprint of 1985a).

Miller, G.J., P. Remeika, J.D. Parks, B. Stout, and V.E. Waters 1988. A preliminary report on half-a-million-year-old cutmarks on mammoth bones from the Anza-Borrego Desert Irvingtonian. Mojave Desert Quaternary Research Symposium, Abstracts of Proceedings, San Bernardino County Museum Association Quarterly 35(3-4):41.

----- 1989a. A preliminary report on half-a-million-year-old cutmarks on mammoth bones from the Anza-Borrego Desert Irvingtonian. Symposium on the Value of the Desert, Anza-Borrego Desert Foundation, Borrego Springs, California, Abstracts p. 8.

----- 1989b. A preliminary report on half-a-million-year-old cutmarks on mammoth bones from the Anza Borrego Desert Irvingtonian. Imperial Valley College Museum Society Occasional Papers 8:1-47.

Miller, G.J., B. Stout, and P. Remeika 1982. Irvingtonian ichnites from the Anza-Borrego Desert, California. Society of Vertebrate Paleontology, Abstracts with Programs 42nd Annual Meeting, Mexico City.

Miller, H.J., and V.R. Maulis 1962. Rockslides in Earthquake Valley quadrangle, Sand Diego County, California. Undergraduate Research Reports, Department of Geology, San Diego State University, California 6(2):1-23.

Miller, K.R. 1980. Petrology, hydrothermal mineralogy, stable isotope geochemistry and fluid inclusion geothermometry of borehole Mesa 31-1, East Mesa geothermal field, Imperial Valley, California. Master's Thesis, University of California, Riverside.

----- 1980. Petrology, hydrothermal mineralology, stable isotope geochemistry and fluid inclusion geothermometry of borehole Mesa 31-1, east Mesa geothermal field, Imperial Valley, California. Institute of Geophysics and Planetary Physics Report 80/1:1- 113.

Miller, R.E. 1977. A Galerskin, finite-element analysis of steady-state flow and heat transport in the shallow hydrothermal system in the East Mesa area, Imperial Valley, California. U.S. Geological Survey, Journal of Research 5:497-508.

Miller, R.H., and M.S. Dockum 1983. Ordovician conodonts from metamorphosed carbonates of the Salton Trough, California. Geology 11:410-412.

Miller, W.J. 1935a. Geologic section across southern Peninsular Ranges of California. California Journal of Mines and Geology 31:115-142.

----- 1935b. Geomorphology of the southern Peninsular range of California. Geological Society of America Bulletin 46:1535-1561.

----- 1944. Geology of the Palm Springs-Blythe strip, Riverside County, California. California Division of Mines and Geology 40:11-72.

Minch, J.A. 1970. Early tertiary paleogeography of a portion of the northern Peninsular Ranges. In Pacific Slope Geology of Northern Baja California and Adjacent Alta California, American Association of Petroleum Geologists, Pacific Section, Fall Field Trip Guidebook p. 83-87.

----- 1972. The late Mesozoic-early Tertiary framework of continental sedimentation of the northern Peninsular Ranges, Baja California, Mexico. Doctoral Dissertation, Department of Geological Sciences, University of California, Riverside.

----- 1979. The late Mesozoic-early Tertiary framework of continental sedimentation of the northern Peninsular Ranges, Baja California, Mexico. In Eocene Depositional Systems, San Diego, California, edited by P.L. Abbott, Society of Economic Paleontologists and Mineralogists, Pacific Section p. 43-67.

Minch, J.A., and P.L. Abbott 1973a. Post batholith geology of the Jacumba area, southeastern San Diego, California. San Diego Society of Natural History Transactions 17:129-136.

----- 1973b. Tertiary stratigraphy of Jacumba Valley, southeastern San Diego County, California. Geological Society of America Abstracts with Programs 5:81-82.

Mitchell, E.D. 1961. A new walrus from the Imperial Pliocene of southern California: with notes on odobenid and otariid humeri. Los Angeles County Museum Contributions in Science 44:1-28.

Mitchell, J.R. 1978a. Field trip; Plaster City; selenite. Rock and Gem 8(10):18-23.

----- 1978b. Imperial Valley concretions. Rock and Gem 8(11):16-19.

----- 1983. Directions to abandoned mines in the Chocolate Mountains. Lost Treasure 8:60-61.

----- 1989. The treasure of Ocotillo. Lapidary Journal 43:71-74.

Mogilner, G.A. 1967. Geology of an area in the western Fish Creek Mountains. Undergraduate Research Reports, Department of Geology, San Diego State University, California 11(4):1-31.

Mohorich, L.M. 1980. Geothermal resources of the

California desert. In Geology and Mineral Wealth of the California Desert, edited by D.L. Fife and A.R. Brown, South Coast Geological Society, Santa Ana, California p. 171-189.

Mooney, W.D., and G.A. McMechan 1982. Synthetic seismogram modeling for the laterally varying structure in the Central Imperial Valley In The Imperial Valley, California, earthquake of October 15, 1979, U.S. Geological Survey Professional Paper 1254:101-107.

Mooney, W.D., W.J. Lutter, J.H. Healey, and G.S. Fuis 1979. Velocity-depth structure in central Imperial Valley, California. EOS, Transactions of the American Geophysical Union 60:876.

Moore, C.V., J.H. Snyder, and P. Sun 1974. Effects of Colorado River water quality and supply on irrigated agriculture. Water Resources Research 10:137-144.

Moore, D.G. 1973. Plate-edge deformation and crustal growth, Gulf of California structural province. Geological Society of America Bulletin 84:1883-1906.

Moore, D.G., and E.C. Buffington 1968. Transform faulting and growth of the Gulf of California since the late Pliocene. Science 161:1238-1241.

Moore, D.G., and J.R. Curray 1982. Geologic and tectonic history of the Gulf of California. In Initial Reports of the Deep Sea Drilling Project 64:1279-1294.

Moore, J.N., and M.C. Adams 1986. Thermal and chemical evolution of the caprock in the Salton Sea geothermal field, California. Geological Society of America Abstracts with Programs 18:699.

----- 1988. Evolution of the thermal cap in two wells from the Salton Sea geothermal system, California. Geothermics 17:695-710.

Moore, R.C. 1967. Geomorphic analysis of some yardangs in the San Felipe Hills Imperial County, California. Undergraduate Research Reports, Geology Department, San Diego State University, California 11(4):38.

Moore, W.G. 1972. Geology of the eastern Fish Creek Mountains, San Diego and Imperial Counties, California. Undergraduate Research Reports, Department of Geology, San Diego State University, California 19:1-18.

Moos, D., R. Kovach, and A. Nur 1981. Measurements of seismic velocity and attenuation at the East Mesa KGRA, Imperial Valley, CA. EOS, Transactions of the American Geophysical Union 62:334.

Mori, J. 1991. Estimates of velocity structure and source depth using multiple P waves from aftershocks of the 1987 Elmore Ranch and Superstition Hills, California, earthquakes. Seismological Society of America Bulletin 81:508-523.

Mori, J., and A. Frankel 1990. Source parameters for earthquakes associated with the 1986 North Palm Springs, California, earthquake determined using empirical green functions. Seismological Society of America Bulletin 80:278-295.

Morgan, J.R. 1964. Provenance of the red, brown and gray porphyritic volcanic clasts from Paleocene conglomerates of southwestern San Diego County, California. Undergraduate Research Reports, Department of Geology, San Diego State University, California 8(4):1-28.

Morgan, T.G. 1971. The paleoecology of the Willis Palms Formation oyster beds. Geology 111B, Department of Geological Sciences, University of California, Riverside 11 p.

Mooris, C.W. and D.A. Campbell 1979. Geothermal reservoir energy recovery; a three-dimensional simulation study of the East Mesa Field. Society of Petroleum Engineers, AIME Annual Fall Technical Conference Exhibit, Paper 54:1-10.

Mooris, R.S. 1987. Tertiary basin formation above middle-crustal shear zones in southern Chocolate Mountains, SE California. Geological Society of America Abstracts with Programs 19:434.

----- 1990. Crustal structure of southeasternmost California: multiple deformational events evident in industry seismic profiles from the Milpitas Wash area, Chocolate Mountains. Master of Science Thesis, San Diego State University, California 139 p.

Morris, R.M., E.A. Keller, and G.L. Meyer 1979. Geomorphology of the Salton Basin, California: selected observations. In Geological Excursions in the Southern California Area, edited by P.L. Abbott, Department of Geological Sciences, San Diego State University, California p. 19-64.

Morris, R.S., D.A. Okaya, E.G. Frost, and P.E. Malin 1986. Base of the Orocopia schist as imaged on seismic reflection data in the Chocolate and Muchacho Mtns. region of SE Calif., and the Sierra Pelona region near Palmdale, Calif. Geological Society of America Abstracts with Programs 18:160.

Morris, R.S., and D.A. Okaya 1986. Crustal geometry of detachment faulting - structural analysis of seismic-reflection data in SE California. Geological Society of America Abstracts with Programs 18:160.

----- 1987. Tertiary overprint and disruption of the Chocolate Mountains thrust system and the Orocopia Schist in the southern Chocolate Mountains, SE California. Geological Society of America Abstracts with Programs 19:779.

Morris, R.S., and E.G. Frost 1985. Geometry of mid-Tertiary detachment faulting overprinted on the Chocolate Mountains thrust system from seismic reflection profiles in the Milpitas Wash region of southeastern California. EOS, Transactions of the American Geophysical Union 66:978-979.

Morley, E.R. Jr. 1963. Geology of the Borrego Mountain quadrangle and the western portion of Shell Reef quadrangle, San Diego County, California. Master of Arts Thesis, University of Southern California, Los Angeles 138 p.

Morse, J.G., and L.D. Thorson 1978. Reservoir engineering study of a portion of the Salton Sea geothermal field. Transactions, Geothermal Resources Council 2:471-474.

Morton, D.M. 1977. Surface deformation in part of the San Jacinto Valley, southern California. U. S. Geological Survey, Journal of Research 5(1):117-124.

Morton, D.M., J.C. Matti, and B.F. Cox 1980. Geologic map of the San Jacinto Wilderness, Riverside County, California. U.S. Geological Survey Map MF-1159-A.

Morton, D.M., J.C. Matti, F.K. Miller, and C.A. Repenning 1986. Pleistocene conglomerate from the San Timoteo Badlands, southern California; constraints on strike-slip displacements on the San Andreas and San Jacinto Faults. Geological Society of America Abstracts with Programs 18(2):161.

Morton, D.M., J.C. Matti, and J.C. Tinsley 1987. Banning Fault, Cottonwood Canyon, San Gorgonio Pass, southern California. In Centennial Field Guide, edited by M.L. Hill, Geological Society of America, Cordilleran Section 490(1):191-192.

Morton, D.M., and J.C. Matti 1993a. Tectonic synopsis of the San Gorgonio Pass and San Timoteo Badlands areas, southern California. San Bernardino County Museum Association Quarterly 40(2):3-14.

----- 1993b. Extension and contraction within an evolving convergent strike-slip fault system: The San Andreas and San Jacinto Fault Zones at their convergence in southern California. In The San Andreas Fault System; Displacement, Palinspastic Reconstruction and Geologic Evolution, edited by R.E. Powell, R.J. Weldon, and J.C. Matti, Geological Society of America Memoir 178: 217-230.

Morton, P.K. 1966. Geologic map of Imperial County, California. California Division of Mines and Geology, Scale 1:125,000.

----- 1977. Geology and Mineral Resources of Imperial County, California. California Division of Mines and Geology County Report 7:1-104.

Mott, W.K. 1980. Analysis of the Elsinore Fault along the southwest flank of the Coyote Mountains. Undergraduate Research Reports, Department of Geology, San Diego State University, California 37(3):1-15.

Mount, J.D. 1974. Molluscan evidence for the age of the Imperial Formation, southern California. Southern California Academy of Sciences Abstracts with Programs p. 9.

----- 1988. Molluscan fauna of the Imperial Formation. In Geologic Field Guide to the Western Salton Basin Area, edited by S.M. Testa and K.E. Green, American Institute of Professional Geologists p. 23-24.

Moyle, W.R., Jr. 1968. Water wells and springs in Borrego, Carrizo, and San Felipe Valley areas, San Diego and Imperial Counties, California. California Department of Water Resources Bulletin 91-15:1-16.

----- 1971. Water wells in the San Luis Rey River valley area, San Diego County, California. California Department of Water Resources Bulletin 91(18):1-11.

----- 1982. Water resources of Borrego Valley and vicinity, California, phase 1 -- definition of geologic and hydrologic characters of the basin. U.S. Geological Survey Open-File Report 82-855:1-39.

Moyle, W.R., Jr., D.J. Downing 1977. Summary of water resources for the Campo, Cuyapaipe, La Posta, and Manzanita Indian Reservations and vicinity, San Diego County, California. U.S. Geological Survey Open-File Report 77-684:1-93.

Mueller, C.S. 1987. The influence of site conditions on near-source high frequency ground motion; case studies from earthquakes in Imperial Valley, CA., Coalinga, CA, and Miramichi, Canada. Doctoral Dissertation, Stanford University, California 239 p.

Mueller, C.S., D.M. Boore, and R.L. Porcella 1982. Detailed study of site amplification at El Centro strong-motion array station No. 6. In Third International Earthquake Microzonation Conference Proceedings, edited by M. Sherif, National Science Foundation, Washington, D.C. 1:413-424.

Mueller, K.J. 1984. Neotectonics, alluvial history and soil chronology of the southwestern margin of the Sierra de Los Cucupas, Baja California Norte. Master of Science Thesis, San Diego State University, California, 363 p.

Mueller, R.F., and K.C. Condie 1964. Stability relations of carbon mineral assemblages in the southern California batholith. Journal of Geology 72:400-411.

Muffler, L.P.J., and B.R. Doe 1968. Composition and mean age of detritus of the Colorado River delta in the Salton Trough, southeastern California. Journal of Sedimentary Petrology 38:384-399.

Muffler, L.P.J., and D.E. White 1968. Origin of $CO_2$ in the Salton Sea geothermal system, southeastern California, U.S.A. Proceedings 23rd International

Geological Congress, Prague, Academia p. 185-194.

Muffler, L.P.J., and D.E. White 1969. Active metamorphism of upper Cenozoic sediments in the Salton Sea geothermal field and the Salton Trough, southeastern California. Geological Society of America Bulletin 80:157-182.

Muir, S.G., and A.E. Fritsche 1981. Three-dimensional model of probable earthquake-induced intrusion structure in Kern Lake sediment compared with sandblow feature formed during the 1979 Imperial Valley Earthquake. Geological Society of America Abstracts with Programs 13:98.

Muir, S.G., and R.F. Scott 1982. Earthquake generated sandblows formed during the main shock. In The Imperial Valley, California, Earthquake of October 15, 1979, U.S. Geological Survey Professional Paper 1254:247-250.

Mullich, J.R. (no date). Geology of a portion of the Coyote Mountains. Manuscript on File, Anza-Borrego Desert State Park, California 4 p.

Muncill, G.E. 1984. Igneous petrologic evolution near a plutonic-metasedimentary contact in the Peninsular Ranges Batholith, southern California. Master of Science Thesis. San Diego State University, California, 112p.

Munguia-Orozco, L. 1983. Strong ground-motion and source mechanism studies for earthquakes in the northern Baja California-southern California region. Doctoral Dissertation, University of California, San Diego 152 p.

Munguia, L., and J.N. Brune 1984. Local magnitude and sediment amplification observations from earthquakes in the northern Baja California - southern California region. Seismological Society of America Bulletin 74:107-119.

Muramoto, F.S., and W.A. Elders 1984. Identification of zones of progressive hydrothermal metamorphism from wire line logs and their correlation to reservoir temperatures in the Salton Sea and Westmorland geothermal systems, Imperial Valley, California. In The Imperial Basin, Tectonics, Sedimentation and Thermal Aspects, edited by C.A. Rigsby, Society of Economic Paleontologists and Mineralogists 40:87-89.

Muramoto, F.S. 1982. Well log study of hydrothermally altered sediments in the Salton Sea geothermal system and Westmorland area, Imperial Valley, California. Master's Thesis, University of California, Riverside.

Muren, C.R. 1980. Quaternary geology and a proposed geologic origin of Fish Creek drainage in San Diego. Undergraduate Research Reports, Department of Geology, San Diego State University, California 36(3):1-16.

Murowchick, J.B. 1979. Preliminary investigations of mineral precipitation from the Salton Sea geothermal brines. Master's Thesis, Pennsylvania State University, University Park.

Murphy, G.P., R.M. Tosdal, J.L. Wooden, J. Kent, and R.B. Vaughn 1990. Chemical and isotopic character of Jurassic granitoids, Cargo Muchacho Mountains, SE California. Geological Society of America Abstracts with Programs 22(3):71.

Murphy, M.A. 1979. California Desert Conservation Area invertebrate paleontological resources study. Bureau of Land Management, Riverside, California.

Murphy, M.A. 1986. The Imperial Formation at Painted Hill, near Whitewater, California. In Geology Around the Margins of the Eastern San Bernardino Mountains, edited by M.A. Kooser and R.E. Reynolds, Inland Geological Society Publications, Redlands, California 1:63-70.

Murray, K.S. 1979. Cenozoic geology of the Palo Verde Mountain volcanic field. Geological Society of America Abstracts with Programs 12:489.

----- 1980. Implications of Cenozoic volcanism in the Palo Verde Mountain volcanic field. In Geology and Mineral Wealth of the California Desert, edited by D.L. Fife and A.R. Brown, South Coast Geological Society, Santa Ana, California p. 256-258.

----- 1981. Tectonic implication of space-time patterns of Cenozoic volcanism in the Palo Verde Mountain volcanic field, southeastern California. Doctoral Dissertation, University of California, Davis 132 p.

----- 1982. Tectonic implications of Cenozoic volcanism in southeastern California. In Mesozoic-Cenozoic Tectonic Evolution of the Colorado River Region, California, Arizona, and Nevada, edited by E.G. Frost and D.L. Martin, Cordilleran Publishers, San Diego, California p. 77-83.

Murray, K.S., J.W. Bell, B.M. Crowe, and D.G. Miller 1980. Geologic structure of the Chocolate Mountains region. In Geology and Mineral Wealth of the California Desert, edited by D.L. Fife and A.R. Brown, South Coast Geological Society, Santa Ana, California p. 121-123.

Murray, K.S., P.R. Butler, and B.W. Troxel 1982. Geology of the Salton Sea 1° x 2° NTMS quadrangle, California, Arizona. Geology and Mineral Resources of the Los Angeles, Needles, Salton Sea, San Bernardino and Trona 1° x 2° NTMS quadrangles Report GJBX-213(82):87-123.

Murray, K.S., and B.M. Crowe 1976. Geology of the northern and central Chocolate Mountains, southeastern, California. Geological Society of America Abstracts with Programs 8:1023-1024.

Naas, J.T. 1983. Aftershocks and energy release associated

with the Imperial Valley Earthquake of October 15, 1979. Undergraduate Research Reports, Department of Geology, San Diego State University, California 42(5):47.

Naeser, C.W., G.A. Izett, and R.E. Wilcox 1973. Zircon fission-track age of Pearlette family ash beds in Meade County, Kansas. Geology 1:187-189.

Nason, R. 1982. Seismic-intensity studies in the Imperial Valley, California, earthquake of October 15, 1979. In The Imperial Valley, California, Earthquake of October 15, 1979, U.S. Geological Survey Professional Paper 1254:259-264.

Needham, P.B., Jr., W.D. Riley, G.R. Conner, and A.P. Murphy 1980. Chemical analyses of brines from four Imperial Valley, Ca., geothermal wells. Society of Petroleum Engineers, AIME Journal 20:105-112.

Negmatullayev, S.K., S.L. Mikhaylov, O.B. Khavroshkin, and V.V. Tsyplakov 1987. Nelineynyye svoystva volnovogo polya zemletryaseniya Imperial Vel'yu 15.10.79. Ixvestiya Akademii Nauk Tadzhikskoy SSSR. Otdeleniye Fiziko Matematicheskikh, Khimicheskikh i Geologicheskikh Nauk 1987(4):78-83.

Nelson, D.C. 1979. Fault patterns and frontal fault tectonics of the southwest Coyote Mountains. Undergraduate Research Reports, Department of Geology, San Diego State University, California 35(3):1-15.

Netherton, R., A.J. Piwinskii, and M. Chan 1977. Viscosity of brine from the Salton Sea geothermal field, California from 25 C to 90 C at 100 kPa. EOS, Transactions of the American Geophysical Union 58:1248.

Neumann, F. 1941. The analysis of the El Centro record of the Imperial Valley Earthquake of May 18, 1940. American Geophysical Union Transactions, 22d Annual Meeting 22:400.

----- 1947. Measurement of permanent ground displacement by geodetic and seismographic methods. Geological Society of America Bulletin 58:1267.

Newberry, J.S. 1861. Geological report. In Report on Colorado River of the West, U.S. 36th Congress, 1st Session, Senate Documents p. 1-154.

Newmark, R.L., P.W. Kasameyer, L.W. Younker, and P.C. Lysne 1985. Thermal gradients in the southern Salton Sea region. EOS, Transactions of the American Geophysical Union 66:365.

----- 1986. Research drilling at the Salton Sea geothermal field, California; the Shallow Thermal Gradient Project. EOS, Transactions of the American Geophysical Union 67:698-707.

----- 1988. Shallow drilling in the Salton Sea region; the thermal anomaly. Journal of Geophysical Research, B, Solid Earth and Planets 93:13,005-13,023.

Niazi, M. 1982a. Coherence of the strong ground motion at neighboring sites of similar surface geology as observed during the 1979 Imperial Valley Earthquake. In Third International Earthquake Microzonation Conference Proceedings, edited by M. Sherif, National Science Foundation, Washington, D.C. 1:425-434.

----- 1982b. Source dynamics of the 1979 Imperial Valley Earthquake from near-source observations (of ground acceleration and velocity). U.S. Geological Survey Open-File Report 82-591:912-934.

----- 1982c. Source dynamics of the 1979 Imperial Valley Earthquake from near-source observations of ground acceleration and velocity. Seismological Society of America Bulletin 72(6):1957-1968.

----- 1985. Spatial coherence of the ground motion produced by the 1979 Imperial Valley Earthquake across El Centro differential array. Physics of the Earth and Planetary Interiors 38:162-173.

Niazi, M., and C.P. Mortgat 1983. Induced rocking and torsional vibrations of a long structure in the near field of the 1979 Imperial Valley Earthquake. EOS, Transactions of the American Geophysical Union 64:766.

Nichols, H.W. 1906. New forms of concretions, sand-calcite concretions from Salton, California. Field Museum, Chicago, Geology Series 3(3):25-31.

Nichols, J.D. 1976. An evaluation of an arid environment utilizing remote sensing; the Borrego Valley, California. Master of Science Thesis, San Diego State University, California 84 p.

Nicholson, C., L. Seeber, P.L. Williams, and L.R. Sykes 1986. Seismic evidence for conjugate slip and block rotation within the San Andreas Fault system, southern California. Tectonics 5(4):629-648.

Nicholson, C., and L. Seeber 1989. Evidence for contemporary block rotation in strike-slip environments: examples from the San Andreas Fault system, southern California. In Paleomagnetic Rotations and Continental Deformation, edited by C. Kissel and C. Laj, Kluwer Academic Publishers: 247-280.

Nicholson, C., and E. Hauksson 1992. The April 1992 M6.1 Joshua Tree Earthquake sequence: seismotectonic analysis and implications. EOS, Transactions of the American Geophysical Union 73:363.

Nicholson, C., and J.M. Lees 1992. Travel-time tomography in the northern Coachella Valley using aftershocks of the 1986 M5.9 North Palm Springs Earthquake. Geophysical Research Letters 19:1-4.

Nielson, J., and G. Kocurek 1984. Migrating zebras of the Algodunes dune field, southeastern California. Society of Economic Paleontologists and Mineralogists, Mid-year Meeting Abstracts 1:60.

Nielson, J., and K.K. Beratan 1990. Tertiary basin

development and tectonic implications, Whipple detachment system, Colorado River extensional corridor, California and Arizona. Journal of Geophysical Research, B, Solid Earth and Planets p. 599-614.

Nishenko, S.P., and L.R. Sykes 1983. Probabilities of occurrence of large plate rupturing earthquakes for the San Andreas, San Jacinto and Imperial Faults, California, 1983-2003. EOS, Transactions of the American Geophysical Union 64:258.

Noble, L.F. 1926. The San Andreas rift and some other active faults in the desert region of southeastern California. Carnegie Institute of Washington Yearbook 25:415-428.

----- 1927. The San Andreas rift and some other active faults in the desert region of southeastern California. Seismological Society of America Bulletin 17:25-39.

----- 1932. The San Andreas rift and some other active faults in the desert region of southeastern California. Carnegie Institute of Washington Yearbook 31:355-363.

Norell, M.A. 1983. Late Neogene lizards from Anza-Borrego Desert, San Diego County, California. Master of Science Thesis, Department of Biology, San Diego State University, California 199 p.

----- 1989. Late Cenozoic lizards of the Anza-Borrego Desert, California. Natural History Museum of Los Angeles County Contributions in Science 414:1-31.

Norris, G.M. 1988. Liquefaction at the Meloland Overcrossing during the Imperial Valley Earthquake of 1979. Bulletin of the Association of Engineering Geologists 25:235-247.

Norris, R.M. 1966. Barchan dunes of Imperial Valley, California. Journal of Geology 74:292-306.

----- 1994. The Indio Hills "Desert Badlands". California Geology 47(5):127-133.

Norris, R.M., and K.S. Norris 1957. Origin of the Algodunes Dunes, Imperial County, California. Geological Society of America Bulletin 68:1838-1839.

----- 1961. Algodones dunes of southeastern California. Geological Society of America Bulletin 72:605-619.

Norris, R.M., E.A. Keller and G.L. Meyer 1979. Geomorphology of the Salton Basin, California: Selected observations. In Geological Excursions in the Southern California area, edited by P.L. Abbott. Department of Geological Sciences, San Diego State University, California p. 19-46.

Norris, R.M., and R.W. Webb 1990. Colorado Desert. In Geology of California, edited by R.M. Norris, and R.W. Webb, John Wiley & Sons, New York p. 250-276.

Nuffer, C.D. 1989. Compositional variation of Holocene sand deposited in a transtensional rift system, Salton Basin, California. Master of Science Thesis, San Diego State University, California 101 p.

Oakes, C.J. 1988. Evidence for replacement of dilute hydrothermal solutions by hot, hypersaline brine in the northeastern part of the Salton Sea geothermal system, California: a fluid inclusion and oxygen isotope study. Master's Thesis, University of California, Riverside 115 p.

O'Connell, M. 1978. Noble gas sampling of the geothermal heat lock at Niland Field, California. Proceedings of the Second workshop on sampling geothermal effluents, Report EPA-600/7-78-121:1-15.

O'Connor, J. 1988. Bar and swale topography analysis. Geosciences 650, University of Arizona, Tucson, Manuscript on File Anza-Borrego Desert State Park, California 7 p.

Odum, H.A. 1982. Metamorphism and depositional environment of the Tule Canyon roof pendant. Undergraduate Research Reports, Department of Geology, San Diego State University, California 39(4):1-41.

Olmsted, F.H., O.J. Loeltz, and B. Irelan 1974. Geohydrology of the Yuma area, Arizona and California. U.S. Geological Survey Professional Paper 486-H:1-227.

Olmsted, F.H., and C.C. MacDonald 1967. Hydrologic studies of the lower Colorado River region. Water Resources Bulletin 3:45-58.

----- 1967. Progress report on geologic investigations of the Yuma and the East Mesa area of Imperial Valley. U.S. Geological Survey Open-File Report 45-58:1-3.

Olmsted, T.L. 1958. Petrology and petrography of an area east of Davies Valley, California. Undergraduate Research Reports, Department of Geology, San Diego State University, California 2(2):1-12.

Olson, A., and R.J. Apsel 1982a. Finite faults and inverse theory with applications to the 1979 Imperial Valley Earthquake. Seismological Society of America Bulletin 72:1969-2001.

----- 1982b. Finite faults and inverse theory with application to the 1979 Imperial Valley Earthquake. U.S. Geological survey Open-file Report 82-0591:877-911.

Olson, A., and J. Orcutt 1980. The effects of realistic Imperial Valley structure on local and near field propagation of seismic waves. EOS, Transactions of the American Geophysical Union 61:108-109.

Olson, E.R. 1976. Oxygen isotope study of diabase, Heber geothermal field, Imperial Valley, California. Geological Society of America Abstracts with Programs 8:1036.

----- 1977. Water-rock ratios and fluid mixing in the Salton

Sea geothermal field. Geological Society of America Abstracts with Programs 9:1119-1120.

----- 1978. Oxygen isotope studies of the Salton Sea geothermal field; new insights. New Zealand Department of Scientific and Industrial Research Bulletin 220:121-126.

O'Malley, M.A. 1969. Petrographic relationships of rocks in a portion of the Santa Ysabel 15' quadrangle, San Diego County, California. Undergraduate Research Reports, Department of Geology, San Diego State University, California 14(2):1-39.

Onken, J.A., and S.L. Rathburn 1988. Fan fluvial system response to base level lowering. Geosciences 650, University of Arizona, Tucson, Manuscript on File Anza-Borrego Desert State Park, California 22 p.

Opdyke, N.D., E.H. Lindsay, N.M. Johnson, and T. Downs 1974. The magnetic polarity stratigraphy of the mammal bearing sedimentary sequence at Anza-Borrego State Park, California. Geological Society of America Abstracts with Programs 6:901.

----- 1977. The paleomagnetism and magnetic polarity stratigraphy of the mammal-bearing section of Anza-Borrego State Park, California. Quaternary Research 7:316-329.

Oquita, R. 1985. Geothermal development in the Imperial Valley. In Geology and Geothermal Energy of the Salton Trough, edited by L.J. Herber, National Association of Geology Teachers, Far Western Section p. 108-111.

Orcutt, C.R. 1890. Geology of the Colorado Desert. California Mining Bureau Annual Report 10:899-919.

----- 1901a. The Colorado Desert. West American Scientist 12(102):2-11.

----- 1901b. The Colorado Desert. West American Scientist 15(128):38-48.

Ordway, A. 1993. Pegmatites in the Chihuahua Valley region, Riverside County, California. In Ashes, Faults and Basins, edited by R.E. Reynolds and J. Reynolds, San Bernardino County Museum Association Special Publication 93-1:97.

O'Rourke, M.J., and R. Dohry 1982. Apparent horizontal propagation velocity for the 1979 Imperial Valley Earthquake. Seismological Society of America Bulletin 72(6, part A):2377-2380.

Orrell, S.E., J.L. Anderson, J.L. Wooden, and J.E. Wright 1987. Proterozoic crustal evolution of the lower Colorado River region; rear-arc orogenesis to an orogenic crustal remobilization. Geological Society of America Abstracts with Programs 19(7):795.

Ortel, W.H. 1977. Geology and reservoir analysis of East Mesa geothermal field, Imperial County, California. American Association of Petroleum Geologists Bulletin 61:1385.

Osborn, W.L. 1989. Formation, diagenesis, and metamorphism of lacustrine sulfates in the Salton Trough. Master's Thesis, University of California, Riverside.

Osterholt, W.R.B. 1934. The Origin of the main physiographical features of Borrego Valley. Master of Science Thesis, University of Southern California, Los Angeles 37 p.

Owen, L.B. 1975. Precipitation of amorphous silica from high-temperature hypersaline geothermal brine. Lawerence Livermore Laboratory, University of California Report 20 p.

----- 1977. Properties of siliceous scale from the Salton Sea Geothermal Field. Transactions, Geothermal Resources Council 1:235-237.

----- 1978. Chemical modification of hypersaline geothermal brine for scale control. Abstracts of Papers of the 144th National Meeting, American Association for the Advancement of Science p. 112.

Owen, L.B., and D. Jackson 1976. Precipitation of amorphous silica from high-temperature-hypersaline geothermal brine. EOS, Transactions of the American Geophysical Union 57:354.

Owens, E., M. Osborne, and L. Kennedy 1988. Metasomatism of the Tumco Formation, Cargo Muchacho Mountains, southeastern California. Geological Society of America Abstracts with Programs 20:219.

Page, L.E., M.A. Chan, and J.D. Tewhey 1979. Cathodoluminescent zonation of anhydrite veins from the Salton Sea geothermal field, California. Geological Society of America Abstracts with Programs 11:121.

Paillet, F.L., and R.H. Morin 1987. Preliminary geophysical well log analysis of the geothermal basin, California. U.S. Geological Survey Circular, Report C-0995:53-54.

----- 1988. Analysis of geophysical well logs obtained in the State 2-14 Borehole, Salton Sea geothermal area, California. Journal of Geophysical Research, B, Solid Earth and Planets 93:12,981-12,994.

Paillet, F.L., editor, R.H. Morin, R.E. Hodges, L.C. Robinson, S.S. Priest, J.H. Sass, J.D. Hendricks, P.W. Kasameyer, G.A. Pawloski, R.C. Carlson, A.G. Duba, J.R. Hearst, and R.L. Newmark 1986. Preliminary report on geophysical well-logging activity on the Salton Sea Scientific Drilling Project, Imperial Valley, California. U.S. Geological Survey Open-File Report 86-544:1-99.

Pajak, A.F. III 1993. The second record of *Tapirus* from the Temecula Valley, southern California, and biostratigraphic implications. Abstracts of Proceedings for the 7th annual Mojave Desert Quaternary Research Symposium and the Desert

Studies Consortium Symposium. San Bernardino County Museum Association Quarterly 40(2):30.

Palmer, A.H. 1916. California earthquakes during 1915. Seismological Society of America Bulletin 6:8-25.

----- 1918. California earthquakes during 1917. Seismological Society of America Bulletin 8:1-12.

Palmer, A.H., J.H. Howard, and D.P. Lande 1975. Geothermal development of the Salton Trough, California and Mexico. National Technical Information Service 45 p.

----- 1976. Geothermal development of the Salton Trough, California and Mexico. In Geothermal World Directory; 1975-1976 Bicentennial Edition, edited by K.F. Meadows, Glendora, California p. 280-317.

Pappajohn, S. 1980. Description of Neogene marine section at Split Mountain, easternmost San Diego County, California. Master of Science Thesis, San Diego State University, California 77 p.

Parcel, R.F, Jr. 1981. Structure and petrology of the Santa Rosa Shear Zone in the Pinyon Flat Area, Riverside County, California. In Geology of the San Jacinto Mountains, edited by A.R. Brown and R.W. Ruff, South Coast Geological Society Annual Field Trip Guidebook, Santa Ana, California 9:139-150.

Pardoen, G.C. 1983. Ambient vibration test results of the Imperial County Services Building. Seismological Society of America Bulletin 73:1895-1902.

Pardoen, G.C., P.J. Moss, and A.J. Carr 1983. Elastic analysis of the Imperial County Services Building. Seismological Society of America Bulletin 73(6, part, A):1903-1916.

Parks, J., B. Stout, G.J. Miller, P. Remeika, and V.E. Waters 1989. A progress report on half-million-year-old marks on mammoth bones from the Anza-Borrego Desert Irvingtonian. Mojave Desert Quaternary Research Symposium, Abstracts of Proceedings, San Bernardino County Museum Association Quarterly 36(2):63.

Pasqualetti, M.J., J.B. Pick, and E.W. Butler 1978. Geothermal energy in Imperial County, California; research conclusions and policy implications. Abstracts of Papers of the 144th National Meeting, American Association for the Advancement of Science p. 111.

----- 1979. Geothermal energy in Imperial County, California; environmental, socio-economic, demographic, and public opinion research conclusions and policy recommendations. Energy 4:67-80.

Payen, L.A., C.H. Rector, E. Ritter, R.E. Taylor, and J.E. Ericson 1978. Comments on the Pleistocene age assignment and associations of a human burial from the Yuha Desert, California. American Antiquity 43:448-452.

Pechmann, J.C., J.B. Minister, and H. Kanamori 1980. Search for seismological precursors to the 1979 Imperial Valley Earthquake. EOS, Transactions of the American Geophysical Union 61:1053.

Pechmann, J.C., and H. Kanamori 1981. Waveform studies of preshocks and aftershocks of the 1979 Imperial Valley Earthquake; evidence for asperities? EOS, Transactions of the American Geophysical Union 62:950-951.

----- 1982. Waveforms and spectra of preshocks and aftershocks of the 1979 Imperial Valley, California, earthquake; evidence for fault heterogeneity? Journal of Geophysical Research 87:10,579-10,597.

Perry, L.E. 1979. Salton Sea concretions. Rock and Gem 9(3):56-58,59.

----- 1981. Mysterious sandstone concretions of the Salton Basin. Lapidary Journal 34:2228-2233.

Peterson, F. 1960. Geology of Alverson Canyon, western Imperial County, California. Undergraduate Research Reports, Department of Geology, San Diego State University, California 4(2):37.

Peterson, G.L. 1970. Distinction between Cretaceous and Eocene conglomerates in the San Diego area. In Pacific Slope Geology of Northern Baja California and Adjacent Alta California, American Association of Petroleum Geologists, Pacific Section, Fall Field Trip Guidebook p. 90-98.

Peterson, G.L., R.G. Gastil, and E.C. Allison 1966. Geology of the Peninsular Ranges. California Division of Mines Bulletin 191:70-73.

----- 1970. Geology of the Peninsular Ranges. California Division of Mines and Geology Mineral Information Service 23:124-127.

Peterson, G.L., R.G. Gastil, J.A. Minch, and C.E. Nordstrom 1968. Clast suites in the late Mesozoic-Cenozoic succession of the western Peninsular Ranges province, southwestern California and northwestern Baja California. Geological Society of America Special Papers 115:177.

Peterson, M.S. 1975. Geology of the Coachella Fanglomerate, edited by J.C. Crowell. In San Andreas Fault in Southern California. California Division of Mines and Geology, Special Report 118:119-126.

Pettinga, J.R. 1991. Structural styles and basin margin evolution adjacent to the San Jacinto Fault Zone, southern California. Geological Society of America Abstracts with Programs, San Diego p. A257.

----- (in prep). Structural styles and basin margin evolution adjacent to the San Jacinto Fault Zone, Borrego Badlands, Anza Borrego Desert State Park, California.

Phelps, P.L., and L.R. Anspaugh, editors 1976. Imperial

Valley Environmental Project; progress report. Lawerence Livermore Laboratory, University of California, Berkeley 50044-76-1:1- 214.

Phillips, M. 1974. A VLF survey of a portion of the Elsinore Fault Zone. Undergraduate Research Reports, Department of Geology, San Diego State University, California 24(2):1-12.

Phillips, T.W. 1977. Turquoise mineralization in the Gold Basin district Imperial County, California. Undergraduate Research Reports, Geology Department, San Diego State University, California 30(2):34.

Pick, J.B., and E.W. Butler 1979. An overview of socioeconomic and environmental studies of geothermal development in the Imperial Valley. In Geology and Geothermics of the Salton Trough, edited by W.A. Elders, University of California, Riverside Campus Museum Contributions 5:77-81.

Pierotti, S.M. 1979. Spatial and temporal analysis of the B-coefficient for the earthquake occurrence relation in southern California. Undergraduate Research Reports, Department of Geology, San Diego State University, California 35(4):24.

Pierson, D.E. 1974. Imperial Valley's proposal to develop a guide for geothermal development within its county. Conference on Research for the Development of Geothermal Energy Resources, California Institute of Technology, Jet Propulsion Laboratory-National Science Foundation, Pasadena p. 160-163.

----- (no date). Geothermal resource development; Imperial County. Annual Conference National Water Supply Improvement Association, Technical Proceedings 5:1-4 p.

Pimentel, K.D., R.R. Ireland, and G.A. Tomkins 1978. Chemical fingerprints to assess the effects of geothermal development on water quality in Imperial Valley. Transactions, Geothermal Resources Council 2(2):527-530.

Pinault, C.T. 1984. Structure, tectonic geomorphology, and neotectonics of the Elsinore Fault Zone between Banner Canyon and the Coyote Mountains, southern California. Master of Science Thesis, San Diego State University, California 231 p.

Pinault, C.T., and T.K. Rockwell 1984. Rates and sense of Holocene faulting on the Elsinore Fault: further constraints on the distribution of dextral shear between the Pacific and North American plates. Geological Society of America Abstracts with Programs 16(6):624.

Piwinskii, A.J., and B.P. Bonner 1978. The effect of diagenetic and hydrothermal alteration on sound propagation in sedimentary rocks from the Dune borehole, Imperial Valley, California. EOS, Transactions of the American Geophysical Union 59:1201.

Piwinskii, A.J., and L. Dengler 1976. Pore structure and mineralogy of core chips from the State of California well No. 1, Salton Sea geothermal field. EOS, Transactions of the American Geophysical Union 57:1017.

Piwinskii, A.J., R. Netherton, and M. Chan 1977. Viscosity of brines from the Salton Sea geothermal field, Imperial Valley, California. University of California, Riverside, Report 52344:1-8.

Pollock, T.C. Geology of Grey Mountain area, Jacumba, California. Undergraduate Research Reports, Department of Geology, San Diego State University, California 39(4):1-12.

Popenoe, F.W. 1959. Geology of the southeastern portion of the Indio Hills, Riverside County, California. Master of Arts Thesis, University of California, Los Angeles.

----- 1963. Coachella Valley, geologic evolution: ocean to grassland to desert. Desert Magazine March 26:12-17.

Porcella, R.L. 1983. Strong-motion program report, January-December 1981. U.S. Geological Survey Circular 914:19.

----- 1984a. Geotechnical investigations at strong-motion stations in the Imperial Valley, California. Earthquake Notes 55:26-27.

----- 1984b. Geotechnical investigations at strong-motion stations in the Imperial Valley, California. U.S. Geological Survey Open-File Report 84-562:1-180.

Porcella, R.L., E.C. Etheredge, R.P. Maley, and J.C. Switzer 1987. Strong-motion data from the Superstition Hills earthquakes of 0514 and 1315 (GMT), November 24, 1987. U.S. Geological Survey Open-File Report 87-672:1-56.

Porcella, R.L., and R.B. Matthiesen 1979. Preliminary summary of U.S. Geological Survey strong-motion records from the October 15, 1979 Imperial Valley Earthquake. U.S. Geological Survey Open-File Report 79-1654:1-42.

----- 1979. Preliminary summary of the U.S. Geological Survey strong-motion records from the October 15, 1979 Imperial Valley Earthquake. Earthquake Engineering Research Institute Newsletter 13:31-37.

Porcella, R.L., R.B. Matthiesen, and R.P. Maley 1982. Strong-motion data recorded in the United States. In the Imperial Valley, California, Earthquake of October 15, 1979. U.S. Geological Survey Professional Paper 1254;289-318.

Porter, L.D. 1982. Data-processing procedures for main-shock motions recorded by the California division of Mines and Geology strong-motion network. In The Imperial Valley, California, Earthquake of October

15, 1979, U.S. Geological Survey Professional Paper 1254:407-432.

----- 1983. Processed data from the strong-motion records of the Imperial Valley Earthquake of October 15, 1979. California Division of Mines and Geology Special Publication 65:317.

Potter, R.M., and G.R. Rossman 1979. The manganese and iron-oxide mineralogy of desert varnish. Chemical Geology 25:79-94.

Powell, C.L., II, 1984a. Bivalve molluscan paleoecology on northern exposures of the marine Neogene Imperial Formation in Riverside County, California. Western Society of Malacologists Annual Report 17:1-32.

----- 1984b. Review of the marine Neogene Imperial Formation, southern California. Society of Economic Paleontologists and Mineralogists, Pacific Section p. 29.

----- 1986. Stratigraphy and bivalve molluscan paleontology of the Neogene Imperial Formation in Riverside County, California. Master of Science Thesis, California State University, San Jose 275 p.

----- 1987a. Correlation between sea level events and deposition of marine sediments in the proto-Gulf of California during the Neogene. Geological Society of America Special Paper 19(7):809.

----- 1987b. Paleogeography of the Imperial Formation of southern California and its molluscan fauna: an overview. Western Society of Malacologists Annual Report 20:11-18.

Powell, J.W. 1891a. The new lake in the desert. Scribner's Magazine 10:463-468.

----- 1891b. The flooding of the Colorado Desert. Engineering and Mining Journal 52:9.

Prescott, W.H., J.C. Savage, and M. Lisowski 1987. Deformation across the Salton Trough, southern California. U.S. Geological Survey Open-File Report 87-374:1-272.

Preston, C. 1966. Geology of a portion of the Sweeney Pass and Carrizo Mountain 7 1/2' quadrangles. Undergraduate Research Reports, Department of Geology, San Diego State University, California 10(4):1-23.

Preston, E.M. 1892. Salton Lake. California State Mining Bureau Report 11:387-393.

Pridmore, C.L., and E.G. Frost 1992. Detachment faults, California's extended past. California Geology January/February:3-17.

Proctor, R.J. 1958. Geology of the Desert Hot Springs area, Little San Bernardino Mountains, California. Master of Arts Thesis, University of California, Los Angeles.

----- 1968. Geology of the Desert Hot Springs-upper Coachella Valley area, California. California Division of Mines and Geology Special Report 64:1-50.

Pruss, D.E., G.W. Olcott, and W.A. Oesterling 1959. Aerial geology of a portion of the Little San Bernardino Mountains, Riverside and San Bernardino Counties, California. Geological Society of America Special Paper 70(12):1741.

Quinn, H.A. 1990. Dynamics stress drop and rupture dynamics of the October 15, 1979 Imperial Valley, California, earthquake. Tectonophysics 175:93-117.

Quinn, H.A., and T.M. Cronin 1984. Micro-paleontology and depositional environments of the Imperial and Palm Spring Formations, Imperial Valley, California. In The Imperial Basin, Tectonics, Sedimentation and Thermal Aspects, edited by C.A. Rigsby, Society of Economic Paleontologists and Mineralogists 40:71-85.

Raleigh, C.B. 1958. Stratigraphy and petrology of a part of the Orocopia schists. Master's Thesis, Claremont Graduate School, California 64 p.

Randall, W. 1977. An analysis of the subsurface structure and stratigraphy of the Salton Sea geothermal anomaly, Imperial Valley, California. Doctoral Dissertation, University of California, Riverside 159 p.

Randall, W., H.T. Meidav, R.W. Rex, and L. Coursey 1968. Electrical resistivity and geochemistry of aquifers in the Durmid dome, Imperial Valley. EOS, Transactions of the American Geophysical Union 49:759.

Rasmussen, G.S. 1982. Historic earthquakes along the San Jacinto Fault Zone, San Jacinto, California. In Neotectonics in Southern California, edited by J.D. Cooper, Geological Society of America, Cordilleran Section p. 115-121.

Raven, P.H., and D.I. Axelrod 1977. Origin and relationships of the California flora. University of California Publications in Botany 72:1-134.

Ravenscroft, A. 1969. Geology of the western Coyote Mountains, San Diego County, California. Undergraduate Research Reports, Department of Geology, San Diego State University, California 14(2):1-31.

Reagor, B.G., C.W. Stover, S.T., Algermissen, K.V. Steinbrugge, P. Hubiak, M.G. Hopper, and L.M. Barnhard 1982. Preliminary evaluation of the distribution of seismic intensities. In The Imperial Valley, California, Earthquake of October 15, 1979, U.S. Geological Survey Professional Paper 1254:251-258.

Real, C.R. 1982. Effects of shaking on residues near the Imperial Fault rupture. In The Imperial Valley, California, Earthquake of October 15, 1979, U.S. Geological Survey Professional Paper 1254:265-271.

Real, C.R., R.D. McJunkin, and E. Leivas 1979. Effects of Imperial Valley Earthquake; 15 October 1979, Imperial County, California. California Geology 32:259-265.

Redway, J.W. 1892. The new lake in the Colorado Desert, California. Royal Geographical Society Proceedings, New Series, London 14:309-314.

----- 1893. Salton Lake. The Geographical Journal 2:170-171.

Reilinger, R. 1983. Geodetic evidence for a seismic slip on the Brawley Fault, southern California, 1972-1974. EOS, Transactions of the American Geophysical Union 64:678.

----- 1984. Coseismic and postseismic vertical movements associated with the 1940 M7.1 Imperial Valley, California, earthquake. Journal of Geophysical Research 89:4531-4537.

Reilinger, R., J. Beaven, L. Gilbert, S. Larsen, K. Hudnut, C.L.V. Aiken, D. Zeigler, B. Strange, M.G. de la Fuente, J.G. Garcia, J. Stowell, W. Young, G. Doyle, and G. Stayner 1990. 1990 Salton Trough-Riverside County GPS network. EOS, Transactions of the American Geophysical Union 71:477.

Reilinger, R., and S. Larsen 1985. Constraints on the age of the present fault configuration in the Imperial Valley, California. EOS, Transactions of the American Geophysical Union 66:1094.

----- 1986. Vertical crustal deformation associated with the 1979 M=6.6 Imperial Valley, California, earthquake--Implications for fault behavior. Journal of Geophysical Research 91:14044-14056.

Reilinger, R., J. Ni, R. Smalley, and M. Bevis 1983. Co-seismic and post-seismic vertical movements associated with the 1940 Imperial Valley, California, earthquake (1931-1972). EOS, Transactions of the American Geophysical Union 64:208.

Reilinger, R.E. 1985. A strain anomaly near the southern end of the San Andreas Fault, Imperial Valley, California. Geophysical Research Letters 12:561-564.

Remeika, P. 1991a. Formational status of the Diablo Redbeds; differentiation between Colorado River affinities and the Palm Spring Formation. Symposium on the Scientific Value of the Desert, Anza-Borrego Foundation, Borrego Springs, California, Abstracts p. 12.

----- 1991b. Additional contributions to the Neogene paleobotany of the Vallecito-Fish Creek Basin and vicinity, Anza-Borrego Desert State Park. Symposium on the Scientific Value of the Desert, Anza-Borrego Foundation, Borrego Springs, California, Abstracts p. 13.

----- 1991c. A preliminary report on calcareous tufa deposits from the Palo Verde Wash area; evidence for the existence of a pre-Lake Cahuilla strandline in the Borrego Badlands, Anza-Borrego Desert State Park, California. Symposium on the Scientific Value of the Desert, Anza-Borrego Foundation, Borrego Springs, California, Abstracts p. 12.

----- 1991d. Paleontology collection management policy. California Department of Parks and Recreation, Colorado Desert District, Anza-Borrego Desert State Park, California 47 p.

----- 1992a. Preliminary report on the stratigraphy and vertebrate fauna of the middle Pleistocene Ocotillo formation, Borrego Badlands, Anza-Borrego Desert State Park, California. Mojave Desert Quaternary Research Symposium, Abstracts of Proceedings, San Bernardino County Museum Association Quarterly 39(2):25-26.

----- 1992b. Paleontological program development at Anza-Borrego Desert State Park, California. In Proceedings of The Third Conference on Fossil Resources in the National Park Service, edited by R. Benton and A. Elder. Abstract with Programs. Fossil Butte National Monument, Wyoming. Natural Resources Report NPS/NRFOBU/NRR 94:14:29.

----- 1993. Achieving paleogeology resource management strategies for general plan development: a progress report on Anza-Borrego Desert State Park, California. Paleontological Resource Management Symposium abstract with programs. Society of Vertebrate Paleontology Fifty-third annual meeting, Albuquerque, New Mexico 13(3):53A.

----- 1994a. Conducting a systematic paleogeology resource inventory of natural features: a case study from Anza-Borrego Desert State Park, California. 4th Conference on Fossil Resources, Colorado Springs, Co., (unpaginated).

----- 1994b. Lower Pliocene angiosperm hardwoods from the Vallecito-Fish Creek Basin, Anza-Borrego Desert State Park, California: deltaic stratigraphy, paleoclimate, paleoenvironment, and phytogeographic significance. Abstracts of Proceedings. 8th Annual Mojave Desert Quaternary Research Center Symposium. San Bernardino County Museum Association Quarterly 41(3):26-27.

Remeika, P., I.W. Fischbein, and S.A. Fischbein 1986. Lower Pliocene petrified wood from the Palm Spring Formation, Anza-Borrego Desert State Park, California. In Geology of the Imperial Valley, California, edited by P.D. Guptil, E.M. Gath and R.W. Ruff, South Coast Geological Society, Annual Field Trip Guidebook, Santa Ana, California 14:65-83.

----- 1988. Lower Pliocene petrified wood from the Palm Spring Formation, Anza-Borrego Desert State Park, California. Review of Palaeobotany and Palynology 56:183-198.

Remeika, P., and G.T. Jefferson 1993. The Borrego Local Fauna: revised basin-margin stratigraphy and paleontology of the western Borrego Badlands, Anza-Borrego Desert State Park, California. In Ashes, Faults and Basins, edited R.E. Reynolds and J. Reynolds. San Bernardino County Museum Association Special Publication 93(1):90-93.

----- 1994. *Mammuthus columbi* and *Mammuthus imperator* from the Borrego Badlands, Anza-Borrego Desert State Park, California. Abstracts of Proceedings. 8th Annual Mojave Desert Quaternary Research Symposium, San Bernardino County Museum Quarterly 41(3):27.

Remeika, P., and L. Lindsay 1993. Geology of Anza-Borrego: Edge of Creation. Sunbelt Publications, Kendall/Hunt Publishing Company Dubuque, Iowa 208 p.

Remeika, P., and J.R. Pettinga 1991. Stratigraphic revision and depositional environments of the middle to late Pleistocene Ocotillo Conglomerate, Borrego Badlands, Anza-Borrego Desert State Park, California. Symposium on the Scientific Value of the Desert, Anza-Borrego Foundation, Borrego Springs, California, Abstracts p. 13.

Rempel, P. 1936. The crescentic dunes of the Salton Sea and their relation to the vegetation. Ecology 17:347-358.

Repenning, C.A. 1992. *Allophaiomys* and the age of the Olyor Suite, Krestovka sections, Yakutia. U.S. Geological Survey Bulletin 2037:1-98.

Rex, R.W. 1966. Heat flow in the Imperial Valley of California. EOS, Transactions of the American Geophysical Union 47:181.

----- 1968. Geochemical water facies in the Imperial Valley of California. EOS, Transactions of the American Geophysical Union 49:758.

----- 1974. Geothermal resources in the Imperial Valley of California. Bulletin Volcanologique 37:461-462.

----- 1983. Salton Sea Scientific Drilling Project deep well Fee no. 7. EOS, Transactions of the American Geophysical Union 64:864.

Rex, R.W., and W. Randall 1969. New thermal anomalies in the Imperial Valley of California; V, present centers. EOS, Transactions of the American Geophysical Union 50:348.

----- 1970. Possible sea-floor spreading of the Imperial Valley of California; V, present centers. EOS, Transactions of the American Geophysical Union 51:422.

Reyes, A., and A. Lopez 1985. Prediccion de la distribucion del movimiento fuerte del terreno en el valle Imperial-Mexicali para dos terremotos de magnitud postulada M-7.0 y 7.4. In Riesgo Sismico en La Baja California, Memoirs of the Sociedad Mexicana de Mecanica de Suelos, A.C., Mexicali p. 15-40.

Reynolds, R.E. 1987. Paleontologic resource assessment, Imperial Irrigation District transmission line, Riverside and Imperial Counties, California. Manuscript on File, San Bernardino County Museum, Redlands, California 28 p.

----- 1989. Paleontologic monitoring and salvage Imperial Irrigation District transmission line Riverside and Imperial Counties, California: final report. Mission Power Engineering Company, Irvine, California, Manuscript on File Anza-Borrego Desert State Park, and San Bernardino County Museum, Redlands, California 169 p.

----- 1993. Landers: earthquakes and aftershocks, compiled by R.E. Reynolds. San Bernardino County Museum Association Quarterly 40(1):1-73.

Reynolds, R.E., L.P. Fay, and R.L. Reynolds 1990. California Oaks Road: an early-late Irvingtonian land mammal age vertebrate fauna from Murrieta, Riverside County, California. Mojave Desert Quaternary Research Symposium, Abstracts of Proceedings, San Bernardino County Museum Association Quarterly 37(2):35.

Reynolds, R.E., and W.A. Reeder 1986. Age and fossil assemblages of the San Timoteo Formation, Riverside County, California. In Geology Around the Margins of the Eastern San Bernardino Mountains, edited by M.A. Kooser and R.E. Reynolds, Inland Geological Society Publications, Redlands, California 1:51-56.

----- 1992. The San Timoteo Formation, Riverside County, California. In Inland Southern California: the Last 70 Million Years, edited by M.O. Woodburne, R.E. Reynolds, and D.P. Whistler, San Bernardino County Museum Association Quarterly 38(3&4):44-48.

Reynolds, R.E., and R.L. Reynolds 1990a. New late Blancan faunal assemblage from Murrieta, Riverside County, California. Mojave Desert Quaternary Research Symposium, Abstracts of Proceedings, San Bernardino County Museum Association Quarterly 37(2):34.

----- 1990b. Irvingtonian? faunas from the Pauba Formation, Temecula, Riverside County, California. Mojave Desert Quaternary Research Symposium, Abstracts of Proceedings, San Bernardino County Museum Association Quarterly 37(2):37.

----- 1993. Rodents and rabbits from the Temecula Arkose. In Ashes, Faults and Basins, edited by R.E. Reynolds and J. Reynolds, San Bernardino County

Museum Association Special Publication 93-1:98-100.

----- 1994. The Victorville Fan and an occurrence of *Sigmodon*. In Off Limits in the Mojave Desert, edited by R.E. Reynolds, San Bernardino County Museum Association Special Publication 94(1):31-33.

Reynolds, R.E., and P. Remeika 1993. Ashes, Faults and Basins: the 1993 Mojave Desert Quaternary Research Center Field Trip. In Ashes, Faults and Basins. edited by R.E. Reynolds and J. Reynolds, San Bernardino County Museum Association Special Publication 93-1:3-33.

Reynolds, S.J., S.M. Richard, G.B. Haxel, R.M. Tosdal, and S.E. Laubach 1988. Geologic setting of Mesozoic and Cenozoic metamorphism in Arizona. In Metamorphism and Crustal Evolution of the Western United States, edited by W.G. Ernst, Prentice Hall, New Jersey, Ruby Volume 7:466-501.

Rial, J.A., V. Pereyra, and G.L. Wojcik 1985. Caustic interpretation of Station 6 Imperial Valley Earthquake record using 3-D ray-tracing. Earthquake Notes 55:7.

----- 1986. An explanation for USGS Station 6 record, 1979 Imperial Valley Earthquake; a caustic induced by a sedimentary wedge. Royal Astronomical Society Geophysical Journal 84:257-278.

Richard, S.M. 1987. Middle to late Mesozoic supracrustal rocks in southeastern California and southwestern Arizona. Geological Society of America Abstracts with Programs 19(7):818-819.

Richards, B., S.A. Schmitz, P. Darling, and C. Vesely (no date). The Arroyo Tapiado mud caves. California Caver 39(2):3-51.

Richardson, C.B. 1976. Cross sections of southern Arizona and adjacent parts of California and New Mexico. Arizona Geological Digest 10:1-5.

Richardson, S.M. 1984. Stratigraphy and depositional environments of a marine-nonmarine Plio-Pleistocene sequence, western Salton Trough, California. Master of Science Thesis, San Diego State University, California 112 p.

Richins, R.L. 1983. Analysis of the geologic constraints and Tertiary overprint on the Vincent Chocolate Mountains thrust system in the southern Chocolate Mountains, Imperial County, California. Undergraduate Research Reports, Department of Geology, San Diego State University, California 43(2):63.

Richter, C.F., C.R. Allen, and J.M. Nordquist 1958. The Desert Hot Springs Earthquakes and their tectonic environment. Seismological Society of America Bulletin 48:315-337.

Rieben, H. 1956. Geology of the Mortmar--Durmid Hills area. (unpublished map).

Rigsby, C.A., editor 1984. The Imperial Basin; tectonics, sedimentation and thermal aspects. Society of Economic Paleontologists and Mineralogists, Pacific Section, Field Trip Guidebook 40:95.

Riley, C.O. 1977. Geochemical analysis of clay seams collected from the La Posta Quartz Diorite, La Posta, San Diego County, California. Master of Science Thesis, San Diego State University, California 150 p.

Rinehart, J.S. 1975. Faulting in geothermal areas. Geothermal Energy 3(12):7-24.

----- 1976. Faulting in geothermal areas. In Geothermal World Directory; 1975-1976 Bicentennial Edition, edited by K.F. Meadows, Glendora, California p. 263-279.

Riney, T.D., J.W. Pritchett, and I.F. Rice 1982. Integrated model of the shallow and deep hydrothermal systems in the East Mesa area, Imperial Valley, California. U.S. Geological Survey Open-File Report 82-980:1-118.

Riney, T.L., J.W. Pritchett, and S.K. Garg 1978. Salton Sea geothermal reservoir simulations. Transactions, Geothermal Resources Council 2(2):571-574.

Riznichenko, Yu.V, T.G. Kondrat'yeva, and S.S. Seyduzova 1977. Fourier spectra and response spectra of seismic oscillations. Physics of the Solid Earth 12:355-361.

Roberts, A.A. 1975. Helium surveys over known geothermal resource areas in the Imperial Valley, California. U.S. Geological Survey Open-File Report 75-427:1-6.

Robinson, J.H. 1965. Environment of the Imperial trough, California during the Quaternary - a paleogeographic problem. Geological Society of America, Abstracts with Programs p. 47.

----- 1966. Environment of the Imperial Valley trough, California, during the Quaternary - A paleogeographic problem. Geological Society of America Special Paper 87:227.

Robinson, J.W. and R.L. Threet 1974. Geology of the Split Mountain area, Anza-Borrego Desert State Park, eastern San Diego County, California. In Recent Geologic and Hydrologic Studies, Eastern San Diego County and Adjacent Areas, edited by M.W. Hart and R.J. Dowlen, San Diego Association of Geologists Field Trip Guidebook, California p. 47-56.

Robinson, P.T., and W.A. Elders 1970. Possible seafloor spreading in the Imperial Valley, California; III, Xenoliths in rhyolite volcanoes as samples of the basement. EOS, American of the Geophysical Union Transactions 51:422.

----- 1971. Late Cenozoic volcanism in the Imperial Valley, California. Geological Society of America Abstracts 3:185.

Robinson, P.T., W.A. Elders, and L.J.P. Muffler 1976. Quaternary volcanism in the Salton Sea geothermal field, Imperial Valley, California. Geological Society of America Bulletin 87:347-360.

Rockwell, T.K. 1989. Behvior of individual fault segments along the Elsinore-Laguna Salada Fault Zone, southern California and northern Baja California: implications for the characteristic earthquake model. In U.S. Geological Survey Redbook on Fault Segmentation and the Controls of Rupture Initiation and Termination. U.S. Geological Survey Open-File Report 89-315:288-308.

----- 1990. Holocene activity of the Elsinore Fault in the Coyote Mountains, southern California. In Western Salton Trough Soils and Tectonics: Friends of the Pleistocene Guidebook, p. 30-42.

Rockwell, T.K., P.M. Merifield, and C.C. Loughman 1986. Holocene activity of the San Jacinto Fault in the Anza seismic gap, southern California. Geological Society of America Special Paper 18(2):77.

Rockwell, T.K., and C.T. Pinault 1986. Holocene slip events on the southern Elsinore Fault, Coyote Mountains, southern California. In Neotectonics and Faulting in Southern California, edited by P.L. Ehlig, Field Trip Number 18, Department of Geology, California State University, Los Angeles p. 193-196.

Rockwell, T.K., and A.G. Sylvester 1979. Neotectonics of the Salton Trough. In Tectonics of the Juncture Between the San Andreas Fault System and the Salton Trough, Southeastern California, edited by J.C. Crowell and A.G. Sylvester, Department of Geological Sciences, University of California Santa Barbara for the Geological Society of America Annual Meeting, San Diego p. 41-52.

Rockwell, T.K., A.P. Thomas, A.L. Stinson, and R. Crisman 1989. Active northeast trending left lateral faults between the Elsinore and San Jacinto Fault Zones, southern California. Geological Society of America Abstracts with Programs 21(5):135.

Rodriguez, C., and M.W. McCann, Jr. 1981. An empirical study of the strong motion during the 1979 Imperial Valley Earthquake. EOS, Transactions of the American Geophysical Union 62:323.

Roedder, E., and K.W. Howard 1988. Fluid inclusions in Salton Sea scientific drilling project core; preliminary results. Journal of Geophysical Research, B, Solid Earth and Planets 93:13,159-13,164.

Rogers, A.F. 1926. Geology of Cormorant Island, Salton Sea, Imperial County, California. Geological Society of America Special Paper 37:219.

----- 1926. Geology of Cormorant Island, Salton Sea, Imperial County, California. Pan-American Geologist 45:249-250.

----- 1934. Salton volcanic domes of Imperial County. Pan-american Geologist 61:372-373.

----- 1935. Salton volcanic domes of Imperial County, California. Geological Society of America Proceedings 1934:328.

----- 1946. Sand fulgarites with enclosed lechatelierite from Riverside County, California. Journal of Geology 54:117-122.

Rogers, J.D. 1985. Earthquake induced bedform distortions--Imperial Valley region. In Geology and Geothermal Energy of the Salton Trough, edited by L.J. Herber, National Association of Geology Teachers, Far Western Section p. 167-171.

Rogers, J.J.W. 1954. Geology of a portion of the Joshua Tree National Monument, Riverside County. In Geology of Southern California, edited by R.H. Jahns, California Division of Mines Bulletin 170:map sheet 24.

----- 1961. Igneous and metamorphic rocks of the western part of Joshua Tree National Monument, Riverside and San Bernardino Counties, California. California Division of Mines and Geology Special Report 68:1-26.

Rogers, T.H. 1965. Geologic map of California, Santa Ana sheet. California Division of Mine and Geology, Scale 1:250,000.

Rojahn, C., editor 1980. Selected papers on the Imperial Valley, California, earthquake of October 15, 1979. U.S. Geological Survey Open-File Report 80-194:1-70.

Rojahn, C., and P.N. Mork 1981. An analysis of strong motion data from a severely damaged structure, the Imperial County Services Building, El Centro, California. U.S. Geological Survey Open-File Report 81-194:1-38.

----- 1982. An analysis of strong motion data from a severely damaged structure, the Imperial County Services Building, El Centro, California. In The Imperial Valley, California, Earthquake of October 15, 1979. U.S. Geological Survey Professional Paper 1254:357-375.

Rojahn, C., and J.T. Ragsdale 1980. Strong-motion records from the Imperial County services building, El Centro. Earthquake Engineering Research Institute, Berkeley p. 173-184.

Rojahn, C., J.T. Ragsdale, and J.D. Gates 1980. October 15, 1979, mainshock strong-motion records from the Meloland Road-Interstate Route 8 overcrossing, Imperial County, California. U.S. Geological Survey Open-File Report 80-1054:1-15.

----- 1982. Mainshock strong-motion records from the Meloland Road-Interstate Highway 8 overcrossing. In The Imperial Valley, California, Earthquake of October 15, 1979. U.S. Geological Survey Professional Paper 1254:377-383.

Rook, S.H., and G.C. Williams 1942. Imperial carbon dioxide gas field. Summary of Operations, California Oil Fields, California Division of Oil and Gas 28:12-33.

Ross, R.J., Jr. 1964. Middle and lower Ordovician formations in southernmost Nevada and adjacent California. U.S. Geological Survey Professional Paper 1180-C:1-101.

Rossbacher, L.A. 1985. The Algodones dune chain, Imperial County, California. Geology and Geothermal Energy of the Salton Trough, California State University, Pomona p. 156-160.

----- 1987. The Algodones Dunes, Imperial Valley, southern California. Geological Society of America, Decade of North American Geology, Centennial Field Guide p. 101-102.

Rowland, R.W. 1972. San Diego County's tungsten boom - 1951 to 1956. Environment Southwest 445:8-10.

Ruff, R.W., P.A. Bogseth and B.M. MacGregor 1981. Geology of the San Jacinto Mountains, Field Trip Road Log. In Geology of the San Jacinto Mountains, edited A.R. Brown and R.W. Ruff, South Coast Geological Society Annual Field Trip Guidebook, Santa Ana, California 9:175-219.

Ruissard, C.I. 1979. Stratigraphy of the Miocene Alverson Formation, Imperial County, California. Master of Science Thesis, San Diego State University, California 125 p.

Rumsey, C.M., and A. McMahan 1986. Mineral resources on the Indian Pass and Picacho Peak study areas, Imperial County, California. U.S. Bureau of Mines Report MLA 18-86:1-28.

Rumsey, C.M., and A. McMahan 1986. Mineral resources on the Indian Pass Wilderness study area, Imperial County, California. U.S. Bureau of Mines Report MLA 39-84:1-31.

Russell, R.J. 1932. Landforms of San Gorgonio Pass, southern California. University of California Publications in Geology 6:23-121.

Rymer, C.M. 1982. Geochemical analysis of spring water in the Lake Henshaw area. Undergraduate Research Reports, Department of Geology, San Diego State University, California 40(5):1-30.

Rymer, M.J. 1989a. New Quaternary age control for strata within the Indio Hills, southern California. In Mojave Desert Quaternary Research Symposium, Abstracts of Proceedings, San Bernardino County Museum Association Quarterly 36(2):64.

----- 1989b. Surface rupture in a fault stepover on the Superstition Hills Fault, California. U.S. Geological Survey Open-File Report 89-315:309-323.

----- 1990. The Bishop Ash in the Coachella Valley: stratigraphic and tectonic implications. Mojave Desert Quaternary Research Symposium, Abstracts of Proceedings, San Bernardino County Museum Association Quarterly 37(2):38.

----- 1991. The Bishop Ash bed in the Mecca Hills. In Geological Excursions in Southern California and Mexico, edited by M.J. Walawender and B.B. Hanan, Geological Society of America Annual Meeting Guidebook, San Diego p. 388-396.

----- 1992. The 1992 Joshua Tree, California, earthquake: tectonic setting and triggered slip. EOS, Transactions of the American Geophysical Union 73:363.

----- 1994. Quaternary fault-normal thrusting in the northwestern Mecca Hills, southern California. In Geological Investigations of an Active Margin, edited by S. F. McGill and T.M. Ross, U.S. Geological Survey, Cordilleran Section Guidebook Trip 9 pp. 325-329.

Rymer, M.J., J. Boley, and R.J. Weldon 1987. Nonuniform rotation (Pleistocene) along the San Andreas Fault in the Indio Hills, southern California. EOS, Transactions of the American Geophysical Union 87:1507.

Rymer, M.J., A.M. Sarna-Wojcicki, C.L. Powell, II, and J.A. Barron 1994. Stratigraphic evidence for late Miocene opening of the Salton Trough in southern California. Cordilleran Section. Geological Society of America 26(2):87.

Rynicki, J.E. 1976. Comparative tectonic modeling of southern California and northern Baja California. Undergraduate Research Reports, Department of Geology, San Diego State University, California 29(2):1-15.

Sabins, F.F., Jr. 1969a. Thermal infrared imagery and its application to structural mapping in southern California. Geological Society of America Bulletin 80:397-404.

----- 1969b. Thermal infrared imagery for geologists. Geological Society of America Special Paper 121:260-261.

Sager, M.S. 1982. The general geology of a portion of the Jacumba Mountains. Undergraduate Research Reports, Department of Geology, San Diego State University, California 41(3):1-16.

Saint, P.K. 1976. Geothermal guide to Mexicali-Imperial rift valley. Geothermal Energy Association, Covina, California map.

Salveson, J.O., and A.M. Cooper 1979a. Exploration and development of the Heber geothermal field, Imperial Valley, California. In Geology and

Geothermics of the Salton Trough, edited by W.A. Elders, University of California, Riverside Campus Museum Contributions 5:82-85.

----- 1979b. Exploration and development of the Heber geothermal field, Imperial Valley, California. Geothermal Resources Council 3:605-608.

----- 1981. Exploration and development of the Heber geothermal field, Imperial Valley, California. New Zealand Geothermal Institute Proceedings p. 3-6.

Sampson, R.J. 1932. Economic mineral deposits of the San Jacinto quadrangle. California Division of Mines Report 28:3-11.

Sampson, R.J., and W.B. Tucker 1931. Feldspar, silica, andalusite and cyanite deposits in California. California Division of Mines Report 27:427-449.

----- 1942. Mineral resources of Imperial County. California Division of Mines Report 38:105-145.

Sandberg, E.C. 1929. The gold quartz veins of the Julian district. Master of Science Thesis, California Institute of Technology, Pasadena 26 p.

Sanders, C.O., K. McNally and H. Kanamori 1981. The state of stress near the Anza seismic gap, San Jacinto Fault Zone, southern California. In Geology of the San Jacinto Mountains, edited by A.R. Brown and R.W. Ruff, South Coast Geological Society Annual Field Trip Guidebook, Santa Ana, California 9:61-67.

Sanders, C.O., and H. Kanamori 1984. A seismotectonic analysis of the Anza seismic gap, San Jacinto Fault Zone, southern California. Journal of Geophysical Research 89:5873-5890.

Sanders, C.O., H. Magistrale, and H. Kanamori 1986. Rupture patterns and preshocks of large earthquakes in the southern San Jacinto Fault Zone. Seismological Society of America Bulletin 76(5):1187-1206.

Sanders, J.E., A.A. Massa, and T.H. Sanders 1982. Recycled immature kerogens from Green River "oil shales" (Eocene) in modern Colorado River sands enhances the prospects for finding petroleum in Gulf of California-Salton Trough. International Congress on Sedimentology 11:151.

Sandusky, A. 1967. Geology of the migmatite exposures near Santa Ysabel, San Diego County, California. Undergraduate Research Reports, Department of Geology, San Diego State University, California 11(5):1-32.

Sanyal, S.K., and H.T. Meidav 1976. Evaluation of the petrophysical characteristics of the East Mesa geothermal anomaly, Imperial Valley, California. Society for Exploration Geophysicists, Abstracts 46:106-107.

Sarna-Wojcicki, A.M., H.R. Bowman, C.E. Meyer, P.C. Rusell, F. Asaro, H. Michael, J.J. Rowe, P.A. Baedecker, and G. McCoy 1980. Chemical analyses, correlations, and ages of late Cenozoic tephra units of east-central and southern California. U.S. Geological Survey Open-File Report 80-231:1-53.

Sarna-Wojcicki, A.M., and M.S. Pringle, Jr. 1992. Laser-fusion 40 Ar/39Ar ages of the tuff of Taylor Canyon and Bishop Tuff, E. California-W. Nevada. EOS Transactions, American Geophysical Union 73(43):241.

Sass, J.H. 1988. Salton Sea Scientific Drilling Program. Earthquakes and Volcanoes 20(4):156-160.

Sass, J.H., S.P. Galanis, Jr., A.H. Lachenburch, B.V. Marshall, and R.J. Moore 1984. Temperature, thermal conductivity, heat flow, and radiogenic heat production from unconsolidated sediments of the Imperial Valley, California. U.S. Geological Survey Open-File Report 84-490:1-40.

Sass, J.H., J.D. Hendricks, S.A. Priest, and L.C. Robinson 1986. Salton Sea Scientific Drilling Program; a progress report. U.S. Geology Survey Circular 974:60-61.

Sass, J.H., S.S. Priest, L.E. Duda, C.C. Carson, J.D. Hendricks, and L.C. Robinson 1988. Thermal regime of the State 2-14 Well, Salton Sea Scientific Drilling Project, Journal of Geophysical Research, B, Solid Earth and Planets 93:12,995-13,004.

Sass, J.H., S.S. Priest, L.C. Robinson, and J.D. Hendricks 1986. Salton Sea Scientific Drilling Project on-site science management. U.S. Geological Survey Open-File Report 86-397:1-24.

Sauer, C. 1929. Landforms in the Peninsular Range of California as developed about Warner's Hot Springs and Mesa Grande. University of California Publications in Geography 3:199-290.

Savage, D.E., and T. Downs 1954. Cenozoic land life of southern California. In Geology of Southern California, edited by R.H. Jahns, California Division of Mines and Geology 170(3):43-58.

Savage, J.C., D.D. Goodreau, and W.H. Prescott 1974. Possible fault slip on the Brawley Fault, Imperial Valley, California. Seismological Society of America Bulletin 64:713-715.

Savage, J.C., M.E. Lisowski, N.E. King, and W.K. Gross 1994. Strain accumulation along the Laguna Salada Fault, Baja California, Mexico. Journal of Geophysical Research 99(B9): 18 109-18 116.

Savage, J.C., and W.H. Prescott 1976. Strain accumulation on the San Jacinto Fault near Riverside, California. Seismological Society of America Bulletin 66(5):1749-1754.

Savage, J.C., W.H. Prescott, M. Lisowski, and N. King 1979a. Deformation across the Salton Trough, California 1973-1977. Journal of Geophysical

Research 84:3069-3079.

----- 1979b. Deformation across the Salton Trough, California 1973-1977. EOS, Transactions of the American Geophysical Union 59:1051.

Savino, J.N., W.L. Rodi, R.C. Goff, and T.H. Jordan 1977. Inversion of combined geophysical data as a geothermal exploration tool; applications to Yellowstone Park and the Imperial Valley. EOS, Transactions of the American Geophysical Union 58:1187.

Savino, J.N., W.L. Rodi, T.H. Jordan, and R.C. Goff 1979. Joint inversion of combined geophysical data as a geothermal exploration tool; applications to Yellowstone Park and the Imperial Valley. Geothermal Resources Council 3:625-628.

Sawicki, D.A. 1978. A structural and petrographic evaluation of the Thing Valley lineament, San Diego County, California. Master of Science Thesis, San Diego State University, California 98 p.

Sawlan, M.G., and J.G. Smith 1984. Petrologic characteristics, age and tectonic setting of Neogene volcanic rocks in northern Baja California Sur, Mexico. In Geology of the Baja California Peninsula, edited by V.A. Frizzell, Jr., Society of Economic Paleontologists and Mineralogists, Pacific Section p. 237-252.

Sawyer, L.M. 1985. Geothermal and mineralogic variations observed in the Holy Sugar 44-P well, of Imperial Valley, California. Master's Thesis, University of California, Los Angeles.

Scaramella, R.J. 1970. Geology of the central portion of the San Felipe Hills, Imperial Valley, California. Undergraduate Research Reports, Department of Geology, San Diego State University, California 15:1-24.

Schaeffer, G.C. 1857. Description of the structure of fossil wood from the Colorado Desert. In 1855 Report of the Explorations and Survey for a Railroad Route from the Mississippi to the Pacific, U.S. Congress, 2nd Session, Senate Executive Document 91, 5(2):338-339.

Scheuing, D.F. 1989. The effect of structure on paleomagnetic data collected in complexly deformed terrains: a case study from the San Jacinto Fault Zone, southern California. Manuscript on File, Lamont-Doherty Geol. Observ., (unpaginated).

Scheuing, D.F., L. Seeber, K.W. Hudnut, and N.L. Bogen 1988. Block rotation in the San Jacinto Fault Zone, southern California. EOS, Transactions of the American Geophysical Union 69:1456.

Scheliga, T.J. Jr. 1963. Geology and water resources of Warner basin, San Diego County, California. Master of Arts Thesis, University of Southern California, Los Angeles 91 p.

Schiffman, P., W.A. Elders, A.E. Williams, S.D. McDowell, and D.K. Bird 1984. Active metasomatism in the Cerro Prieto geothermal system, Baja California, Mexico: a telescoped low-pressure, low-temperature metamorphic facies series. Geology 12:12-15.

Schremp, L.A. 1981. Archaeogastropoda from the Pliocene Imperial Formation of California. Journal of Paleontology 55(5):1123-1136.

Schmidt, N. 1990. Plate tectonics and the Gulf of California region. California Geology 43:252-256.

Schnabel, P., H.B. Seed, and, J. Lysmer 1972. Modification of seismograph records for effect of local soil conditions. Seismological Society of America Bulletin 62:1649-1664.

Schnapp, M., and G. Fuis 1976. Preliminary catalogue of earthquakes in the northern Imperial Valley, California, October 1, 1976-December 31, 1976. U.S. Geological Open-File Report 77-431:1-19.

----- 1977. Preliminary catalogue of earthquakes in the northern Imperial Valley, California, January 1, 1977-March 31, 1977. U.S. Geological Open-File Report 78-74:1-15.

Schroeder, R.C. 1976. Reservoir engineering report for the Magma-SDG&E geothermal experimental site near the Salton Sea, California. University of California Riverside Report 52094:1-62.

Schultejann, P.A. 1984. The Yaqui Ridge antiform and detachment fault: mid-Cenozoic extensional terrain west of the San Andreas Fault. Tectonics 3:677-691.

----- 1985. Structural trends in Borrego Valley, California: interpretations from SIR-A and SESAT SAR. Photogrammetric Engineering and Remote Sensing 51:(10):1615-1624.

Schultze, L.E., and J.D. Bauer 1982a. Comparison of methods for recovering metal values from Salton Sea KGRA brines. Geothermal Resources Council 6:111-113.

----- 1982b. Operation of a mineral recovery unit on brine from Salton Sea known geothermal resource area. U.S. Bureau of Mines Report of Investigation 8680:1-25.

Scott, E. 1992. First record of *Arctodus simus* (Cope), 1879 (Mammalia; Carnivora; Ursidae) from Riverside County, California. Mojave Desert Quaternary Research Symposium, Abstracts of Proceedings, San Bernardino County Museum Association Quarterly 39(2):27.

Scott, J.S. 1992. Microearthquake studies in the Anza seismic gap. Doctoral Dissertation, University of California, San Diego.

Scott, S., and S.E. Wood 1971-1972. California's bright and geothermal future. Cry California 7:10-23.

----- 1975. California's bright and geothermal future. In Energy and Man; Technical and Social Aspects of Energy, edited by M.G. Morgan, IEEE Press, New York p. 185-197.

Scrivner, P.J., and D.J. Bottjer 1986. Neogene avian and mammalian tracks from Death Valley National Monument, California: their context, classification and preservation. Palaeogeography, Palaeoclimatology, Palaeoecology 57:285-331.

Seamount, D.T. 1981. Well log study of the hydrothermally altered sediments of the Cerro Prieto geothermal field, Baja California, Mexico. Master of Science Thesis, Department of Geological Sciences, University of California, Riverside 165 p.

Seeber, L., and N.L. Bogen 1985. Block rotation along the southern San Jacinto Fault Zone. Abstract. Eos (Transactions, American Geophysical Union) 66:953.

Seed, H.B., I.M. Idriss, and F.W. Kiefer 1969. Characteristics of rock motions during earthquakes. American Society of Civil Engineers, Journal of Soil Mechanics and Foundation Division 95(6783):1199-1218.

Segall, P., and D.D. Pollard 1980. Mechanics of discontinuous faults. Journal of Geophysical Research 85:4337-4350.

Sentianin, B.A. 1989. Petrology and provenance of sandstones in the Pleistocene Borrego Formation, Salton Basin, California. Master of Science Thesis, San Diego State University, California 79 p.

Severson, L.K., and McEvilly, T.V. 1987. Analysis of seismic reflection data from the Imperial Valley, California. Geological Society of America Abstracts with Programs 19:449-450.

Shafiquallah, M., E.G. Frost, D.L. Frost, and P.E. Damon 1990. Regional extension and gold mineralization in the southern Chocolate Mountains, southeastern California- K-Ar constraints from fault rocks. Geological Society of America Abstracts with Programs 22(3):83.

Shakal, A.F. 1977. Source, path and side effects on strong motion records from the Kern County, Borrego Mountain and San Fernando earthquakes. EOS, Transactions of the American Geophysical Union 58:445.

Shannon, D. 1977. Geological reconnaissance study of the northern portion of the Midway Mountains Imperial County, California. Undergraduate Research Reports, Department of Geology, San Diego State University, California 31(2):1-40.

Shannon, Wilson, and Agbabian Associates 1976. Geotechnical and strong motion earthquake data from the U.S. accelerograph stations; Ferndale, Chalome, and El Centro, California. Report 1 NUREG-29.

----- 1980. Site-dependent response at El Centro, California accelerograph station including soil/structure interaction effects. Report NUREG/CR-1642:1-149.

Sharp, R.P. 1964. Wind-driven sand in Coachella Valley. Geological Society of America Bulletin 75:785-804.

----- 1978. Geological aspects of the eastern Mojave Desert and Salton Trough. In Aeolian Features of Southern California; a Comparative Planetary Geology Guidebook, edited by R. Greeley, et al., National Aeronautics and Space Administration p. 1-7.

----- 1979a. Intradune flats of the Algodones chain, Imperial Valley, California. California Institute of Technology, Pasadena, Division of Geological and Planetary Science Publication 3195:1-9.

----- 1979b. Intradune flats of the Algodones chain, Imperial Valley, California. Geological Society of America Bulletin 90:1908-1916.

Sharp, R.V. 1965. Geology of the San Jacinto Fault Zone in the Peninsular Ranges of southern California. Doctoral Dissertation, California Institute of Technology, Pasadena 184 p.

----- 1967a. Ancient mylonite zone and fault displacements in the Peninsular Ranges of southern California. Geological Society of America Special Papers 101:333.

----- 1967b. San Jacinto Fault Zone in the Peninsular Ranges of southern California. Geological Society of America Bulletin 78(6):705-730.

----- 1972a. The Borrego Mountain Earthquake of April 9, 1968, introduction. In The Borrego Mountain Earthquake of April 9, 1968, U.S. Geological Survey Professional Paper 787:1-2.

----- 1972b. Tectonic setting of the Salton Trough. In The Borrego Mountain Earthquake of April 9, 1968, U.S. Geological Survey Professional Paper 787:3-15.

----- 1972c. Map showing the recently active branches along the San Jacinto Fault Zone between the San Bernardino area and Borrego Valley, California. U.S. Miscellaneous Investigations Map I-675.

----- 1975. En echelon fault patterns of the San Jacinto Fault Zone. In San Andreas Fault in Southern California, edited by J.C. Crowell, California Division of Mines and Geology, Special Report 118:147-152.

----- 1976. Surface faulting in Imperial Valley during the earthquake swarm of January-February, 1975. Seismological Society of America Bulletin 66:1145-1154.

----- 1977. Map showing the Holocene surface expression of the Brawley Fault, Imperial County, California. U.S. Geological Survey Miscellaneous Field Study Map MF-838.

----- 1979. Some characteristics of the eastern Peninsular

Ranges mylonite zone. In Proceedings, Conference VIII, Analysis of Actual Fault Zones in Bedrock, U.S. Geological Survey Open-File Report 79-1239:258-267.

----- 1981. Variable rates of late Quaternary strike slip on the San Jacinto Fault Zone, southern California. Journal of Geophysical Research 86:1754-1762.

----- 1982a. Tectonic setting of the Imperial Valley region. In The Imperial Valley, California, Earthquake of October 15, 1979. U.S. Geological Survey Professional Paper 1254:5-14.

----- 1982b. Comparison of the 1979 surface faulting with earlier displacements in the Imperial Valley. In The Imperial Valley, California, Earthquake of October 15, 1979. U.S. Geological Survey Professional paper 1254:213-221.

----- 1986. Holocene sedimentation in the Imperial Valley, California. Geological Society of America Abstracts with Programs 18:183.

----- 1989. Pre-earthquake displacement and triggered displacement on the Imperial Fault associated with the Superstition Hills Earthquake of 24 November 1987. Seismological Society of America Bulletin 79:466-479.

Sharp, R.V., C.R. Allen, and M.F. Meier 1959. Pleistocene glaciers on southern California mountains. American Journal of Science 257:81-94.

Sharp, R.V., K.E. Budding, J. Boatwright, M.J. Ader, M.G. Bonilla, M.M.Clark, T.E. Fumel, K.K. Harms, J.J. Lienkaemper, D.M. Morton, B.J. O'Neill, C.L. Ostergren, D.J. Ponti, M.H. Rymer, J.O. Saxton, and J.D. Sims 1989. Surface faulting along the Superstition Hills Fault Zone and nearby faults associated with the earthquakes of 24 November 1987. Seismological Society of America Bulletin 79:252-281.

Sharp, R.V., and M.M. Clark 1972. Geologic evidence of previous faulting near the 1968 rupture of the Coyote Creek Fault. U.S. Geological Survey Professional Paper 787:131-140.

Sharp, R.V., and J.J. Lienkaemper 1982. Preearthquake and postearthquake near-field leveling across the Imperial Fault and the Brawley Fault Zone. In The Imperial Valley, California, Earthquake of October 15, 1979. U.S. Geological Survey Professional Paper 1254:169-182.

Sharp, R.V., J.J. Lienkaemper, M.G. Bonilla, D.B. Burke, B.F. Fox, D.G. Herd, D.M. Miller, D.M. Morton, D.J. Ponti, M.J. Rymer, J.C. Tinsley, J.C. Yount, J.E. Kale, E.W. Hart, and K.E. Sieh 1982. Surface faulting in the central Imperial Valley. In The Imperial Valley, California, Earthquake of October 15, 1979. U.S. Geological Survey Professional Paper 1254:119-143.

Sharp, R.V., J.J. Lienkaemper, and M.J. Rymer 1982. Surface displacement on the Imperial and Superstition Hills Faults triggered by the Westmorland, California, earthquake of 26 April 1981. U.S. Geological Survey Open-File Report 82-282.

Sharp, R.V., M.J. Rymer, and J.J. Lienkaemper 1982. Surface displacement on the Imperial and Superstition Hills Faults triggered by the Westmorland, California, earthquake of 26 April 1981. Seismological Society of America Bulletin 76:949-965.

Sharp, R.V., and J.L. Saxton 1989. High precision three-dimensional records of 1987 surface displacement on the Superstition Fault. Seismological Society of America Bulletin 79:376-389.

Sharpe, R.D., and G.G. Cork 1995. Geology and mining of the Miocene Fish Creek Gypsum in Imperial County, California. In 29th Forum on the Geology of Industrial Minerals: Proceedings, edited by M. Tabilio and D.L. Dupras. California Department of Conservation, Division of Mines and Geology, Special Publication 110:169-180.

Shaw, C.A. 1981. The middle Pleistocene El Golfo local fauna from northwestern Sonora, Mexico. Master of Science Thesis, Department of Biology, California State University, Long Beach 141 p.

Shaw, C.A., and H.G. McDonald 1987. First record of giant anteater (Xenarthra; Myrmecophagidae) in North America. Science 236:186-188.

Shaw, K. 1978. Magnetic survey, southern portion of the Elsinore Fault, Coyote Mountains, Imperial County. Undergraduate Research Reports, Department of Geology, San Diego State University, California 32(3):1-31.

Shaw, S.E., J.A. Cooper, J.R. O'Neil, V.R. Todd, and J.L. Wooden 1986. Strontium, oxygen, and lead isotope variations across a segment of the Peninsular Ranges Batholith, San Diego County, California. Geological Society of America abstracts with programs 18:183.

Shearer, C.K., J.J. Papike, S.B. Simon, B.L. Davis, and J.C. Laul 1988. Mineral reactions in altered sediments from the California State 2-14 Well; variations in the modal mineralogy, mineral chemistry, and bulk composition of the Salton Sea Scientific Drilling core. Journal of Geophysical Research, B, Solid Earth and Planets 93:13,104-13,122.

Sheehan, J.R. 1986. A tectonic model of the Salton Trough. In Geology of the Imperial Valley, California, edited by P.D. Guptil E.M. Gath and R.W. Ruff, South Coast Geological Society, Annual Field Trip Guidebook, Santa Ana, California 14:165-172.

----- 1988. Tectonic model of the Salton Trough. In

Geologic Field Guide to the Western Salton Basin Area, edited by S.M. Testa and K.E. Green, American Institute of Professional Geologists p. 7-14.

Sheldon, G.W. 1959. Julian gold mining days. Master of Science Thesis, San Diego State University, California 277 p.

Shelton, C.E. 1963. A guide to Coachella Valley. Desert Magazine, Palm Desert, California March:28-36.

----- 1966. Geology illustrated. W.H. Freeman and Company, San Francisco, California 434 p.

Shepard, F.P. 1964. Sea-floor valleys of the Gulf of California. In Marine geology of the Gulf of California, edited by T.H. van Andel, and G.G. Shor. American Association of Petroleum Geologists Memoir 3:157-192.

Sheridan, J.M., J.R. Weldon, II, and C.A. Thorton 1994. Stratigraphy and deformational history of the Mecca Hills, southern California. In Geological Investigations of an Active Margin, edited by S.F. McGill and T.M. Ross, U.S. Geological Survey, Cordilleran Section Guidebook Trip 15 pp. 319-325.

Sheridan, J.M., and J.R. Weldon, II 1994. Accomodation of compression in the Mecca Hills, California. In Geological Investigations of an Active Margin, edited by S.F. McGill and T.M. Ross, U.S. Geological Survey, Cordilleran Section Guidebook Trip 15 pp. 330-336.

Shifflett, H., P. Nicolay, E. Wiebe and R. Crisman 1986. Holocene offsets, San Andreas Fault, Mecca Hills, California. In Geology of the Imperial Valley, California, edited by P.D. Guptil, E. M. Gath and R. W. Ruff, South Coast Geological Society Annual Field Trip Guidebook, Santa Ana, California 14:159-164.

Shirley, J.H. 1988. Forces based on tidal correlations of historic M 6 earthquakes of the Colorado Delta-Imperial Valley and San Jacinto Fault Zone, California. EOS, Transactions of the American Geophysical Union 69:1300.

Shremp, L.A. 1981. Archaeogastropoda from the Pliocene Imperial Formation of California. Journal of Paleontology 55:1123-1136.

Sieh, K.E. 1978. Slip along the San Andreas Fault associated with the great 1857 earthquake. Seismological Society of America Bulletin 68:1421-1448.

----- 1982. Slip along the San Andreas Fault associated with the earthquake. In The Imperial Valley, California, Earthquake of October 15, 1979. U. S. Geological Survey Professional Paper 1254:155-160.

----- 1986. Slip rate across the southern San Andreas and prehistoric earthquakes at Indio, California. EOS, Transactions of the American Geophysical Union 67:1200.

Signorotti, V.J. 1991. Imperial Valley's geothermal resources come of age. In The Diversity of Mineral and Energy Resources of Southern California, edited by M.A. McKibben. Society of Economic Geologists 12:117-121.

Sigurdson, D.R., H.T. Meidav, and R.V. Sharp 1971. Structure of sediments under the Salton Sea. Geological Society of America Abstracts 3:192-193.

Silver, L.T., and B.W. Chappell 1988. The Peninsular Ranges Batholith: an insight into the evolution of the Cordilleran batholiths of southwestern North America. Transactions of the Royal Society of Edinburgh, Earth Sciences 79:105-121.

Silver, L.T., H.P. Taylor, Jr., and B. Campbell 1979. Some petrological, geochemical and geochronological observations of the Peninsular Ranges Batholith near the international border of the U.S.A. and Mexico. In Mesozoic Crystalline Rocks: Peninsular Ranges Batholith and Pegmatites, Point Sal Ophiolite, edited by P.L. Abbott and V.R. Todd, Department of Geological Sciences, San Diego State University, California p. 83-110.

Silver, L.T., H.P. Taylor, Jr., and B. Chappell 1979. Some petrological, geochemical and geochronological observations of the Peninsular Ranges Batholith near the international border of the U.S.A. and Mexico. In Mesozoic crystalline rocks, edited by P.L. Abbott and V.R. Todd. Department of Geological Sciences, San Diego State University p. 83-110.

Silver, P.G. 1983. Retrieval of source-extent parameters for the Imperial Valley Earthquake of October 15, 1979. Carnegie Institute of Washington Yearbook 82:522-529.

Silver, P.G., and T. Masuda 1985. A source extent analysis of the Imperial Valley Earthquake of October 15, 1979 and the Victoria Earthquake of June 9, 1980. Journal of Geophysical Research 90:7639-7651.

Simoni, T.R., Jr. 1980. Geophysical and lithologic data from a test well on Clark Lake, San Diego County, California. U.S. Geological Survey Open-File Report 80-1302:1.

Simpson, C. 1984. Borrego Springs-Santa Rosa mylonite zone: a late Cretaceous west-directed thrust in southern California. Geology 12:8-11.

----- 1985. Deformation of granitic rocks across the brittle-ductile transition. Journal of Structural Geology 7:503-511.

----- 1986. Microstructural evidence for northeastward movement on the Vincent-Chocolate Mountains thrust system. Geological Society of America Abstracts with Programs 18:185.

----- 1990. Microstructural evidence for northeastward movement on the Chocolate Mountains Fault Zone, southeastern California. Journal of Geophysical Research, B, Solid Earth and Planets 95:529-537.

Sims, J.D. 1975. Determining earthquake recurrence intervals from deformational structures in young lacustrine sediments. Tectonophysics 29:141-152.

Sims, S.J. 1960. Geology of part of the Santa Rosa Mountains, Riverside County, California. Doctoral Dissertation, Stanford University, Menlo Park, California.

Sinfeld, L.D. 1983. The Geology of the eastern facies of the La Posta pluton. Undergraduate Research Reports, Department of Geology, San Diego State University, California 43(2):1-44.

Singh, J.P. 1988. 1979 Imperial Valley Earthquake; its immediate, short- and long-term impact on seismic design practice. U.S. Geological Survey Open-File Report 88-0013-A:141-144.

Singh, K., J. Fried, R. Apsel, and J. Brune 1981. Spectral attenuation of SH-wave along the Imperial Fault and a preliminary model of Q in the region. EOS, Transactions of the American Geophysical Union 62:969.

----- 1982a. Spectral attenuation of SH-wave along the Imperial Fault and a preliminary model of Q in the region. U.S. Geological Survey Open-File Report 82-0591:955-976.

----- 1982b. Spectral attenuation of SH-wave along the Imperial Fault. Seismological Society of America Bulletin 72:2003-2016.

Sipkin, S.A. 1989. Moment-tensor solutions for the 24 November 1987 Superstition Hills, California, earthquakes. Seismological Society of America Bulletin 79:493-499.

Skinner, B.J. 1963. Sulfides deposited by the Salton Sea geothermal brine. Mining Engineering 15:60-61.

Skinner, B.J., D.E. White, H.J. Rose, and R.E. Mays 1967. Sulfides associated with the Salton Sea geothermal brine. Economic Geology 62:316-330.

Skrivan, J.A. 1977. Digital-model evaluation of the ground-water resources in the Ocotillo-Coyote Wells basin, Imperial County, California. California Water Resources Investigation 77-30:1-50.

Slade, M.A., G.A. Lysenga, and A. Raefsky 1984. Modeling of the surface static displacements and fault plane slip for the 1979 Imperial Valley Earthquake. Seismological Society of America Bulletin 74:2413-2433.

Slade, M.A., G.A. Lysenga, A. Raefsky, S. Hartzell, and R. Scott 1982. Finite element modeling of the 1979 Imperial Valley Earthquake. EOS, Transactions of the American Geophysical Union 63:1119.

Small, M.A. 1961. Settlement studies by means of precision leveling. Bulletin Geodesique 62:317-325.

----- 1973. Settlement studies by means of precision leveling. In Report on geodetic measurements of crustal movement, 1906-71, National Oceanic and Atmospheric Administration-National Ocean Survey National Geodetic Society 22 p.

Smith, D.B, B.R. Berger, and R.W. Tosdal 1987. Geochemical studies in the Indian Pass and Picacho Peak Bureau of Land Management Wilderness Study Area, Imperial County, southern California. Journal of Geochemical Exploration 28:479-494.

Smith, D.B, B.R. Berger, R.W. Tosdal, D.R. Sherrod, G.L. Raines, A. Griscom, MG. Helferty, C.M. Rumsey, and A.B. McMahan 1987. Mineral resources of the Indian Pass and Picacho Peak Wilderness Study Area, Imperial County, California. Geological Survey Bulletin, Report B 1711-A:A1-A21.

Smith, D.D. 1962. Geology of the northeast quarter of the Carrizo Mountain quadrangle, Imperial County, California. Master of Science Thesis, University of Southern California, Los Angeles 89 p.

Smith, D.R. 1972. Geology of an area east of Fish Creek gorge. Undergraduate Research Reports, Department of Geology, San Diego State University, California 20(3):1-22.

Smith, G. 1984. *E. equus*: immigrant or emigrant? Science Magazine, Crosscurrents 84:76-77.

Smith, J.L. 1979. Geology and commercial development of the East Mesa geothermal field, Imperial Valley, California. In Geology and Geothermics of the Salton Trough, edited by W.A. Elders, University of California, Riverside Campus Museum Contributions 5:86-94.

Smith, J.T. 1988. Sources of oceanographic, geologic, and molluscan data for the northern Gulf of California and the Salton Trough of southern California. Western Society of Malacologists Annual Report 20:5-10.

Smith, P.B. 1970. New evidence for a marine embayment along the lower Colorado River area, California and Arizona. Geological Society of America Bulletin 81:1411-1420.

Smith, R.S.U. 1972. Barchan dunes in a seasonally reversing wind regime, southern Imperial County, California. Geological Society of America Abstracts with Programs 4:240-241.

----- 1977. Barchan dunes; development persistence and growth in a multidirectional wind regime, southeastern Imperial County, California. Geological Society of America Abstracts with Programs 9:502.

----- 1978. Guide to selected features of aeolian geomorphology in the Algodones dune chain, Imperial County, California. In Aeolian Features of

Southern California; a Comparative Planetary Geoogy Guidebook, edited by R. Greeley, and others. National Aeronautics and Space Administration p. 73-98.

----- 1980. Maintenance of barchan size in the southern Algodones dune chain, Imperial County, California. U.S. National Aeronautics and Space Administration Technical Memorandum 81776:253-254.

Smith, R.S.U., W.E. Yeend, J.C. Dohrenwend, and D.D. Gese 1984. Mineral resources of the Algodones Dunes Wilderness Study Area (CDCA-360), Imperial County, California. U.S. Geological Survey Open-File Report 84-630:1-13.

Snay, R.A., M.W. Cline, and E.L. Timmerman 1982a. Horizontal deformation in the Imperial Valley, California, between 1934 and 1980. Journal of Geophysical Research 87:3959-3968.

----- 1982b. Regional deformation of the earth model for the San Diego - El Centro region. Fourth Annual Conference on the NASA Geodynamics Program; Crustal Dynamics Project, Washington D.C. p. 72.

Snay, R.A., A.R. Drew, W.E. Strange, H.C. Neugbauer, R.E. Reilinger, and W. Thatcher 1988. Combining GPS and classical geodetic surveys for crustal deformation in the Imperial Valley, California. EOS Transactions of the American Geophysical Union 69:326.

Sosky, J.L., and V.C. Kelley 1935. Wave-built pumice deposits and Salton rhyolite hills. Pan-Pacific Geologist 63:319-320.

----- 1936. Wave-built pumice deposits and Salton rhyolite hills. Geological Society of America Proceedings p. 341.

Southard, G. 1972. A magnetic survey of a portion of the Elsinore Fault through Vallecito Valley. Undergraduate Research Reports, Department of Geology, San Diego State University, California 20(3):1-19.

Spencer, J.E., and S.J. Reynolds 1990. Relationship between Mesozoic and Cenozoic tectonic features in west central Arizona and adjacent southeastern California. Journal of Geophysical Research, B, Solid Earth and Planets 95:539-555.

Spudich, P., and E. Cranswick 1984a. Direct observation of rupture propagation during the 1979 Imperial Valley Earthquake using a short baseline accelerometer array. Seismological Society of America Bulletin 74:2083-2114.

----- 1984b. Soil strains and horizontal propagation velocities of strong ground motions observed during the 1979 Imperial Valley, California, earthquake. Proceedings of the World Conference on Earthquake Engineering 8(2):231-238.

Stanley, G.M. 1962. Prehistoric lakes of the Salton Sea Basin. Geological Society of America Abstracts with Programs 73:249-250.

----- 1965. Deformation of Pleistocene Lake Cahuilla shoreline, Salton Basin, California. Geological Society of America Special Paper Abstracts with Programs 87:165.

Stearns, R.E.C. 1879. Remarks on fossil shells from the Colorado Desert. American Naturalist 13:141-154.

----- 1901. The fossil fresh water shells of the Colorado Desert, their distribution, environment and variation. Proceedings of the U.S. National Museum 24(1256):271-299.

Steer, B.L., and P.L. Abbott 1981. Paleohydrology of the Eocene Ballena River, Sonora to San Diego. Geological Society of America Abstracts with Programs 13:108.

Stensrud, H.L., and G.R. Gastil 1978. Spotted tonalite dikes from the eastern margin of the southern California batholith. Geological Society of America abstracts with programs 10(3):148.

Stephen, M.F., and D.S. Gorsline 1975. Sedimentary aspects of the New River Delta, Salton Sea, Imperial County, California. In Deltas, Models for Exploration, edited by M.L. Broussard, Houston Geological Society, Texas p. 267, 282.

Stewart, J.D. and M.A. Roeder 1993. Razorback sucker (*Xyrauchen*) fossils from the Anza-Borrego Desert and the ancestral Colorado River. In Ashes, Faults and Basins, edited by R.E. Reynolds and J. Reynolds, San Bernardino County Museum Association Special Publication 93-1:94-96.

Stewart, R.M. 1956. Geology and mineral resources of Anza-Borrego Desert State Park. California Division of Mines and Geology, Manuscript on File Anza-Borrego Desert State Park, California 20 p.

----- 1958. Mines and mineral resources of Santa Ysabel quadrangle, San Diego County, California. California Division of Mines Bulletin 117:21-38.

Stinson, A.L. 1990. Structural deformation within the Pinyon Mountains, San Diego County, California. Master of Science Thesis, San Diego State University, California, 133 p.

Stock, J., M.B. Arturo, S.V. Francisco, and M.M. Miller 1991. Miocene to Holocene extensional volcanic stratigraphy of northeast Baja California, Mexico. In Geological Excursions in Southern California and Mexico, edited by M.J. Walawender and B.B. Hanan, Geological Society of America Annual Meeting Guidebook, San Diego p. 44-67.

Stock, J.M., and K.V. Hodges 1989. Pre-Pliocene extension around the Gulf of California and the transfer of Baja California to the Pacific Plate. Tectonics 8(1):99-115.

Stokes, W.C., and K.C. Condie 1961. Pleistocene bighorn sheep from the Great Basin. Journal of Paleontology 35:598-609.

Stone, W.E., and P.J. MacLean 1987. Experimental metamorphic study of sediment from a $CO_2$ mudpool in the Salton Trough, California, implications for metamorphism of Archean gold ores in Canada. Geological Association of Canada, Mineralogical Association of Canada, Canadian Geophysical Union Abstracts 12:92.

Storms, W.H., and H.W. Fairbanks 1965 (originally published 1893). Old mines of southern California - desert - mountain - coastal areas including the Calico-Salton Sea-Colorado River districts and southern counties. Frontier Book Company, Toyahvale, Texas 96 p.

Stotts, J.L. 1965. Stratigraphy and structure of the northwest Indio Hills, Riverside County, California. Master of Arts Thesis, Department of Geological Sciences, University of California, Riverside.

Stout, B.W., G.J. Miller, and P. Remeika 1987. Neogene mega-vertebrate ichnites from the Vallecito Basin, Anza-Borrego Desert State Park, California. Symposium on the Scientific Value of the Desert, Anza-Borrego Foundation, Borrego Springs, California p. 5.

Stout, B.W., and P. Remeika 1991. Status report on three major camelid tracksites in the lower Pliocene delta sequence, Vallecito-Fish Creek Basin, Anza-Borrego Desert State Park, California. Symposium on the Scientific Value of the Desert, Anza-Borrego Foundation, Borrego Springs, California p. 9.

Strahorn, A.T., E.B. Watson, A.E. Kocher, E.C. Eckmann, and J.B. Hammon 1918. Soil survey of the El Centro area, California. U.S. Department of Agriculture, Bureau of Soils, Washington D.C. 59 p.

Strand, C.L. 1980. Mud volcanoes, faults and earthquakes of the Colorado delta before the twentieth century. San Diego Historical Society Journal of San Diego History, p. 43-63.

Strand, C.L., R.G. Gastil, and M. Marshall 1981. Determination of epicenter and magnitude for the southern California-Baja California earthquake of February 23, 1892. Geological Society of America Abstracts with Programs 13:108.

Strand, R.G. 1962. Geologic map of California, San Diego-El Centro sheet. California Division of Mines and Geology, Scale 1:250,000.

Strand, R.G. 1982. San Diego - El Centro Sheet; in the collection of Bouguer gravity map of California. California Division of Mines and Geology.

Strange, W.E. 1982. Vertical motions associated with earthquakes on the Imperial Valley Fault. EOS, Transactions of the American Geophysical Union 63:906.

----- 1986. The relation of vertical motions to faulting and seismic limitations in the Salton Trough. EOS, Transactions of the American Geophysical Union 67:369.

Streiff, D. 1971. Geology of the Yuha Basin, southwestern Imperial County, California. Undergraduate Research Reports, Department of Geology, San Diego State University, California 18:1-30.

Stuart, C.J. 1963. Geology of the Big Four Mine, Julian, California. Undergraduate Research Reports, Department of Geology, San Diego State University, California 7(3):1-9.

Stump. T.E. 1970. Stratigraphy and paleoecology of the Imperial Formation in the western Colorado Desert. Undergraduate Research Reports, Department of Geology, San Diego State University, California 16(2):1-118.

----- 1972. Stratigraphy and paleoecology of the Imperial Formation in the western Colorado Desert. Master of Science Thesis, San Diego State University, California 132 p.

Stump. T.E., and J.D. Stump 1972. Age, stratigraphy, paleoecology and Caribbean affinities of the Imperial fauna of the Gulf of California depression. Geological Society of America Special Paper 4(3):243.

Sturrock, A.M., Jr. 1977. Evaporation and radiation measurements at Salton Sea, California. U.S. Geological Survey Open-File Report 77-74:1-42.

----- 1978. Evaporation and radiation measurements at Salton Sea, California. U.S. Geological Survey Water-Supply Paper 2053-1-26.

Sturz, A. 1974. A study of the Na-K variation in the feldspars of the Rattlesnake Granite, Cuyamaca Peak quadrangle, southern California. Undergraduate Research Reports, Department of Geology, San Diego State University, California 24(2):1-44.

----- 1989. Low-temperature hydrothermal alteration in near-surface sediments, Salton Sea geothermal area. Journal of Geophysical Research, B, Solid Earth and Planets 94:4015-4025.

Sumner, J.R. 1971. Tectonic significance of gravity and aeromagnetic investigations in Sonora and Arizona at the head of the Gulf of California. Geological Society of America Abstracts 3:204-205.

Suter, V.E. 1979. Geothermal development in California's Imperial Valley. Interstate Oil Compact Commission, Committee Bulletin 21:17-21.

Swajian, A., G.L. Morris, and K.D. Pimentel 1978. Water quality management for Imperial Valley geothermal development. Transactions of the Geothermal

Resources Council 2(2):631-634.
Swanberg, C.A. 1973. Preliminary results of geothermal well Mesa 6-2, Mesa anomaly, Imperial Valley, California. EOS, Transactions of the American Geophysical Union 54:1215-1216.

----- 1974a. Heat flow and geothermal potential of the East Meas KGRA, Imperial Valley, California. Conference on Research for the Development of Geothermal Energy Resources, California Institute of Technology, Jet Propulsion Laboratory-National Science Foundation, Pasadena p. 85-97.

----- 1974b. The Mesa geothermal anomaly; a comparison of results from surface geophysics and deep drilling. EOS, Transactions of the American Geophysical Union 56:1102.

----- 1975. The Mesa geothermal anomaly, Imperial Valley, California; a comparison and evaluation of results obtained from surface geophysics and deep drilling. United Nations Symposium on the Development and Uses of Geothermal Resources Abstracts 2.

----- 1981. Geothermal resources of rifts; a comparison of the Rio Grande Rift and the Salton Trough. Tectonophysics 94:659-678.

Swanberg, C.A., and S. Alexander 1978. The use of random groundwater data from WATSTORE in geothermal exploration; an example from the Imperial Valley, California. Transactions of the Geothermal Resources Council 2(2):635-638.

----- 1979. Use of water quality file WATSTORE in geothermal exploration; an example from the Imperial Valley, California. Geology 7:108-111.

Sweet, J., R.A. Wiggins, and G.A. Frazer 1977. Modeling the source mechanism of the Imperial Valley, California, earthquake of 1940. EOS, Transactions of the American Geophysical Union 58:1193.

Sweet, M.L., and G. Kocurek 1987. Preliminary observations of airflow over a complex crescentic dune (DRAA) in the Algodones Dune field, California. Society of Economic Paleontologists and Mineralogists 4:82.

----- 1988. Aerodynamic wake effects in the lee of aeolian dunes; implications for dune spacing. Geological Society of America Abstracts with Programs 20(7):176-177.

Sweet. M.L., J. Nelson, K. Havholm, and J. Farrelly 1987. Deposits and migration of the Algodones Dune field, southeastern California. Geological Society of America Abstracts with Programs 19(7):862.

Sweet., M.L., J. Wielson, K. Havholm, and J. Farrelly 1988. Algodones Dune field of southeastern California; case history of a migrating modern dune field. Sedimentology 35:939-952.

Swenson, G.A., III 1981. The ground water hydrology of Jacumba Valley, California and Baja California. Master of Science Thesis, San Diego State University, California 264 p.

Sykes, G.G. 1914. Geographical features of the Cahuilla basin. In The Salton Sea: A Case Study of the Geography, the Geology, the Floristics and the Ecology of a Desert Basin, edited by D.T. MacDougal. Carnegie Institute of Washington Publication 193:13-20.

----- 1926. The delta and estuary of the Colorado River. Geographical Review 16:232-255.

----- 1937a. The Colorado delta. American Geographic Society Special Publication 19:1-193.

----- 1937b. The Colorado River delta. Carnegie Institute of Washington Publication 460:1-193.

----- 1937c. Delta, estuary, and lower portion of the channel of the Colorado River 1933 to 1935. Carnegie Institute of Washington Publication 480:1-70.

Sykes, R.L., and S.P. Nishenko 1984. Probabilities of occurrence of large plate rupturing earthquakes for the San Andreas, San Jacinto, and Imperial Faults, California, 1983-2003. Journal of Geophysical Research 89B:5905-5927.

Sylvester, A.G. 1979. Earthquake damage in Imperial Valley, California, May 18, 1940, as reported by T.A. Clark. Seismological Society of America Bulletin 69:547-568.

----- 1991. Palm Tree Structure in the Central Mecca Hills. In Geological Excursions in Southern California and Mexico, edited by Michael J. Walawender and Barry B. Hanan, Geological Society of America Annual Meeting Guidebook, San Diego p. 378-387.

Sylvester, A.G., and M. Bonkowski 1979. Basement rocks of the Salton Trough region. In Tectonics of the Juncture Between the San Andreas Fault System and the Salton Trough, Southeastern California, edited by J.C. Crowell and A.G. Sylvester, Department of Geological Sciences, University of California, Santa Barbara p. 65-76.

----- 1991. Structure section in Painted Canyon, Mecca Hills, southern California. In Geological Excursions in Southern California and Mexico, edited by Michael J. Walawender and Barry B. Hanan, Geological Society of America Annual Meeting Guidebook, San Diego p. 372-377.

Sylvester, A.G., and R.R. Smith 1975. Structure section across the San Andreas Fault Zone, Mecca Hills. In San Andreas Fault in Southern California. A guide to San Andreas Fault from Mexico to Carrizo Plain, edited by J.C. Crowell, California Division of Mines and Geology Special Report 118:111-118.

----- 1976. Tectonic transpression and basement-controlled deformation in the San Andreas Fault Zone, Salton Trough, California. American Association of Petroleum Geologists Bulletin 60:2012-2081.

Tansev, E.O., and M.L. Wasserman 1978. Modeling the Heber geothermal reservoir. Transactions of the Geothermal Resources Council 2(2):654-648.

Tarbet, L.A. 1951. Imperial Valley. American Association of Petroleum Geologists Bulletin 35:260-263.

Tarbet, L.A., and W.H. Holman 1944. Stratigraphy and micropaleontology of the west side of Imperial Valley. American Association of Petroleum Geologists Bulletin 28:1781-1782.

Tarif, P.A., R.H. Wilkens, C.H. Cheng, and F.L. Paillet 1988. Laboratory studies of acoustic properties of samples from Salton Sea Scientific Drilling Project and their relation to microstructure and field measurements. Journal of Geophysical Research, B, Solid Earth and Planets 93:13,057-13,067.

Taylor, D.W. 1966. Summary of North American Blancan nonmarine molluscs. Malacologica 4:1-172.

----- 1975. Index and bibliography of late Cenozoic freshwater Mollusca of western North America. Museum of Paleontology University of Michigan, Ann Arbor, Claude W. Hibbard Memorial Volume 1:1-384.

----- 1981. Freshwater molluscs of California: a distributional checklist. California Department of Fish and Game 67(3):140-163.

Teilman, M., R.B. Weiss, and H.T. Meidav 1977. Imperial Valley Fault compilation. Transactions of the Geothermal Resources Council 1:289-290.

Teissere, R.F. 1968. Paleomagnetic investigation of the San Marcos Gabbro, southern California. Master of Science Thesis, University of California, Riverside 57 p.

Teissere, R.F., and M.E. Beck 1973. Divergent Cretaceous paleomagnetic pole position for the southern California batholith, U.S.A. Earth and Planetary Science Letters 18:296-300.

Terpstra, P. 1980. A computer simulation of the temporal development of geothermal energy systems with implications for the Salton Sea geothermal field of California. Master's Thesis, University of Illinois, Chicago.

Terres, R., and J.C. Crowell 1979. Plate tectonic framework of the San Andreas-Salton Trough juncture. In Tectonics of the Juncture Between the San Andreas Fault System and the Salton Trough, Southeastern California, edited by J.C. Crowell and A.G. Sylvester, Geological Society of America Annual Meeting Guidebook p. 15-25.

Tewhey, J.D. 1977a. The effect of hydrothermal alteration on porosity and permeability in the Salton Sea geothermal field, California. Geological Society of America Abstracts with Programs 9:1197-1198.

----- 1977b. Geologic characteristics of a portion of the Salton Sea geothermal field. Transactions of the Geothermal Resources Council 1:291-293.

Thatcher, W., J.N. Brune, and N.D. Clay 1971. Seismic evidence on the crustal structure of the Imperial Valley region. Geological Society of America Abstracts 3:208.

Thatcher, W., and R.M. Hamilton 1971. Spacial distribution and source parameters of the Coyote Mountain aftershock sequence, San Jacinto Fault Zone. Geological Society of America Abstracts with Programs 3:208-209.

----- 1973. Aftershocks and source characteristics of the 1969 Coyote Mountain Earthquake, San Jacinto Fault Zone, California. Bulletin of the Seismological Society of America 63:647-661.

Thatcher, W., J.A. Hileman, and T.C. Hanks 1975. Seismic slip distribution along the San Jacinto Fault Zone, southern California, and its implications. Geological Society of America Bulletin 86:1140-1146.

Theilege, E., M.B. Womer, and R. Papson 1978. Geological field guide to the Salton Trough. In Aeolian Features of Southern California: a Comparative Planetary Geology Guidebook, edited by R. Greely, et al. National Aeronautics and Space Administration Office of Planetary Geology p. 100-159.

Theodore, T.G. 1967a. Structure and petrology of the gneisses and mylonites at Coyote Mountain, Borrego Springs, California. Doctoral Dissertation, University of California, Los Angeles 268 p.

----- 1967b. Structure and petrology of the gneisses and mylonites at Coyote Mountain, Borrego Springs, California. Geological Society of America Abstracts with Programs 28:954B.

----- 1968. High-grade mylonite zone in southern California. Geological Society of America Special Paper 101:220.

----- 1970. Petrogenesis of mylonites of high metamorphic grade in the Peninsular Ranges of southern California. Geological Society of America Bulletin 81:435-449.

Theodore, T.G., and R.V. Sharp 1975. Geologic map of the Clark Lake quadrangle, San Diego County, California. U.S. Geological Survey Miscellaneous Field Study Map MF-644.

Thesken, R.S. 1977. A geochemical study of a hot spring and well waters in Jacumba, California. Undergraduate Research Reports, Department of Geology, San Diego State University, California 31(2):1-30.

Thilakaratne, V., and M. Vucetic 1990. Analysis of the seismic response at the Imperial wildlife liquefaction array in 1987. Proceedings of the U.S. National Conference on Earthquake Engineering 4(3):773-782.

Thomas, A., H. Magistrale, and S. Day 1991. Seismicity associated with a northwestern extension of the Cerro Prieto Fault, northern Baja California, Mexico. Geological Society of America Abstracts with Programs p. A91.

Thomas, H.W., and L.G. Barnes 1993. Discoveries of fossil whales in the Imperial Formation, Riverside County, California: possible further evidence of the northern extent of the proto-Gulf of California. In Ashes, Faults and Basins, edited by R.E. Reynolds and J. Reynolds, San Bernardino County Museum Association Special Publication 93-1:34-36.

Thomas, K. 1983. Analysis of the aftershocks of the Borrego Mountain Earthquake, April 9, 1968. Undergraduate Research Reports, Department of Geology, San Diego State University, California 42(6):1-8.

Thomas, R.G. 1963. The late Pleistocene 150-foot fresh water beach line of the Salton Sea area. Southern California Academy of Sciences 62(1):9-17.

Thompson, D.G. 1929. Mojave Desert region. U.S. Geological Survey Water-Supply Paper 578:1-759.

Thompson, D.G., and J.I. Mead 1982. Late Quaternary environments and biogeography in the Great Basin. Quaternary Research 17:39-55.

Thompson, J.M., and R.O. Fournier 1988. Chemistry and geothermometry of brine produced from the Salton Sea scientific drill hole, Imperial Valley, California. Journal of Geophysical REsearch, B, Solid Earth and Planets 93:13,165-13,173.

Thompson, R.S. 1991. Pliocene embayment and climates in the western United States. In Pliocene Climates, edited by T.M. Cronin and H.J. Dowsett, Quaternary Science Reviews 10(2/3):115-132.

Thompson, R.W. 1968. Tidal flat sedimentation of the Colorado River delta, northwestern Gulf of California. Geological Society of America Memoir 107:1-133.

----- 1975. Tidal flat sediments of the Colorado delta, northwestern Gulf of California. In Tidal deposits, edited by R.N. Ginsburg, Springer-Verlag, New York p. 57-65.

Tomsen, D.E. 1974. Power from the Salton Trough. Science News 106(2):28-29.

Threet, R.L. 1974. Alternative interpretations for the southern portion of the San Jacinto Fault Zone, San Diego County, California. In Recent Geologic and Hydrologic Studies, Eastern San Diego County and Adjacent Areas, edited by M.W. Hart and R.J. Dowlen, San Diego Association of Geologists Field Trip Guidebook p. 36-46.

----- 1978. Features of the San Jacinto Fault Zone and the Borrego Mountain earthquake. National Association of Geology Teachers Guidebook for Fieldtrip No. 2, 25 p.

----- 1989. Field guide for two trips in the Imperial Valley, California; geomorphology of western Salton Sink, and San Jacinto Fault Zone and Borrego Mountain Earthquake of April 16, 1978. National Association of Geology Teachers Guidebook, Far Western Section 42 p.

Timm, W.C. 1967. Geologic reconnaissance of the central portion of the Palo Verde Mountains, Imperial County, California. Undergraduate Research Reports, Department of Geology, San Diego State University, California 11(5):1-32.

Todd, V.R. 1977. Geologic map of the Agua Caliente Springs quadrangle, San Diego, California. U.S. Geological Survey Open-File Report 77-742:1-17.

----- 1977a. Geologic map of the Cuyamaca Peak 7 1/2-minute quadrangle, San Diego County, California. U.S. Geological Survey Open-File Report 77-405:1-17.

----- 1977b. Mafic plutonic rocks of two ages in the Peninsular Ranges Batholith, south-central San Diego County, California. Geological Society of America Abstracts with Programs 9:516.

----- 1978. Geologic map of the Monument Peak quadrangle, San Diego, California. U.S. Geological Survey Open-File Report 78-697:1-45.

----- 1979a. Itinerary for field trip to the Peninsular Ranges Batholith, San Diego - road log for Friday, November 9, 1979. In Mesozoic Crystalline Rocks, Peninsular Ranges Batholith and Pegmatites, Point Sal Ophiolite, edited by P.L. Abbott and V.R. Todd, Department of Geological Sciences, San Diego State University, California p. 233-267.

----- 1979b. Geologic map of the Mount Laguna quadrangle, San Diego County, California. U.S. Geological Survey Open-File Report 79-862:1-40.

----- 1979c. Vertical tectonics in the Elsinore Fault zone south of 33°7'30". Geological Society of America Abstracts with Programs 11:528.

----- 1980. Quaternary deformation in southern part of Elsinore Fault Zone. U.S. Geological Survey Professional Paper 1157:1-101.

----- 1982. Geologic map of the Tule Springs quadrangle, San Diego County, California. U.S. Geological Survey Open-File Report 82-0221:-132.

Todd, V.R., D.E. Detra, J.E. Kilburn, A. Griscom, F.A. Kruse, and E.L. McHugh 1987. Mineral resources of the Fish Creek Mountains Wilderness Study Area, Imperial County, California. U.S. Geological Survey Bulletin, Report B1171-C:C1-C14.

Todd, V.R., B.G. Erskin, and D.M. Morton 1988. Metamorphic and tectonic evolution of the northern Peninsular Ranges Batholith, southern California. In Metamorphism and crustal evolution of the western

United States, edited by W.G. Ernst. Rubey Volume 7. Prentice-Hall, Englewood Cliffs p. 894-937.

Todd, V.R., and W.C. Hoggatt 1976. The Elsinore Fault Zone in the Tierra Blanca Mountains, eastern San Diego County, California. Geological Society of America Abstracts with Programs 8:416-417.

Todd, V.R., R.C. Jachens, D.L. Kimbrough, G.H. Girty, and J.M. Hammarstrom 1991. The El Cajon 30' X 60' quadrangle, southern California: a geologic history from early Mesozoic continental growth to modern rift-transform tectonics. Abstract with programs. Geological Society of America annual meeting. San Diego p. A479.

Todd, V.R., J.E. Kilburn, D.E. Detra, A. Griscom, F.A. Kruse, and E.L. McHugh 1987. Mineral resources of the Jacumba (In-Ko-pah) Wilderness Study Area, Imperial County, California. U.S. Geological Survey Bulletin, Report B1171-D:D1-D18.

Todd, V.R., D.L. Kimbrough, and C.T. Herzig 1994. The Peninsular Rangers batholith from western volcanic arc to eastern mid-crustal intrusive and metamorphic rocks, San Diego County, California. In Geological Investigations of an Active Margin, edited by S.F. McGill and T.M. Ross, U.S. Geological Survey, Cordilleran Section Guidebook Trip 9, p. 227-324.

Todd, V.R., R.E. Learned, T.J. Peters, and R.T. Mayerel 1983. Mineral resource potential of the Sill Hill, Hauser, and Caliente roadless areas, San Diego County, California. U.S. Geological survey Miscellaneous Field Studies Map MF-1547-A:1-12.

Todd, V.R., and T.J. Peters 1984. Sill Hill, Hauser, and Caliente roadless areas, California. U.S. Geological Survey Professional Paper 1300:371-372.

Todd, V.R., and S.E. Shaw 1979. Structural, metamorphic and intrusive framework of the Peninsular Ranges Batholith in southern San Diego County, California. In Mesozoic Crystalline Rocks: Peninsular Ranges Batholith and Pegmatites, Point Sal Ophiolite, edited by P.L. Abbott and V.R. Todd, Department of Geological Sciences, San Diego State University, California p. 177-231.

----- 1980. Intrusive, metamorphic, and structural framework of the Peninsular Ranges Batholith in southern San Diego County (latitude 32°45'33"), California. Geological Society of America Abstracts with Programs 12:156.

----- 1981. Deformation in the Peninsular Range Batholith, San Diego County, California. In Tectonic Framework of the Mojave and Sonoran Deserts, California and Arizona, edited by K.A. Howard, U.S. Geological Survey Open-File Report 81-0503:110-112.

----- 1985. S-type granitoids and an I-S line in the Peninsular Ranges Batholith, southern California. Geology 13:231-233.

Todd, V.R., G.H. Girty, S.E. Shaw, and R.C. Jachens 1991. Geochemical geochronologic and structural characteristics of Jurassic plutonic rocks, Peninsular Ranges, California. Geological Society of America Abstracts with Programs p. A249.

Toms, R.S.H. 1983. Geothermal research in the deep hole, Salton Sea Scientific Drilling Project. EOS, Transactions of the American Geophysical Union 64:864.

Tooms, J.S. 1970. Review of the knowledge of metalliferous brines and related deposits. Institute of Mining and Metallurgy Transactions 79:B116-B126.

Toppozada, T.R. 1993. The Landers-Big Bear Earthquake sequence and its felt effects. California Geology 46(1):3-9.

Torres, S.G. 1982. The geology of the Blue Lady pegmatite dyke of Chihuahua Valley, San Diego County, California. Undergraduate Research Reports, Department of Geology, San Diego State University, California 40(7):1-60.

Tosdal, R.M. 1983. Mesozoic crystalline rocks of the Mule Mountains and Vincent-Chocolate Mountain thrusts, southern California and southwest Arizona; juxtaposed pieces of the Jurassic magnetic arc of southwest North America. In Proceedings of the Circum-Pacific Terrains Conference, edited by D.G. Howell, D.L. Jones, A. Cox, and A.M. Nur, Stanford University Publications, Geological Sciences 18:193-194.

----- 1986. Gneissic host rocks to gold mineralization in the Cargo Muchacho Mountains, southeastern California. In Geological Setting of Gold and Silver Mineralization in Southeastern California and Southwestern Arizona, Geological Society of America Field Guide 5(6):139-142.

----- 1988. Mesozoic rock units along the late Cretaceous Mule Mountains thrust system, southeastern California and southwestern Arizona. Doctoral Dissertation, University of California, Santa Barbara 394 p.

----- 1990a. Constraints on the tectonics of the Mule Mountains thrust system, southeast California and southwest Arizona, Journal of Geophysical Research, B, Solid Earth and Planets 95(12):20,025-20,048.

----- 1990b. Jurassic low-angle ductile shear zones, SE California and SW Arizona: thrust faults, extensional faults, or rotated high-angle faults? Geological Society of America Abstracts with Programs 22(3):89.

Tosdal, R.M., and G.B. Haxel 1982. Two belts of late Cretaceous to early Tertiary crystalline thrust faults in southwest Arizona and southeastern California. Geological Society of America Abstracts with Programs 14:240.

----- 1987. Late Jurassic regional metamorphism and deformation, southeastern California and southwestern Arizona. Geological Society of America Abstracts with Programs 19(7):870-871.

Tosdal, R.M., G.B. Haxel, and J.T. Dillon 1986. Evidence for and tectonic implications of northwestward movement on the Chocolate Mountains thrust, SE Calif. Geological Society of America Abstracts with Programs 18:193.

Tosdal, R.M., and D.R. Sherrod 1985. Geometry of Miocene extensional deformation, lower Colorado River region, southeastern California and southwestern Arizona; evidence for the presence of a regional low-angle normal fault. Lunar and Planetary Institute, Houston, Texas, Contribution 575:147-151.

Towse, D. 1975a. An estimate of the geothermal energy resource in the Salton Trough, California. University of California Riverside Report 51851:1-22.

----- 1975b. Estimating the geothermal resources; the Salton Trough, California. United Nations Symposium on the Development and Uses of Geothermal Resources Abstracts 2.

----- 1975c. Geothermal resources of the Salton Trough, California. American Association of Petroleum Geologists, Society of Economic Paleontologists and Mineralogists Abstract 2:75.

Tracy, S.A. 1984. Magnetic survey across the Elsinore Fault, southwestern Imperial Valley, California. Undergraduate Research Reports, Department of Geology, San Diego State University, California 44(2):1-23.

Trent, D.D. 1985. Geology of Joshua Tree National Monument. California Geology 37:75-85.

Trifunac, M.D. 1969. Investigation of strong earthquake ground motion. Doctoral Dissertation, California Institute of Technology, Pasadena 153 p.

----- 1972. Tectonic stress and the source mechanism of the Imperial Valley, California, earthquake of 1940. Seismological Society of America Bulletin 62:1283-1302.

Trifunac, M.D., and J.N. Brune 1970. Complexity of energy release during the Imperial Valley, California, earthquake of 1940. Seismological Society of America Bulletin 60:137-160.

Truschel, J.P. 1980. A petrographic study of the Julian Schist. Undergraduate Research Reports, Department of Geology, San Diego State University, California 37(3):1-19.

Tsang, C.F., D.S. Mangold, and M.J. Lippmann 1980. Simulation of reinjection at Cerro Prieto using an idealized two-reservoir geological model. Energy and Environment Division, Lawrence Berkeley Laboratory, California, Report 10686:52-57.

Tucker, A.B., R.M. Feldman, and C.L. Powell, II 1994. *Speocarcinus berglundi* n. sp. (Decapoda: Brachyura), a new crab from the Imperial Formation (late Miocene-late Pliocene) of southern California. Journal of Paleontology 68(4):800-807.

Tucker, W.B., and R.J. Sampson 1945. Mineral resources of Riverside County. California Division of Mines Report 42:121-182.

Tucker, W.B., and G.H. Reed 1939. Mineral resources of San Diego County. California Division of Mines Report 35:8-55.

Turley, T.J. 1972. Mud volcanoes; an unusual natural phenomena. Rocks and Minerals 47:200-201.

Twiss, R. J. Sidener, G. Bingham, J.E. Burke, and C.H. Hall 1980. Potential impacts of geothermal development on outdoor recreational use of the Salton Sea. University of California, Riverside Report 136897:1-61.

Udwadia, F.E., and M.D. Trifunac 1972. Studies of strong earthquake motions and microtremor process. Microzonation Conference, University of Washington, Seattle p. 319-325.

----- 1973. Comparison of earthquake and microtremor ground motion in El Centro, California. Seismological Society of America Bulletin 63:1227-1253.

----- 1974. (Comments on: Comparison of earthquake and microtremor ground motion in El Centro, California.) a reply. Seismological Society of America Bulletin 64:497.

Ulrich, F.P. 1941. The Imperial Valley Earthquake of 1940. Seismological Society of America Bulletin 31(2):13-31.

U.S. Bureau of Reclamation (prior to late 1940's). Unpublished geologic reports on the proposed routes of the Coachella branch of the All American Canal. Headquarters, Denver, Colorado.

U.S. Department of Energy, Geothermal Energy Division 1980. Environmental assessment, geothermal energy, Heber geothermal binary-cycle demonstration project; Imperial County, California. Report DOE/EA-0119:1-298.

U.S. Energy Research and Development Administration 1975. Problems in identifying, developing, and using geothermal resources. Report RED-75-330:1-71.

U.S. Geological Survey 1968. Water resources data for California; volume I, Colorado River basin, southern Great Basin from Mexican border to Mono Lake basin, and Pacific slope basins from Tijuana

River to Santa Maria River. Water Data Reports.
----- 1972. The Borrego Mountain Earthquake of April 9, 1968. Professional Paper 787:1-207.
----- 1979. Land use and land cover and associated maps for El Centro, California, Arizona. Open-File Report 79-1175.
----- 1980. Land use and land cover and associated maps for Salton Sea, California, Arizona. Open-File Report 80-857.
----- 1982. The Imperial Valley, California, earthquake of October 15, 1979. Professional Paper 1254:1-451.
----- 1983. Aeromagnetic map of the Salton Sea area, California. Open-File Report 83-0664.
----- 1988. Geothermal energy resources. Bulletin 1850:71-72.
U.S. National Oceanic and Atmospheric Administration 1972. Seismological field survey intensity distribution and field effects, strong-motion seismograph records, and response spectra. In The Borrego Mountain Earthquake of April 9, 1968, U.S. Geological Survey Professional Paper 787:141-157.
Urban, T.C., W.H. Diment, and M. Nathenson 1978. East Mesa geothermal anomaly, Imperial County, California; significance of temperatures in a deep drill hole near thermal equilibrium. Transactions of the Geothermal Resources Council 2(2):667-670.
Vallette-Silver, N.J., F. Tera, M.J. Pavich, J. Klein, and R. Middelton 1988. 10Be-9Be in the Salton Sea geothermal system. Transactions of the Geothermal Resources Council 12:143-150.
Van Andel, T.H. 1964. Recent marine sediments of Gulf of California. In Marine geology of the Gulf of California, edited by T. H. van Andel, and G.G. Shor. American Association of Petroleum Geologists memoir 3:216-310.
Van Andel, T.H., and G.G. Shor, editors 1964. Marine geology of the Gulf of California. American Association of Petroleum Geologists Memoir 3:1-408.
Van Buskirk, M.C., and M.A. McKibben 1993. The Modoc fossil hot spring deposit. In Ashes, Faults and Basins, edited by R.E. Reynolds and J. Reynolds, San Bernardino County Museum Association Special Publication 93-1:81.
Van de Kamp, P.C. 1973. Holocene continental sedimentation in the Salton Basin, California, a reconnaissance. Geological Society of America Bulletin 84:827-848.
Van de Kamp, P.C., B.E. Leake, and A. Senior 1976. The petrology and geochemistry of some Californian arkoses with application to identifying gneisses of metasedimentary origin. Journal of Geology 84:195-212.

Van De Verg, P.E. 1976. Seismic refraction investigation of the Dunes thermal anomaly, Imperial County, California. Master's Thesis, University of California, Riverside.
Van Nort, S.D., and M. Harris 1984. Geology and mineralization of the Picacho gold prospect Imperial County, California. Arizona Geological Society Digest 15:175-183.
Van Wagenen, L.G. 1979. San Diego Gas and Electric Company geothermal activities. In Geology and Geothermics of the Salton Trough, edited by W.A. Elders, University of California, Riverside Campus Museum Contributions 5:101-103.
Vaughan, F.E. 1918. Evidence in the San Gorgonio Pass, Riverside County, California, of a late Pliocene extension of the Gulf of California. Geological Society of America Bulletin 29(1):164-165.
----- 1922. Geology of the San Bernardino Mountains north of San Gorgonio Pass. University of California Department of Geological Sciences Bulletin 13:319-411.
Vaughan, T.W. 1904. A California Tertiary coral reef and its bearing on American Recent coral faunas. Science 19:503.
----- 1917a. The reef-coral fauna of Carrizo Creek, Imperial County, California and its significance. U.S. Geological Survey Professional Paper 98T:355-387.
----- 1917b. Significance of reef coral fauna at Carrizo Creek, Imperial Co., Calif. Washington Academy of Science Journal 7:194.
Vaux, H.T., Jr. 1975. Desalting Imperial Valley geothermal brines; some economic and policy considerations. California University Water Resources Center Report 33:135-145.
Veatch, J.A. 1858. Notes on a visit to the mud volcanoes in the Colorado Desert in the month of July, 1857. American Journal of Science and Arts, 2nd Series, 26:288-295.
Ver Planck, W.E. 1952. Gypsum in California: Fish Creek Mountains deposit, Imperial and San Diego Counties. California Division of Mines Bulletin 163:28-35.
----- 1954. Salines in southern California. In Geology of Southern California, edited by R.H. Jahns, California Division of Mines Bulletin 170(8):6-9.
----- 1957. Salines and salt, mineral commodities of California. California Division of Mines Bulletin 176:476-477,486,494.
----- 1958. Salt in California. California Division of Mines Bulletin 175:1-168.
Vetter, O., D.A. Campbell, and M.J. Walker 1978. Geothermal fluid investigations at Republic's East Mesa test site. Transactions of the Geothermal Resources Council 2(2):693-695.

Veysey, V.V. and others 1973. National Conference on Geothermal Energy, Proceedings. University of California, Riverside 301 p.

Vialov, O.S. 1966. The traces of the vital activity of organisms and their paleontological significance. Ukrainian Academy of Science, Soviet Socialist Republics 219 p.

Vonder Haar, S.P. and J.H. Howard 1978. Geology and geothermal resources of the Salton Trough, California, in relation to rift zone tectonics. Conference Proceedings, Los Alamos Scientific Laboratory, New Mexico LA-C 7478:97-98.

----- 1980. Regional geology setting of the Cerro Prieto geothermal field. Energy and Environment Division, Lawrence Berkeley Laboratory, California 10686:39-38.

Vonder Haar, S.P., J. Nobel, M.T. O'Brien, and J.H. Howard 1980. Subsurface geology of the Cerro Prieto geothermal field. Energy and Environment Division, Lawrence Berkeley Laboratory, California 10686:39-43.

Vrla, J.G. 1982. Geology of the Jojoba Wash area, northern Jacumba Mountains, California. Undergraduate Research Reports, Department of Geology, San Diego State University, California 41(4):1-17.

Vucetic, M., R. Dobry, K.H. Stokoe II, R.S. Ladd, and T.L. Youd 1986. Evaluation of a liquefaction case history; Heber Road site, 1979 Imperial Valley Earthquake. European Conference on Earthquake Engineering 8(2):5.3/57-5.3/64.

Waananen, A.O., and W.R. Moyle, Jr. 1972. Water resources effects. In The Borrego Mountain Earthquake of April 9, 1968, U.S. Geological Survey Professional Paper 787:183-189.

Wagoner, J.L. 1977. Stratigraphy and sedimentation of the Pleistocene Brawley and Borrego Formations in the San Felipe Hills area, Imperial County, California. Master of Science Thesis, Department of Geological Sciences, University of California, Riverside 128 p.

----- 1978. The stratigraphy and sedimentology of the Pleistocene Brawley and Borrego Formations in the San Felipe Hills area, Imperial Valley, California. Geological Society of America Abstracts with Programs 12:152.

----- 1980. Grain-size distribution analysis of non-marine sandstone, Imperial Valley, California, U.S.A. Geological Society of America Abstracts with Programs 12:158.

Walawender, M.J. 1979. Basic plutons of the Peninsular Ranges Batholith, southern California. In Mesozoic crystalline rocks, edited by P.L. Abbott, and V.R. Todd. Department of Geological Sciences, San Diego State University p. 151-162.

Walawender, M.J., R.G. Gastil, J.P. Clinkenbeard, W.V. McCormick, B.G. Eastman, R.S. Wernicke, M.S. Wardlaw, S.H. Gunn, and B.M. Smith 1990. Origin and evolution of the zoned La Posta-type plutons, eastern Peninsular Ranges Batholith, southern and Baja California. Geological Society of America Memoir 174:1-18.

Walawender, M.J., and W.W. Gross 1986. Geochemistry and origin of mafic inclusions in the grantic rocks of the Peninsular Ranges Batholith, southern California. Geological Society of America abstracts with programs 18:194.

Walawender, M.J., and T.E. Smith 1980. Geochemical and petrologic evolution of the basic plutons of the Peninsular Ranges Batholith, southern California. Journal of Geology 88:233-242.

Wald, D.J., D.V. Helmberger, and S.H. Hartzell 1990. Rupture process of the 1987 Superstition Hills Earthquake from inversion of strong-motion data. Seismological Society of America Bulletin 80:1079-1098.

Wald, D.J., and P.G. Sommerville 1987. Semi-empirical modeling of recorded accelerations from the 1979 Imperial Valley Earthquake. EOS, Transactions of the American Geophysical Union 68:1355.

Wald, D.J., P.G. Sommerville, and D.V. Helmberger 1987. Compatibility of accelerograms of the 1979 Imperial Valley Earthquake with slip-distribution asperity models. Seismological Research Letters 58:10.

Walker, G.W. 1953. California district. U.S. Geological Survey Report TEI-330:224-225.

Walker, G.W., T.G. Lovering, and S.G. Hal 1956. Radioactive deposits in California. California Division of Mines Special Report 49:1-38.

Walker, J.R., and G.R. Thompson 1989. Structural variations in chlorite and illite in a diagenetic sequence from the Imperial Valley, California. Clay Minerals Society, Annual Clay Minerals Conference 26:74.

----- 1990. Structural variations in chlorite and illite in a diagenetic sequence from the Imperial Valley, California. Clays and Clay Minerals 38:315-321.

Walker, M. 1981. Gravity survey of the southern part of the Elsinore Fault. Undergraduate Research Reports, Department of Geology, San Diego State University, California 38(6):1-40.

Wallace, R.D. 1982. Evaluation of possible detachment faulting west of the San Andreas, southern Santa Rosa Mountains, California. Master of Science, San Diego State University, California 77 p.

Wallace, R.D., and D.J. English 1982. Evaluation of possible detachment faulting west of the San Andreas, southern Santa Rosa Mountains, California. In Mesozoic-Cenozoic Tectonic

Evolution of the Colorado River Region, California, Arizona, and Nevada, edited by E.G. Frost and D.L. Martin, San Diego Cordilleran Publishers, California p. 503-509.

Walter, D.L. 1984. Formation of a tungsten mine. Undergraduate Research Reports, Department of Geology, San Diego State University, California 44(2):1-16.

Ware, G.C., Jr. 1958. The geology of a portion of the Mecca Hills, Riverside County, California. Master of Arts Thesis, University of California, Los Angeles.

Warn, F. 1966. Sinkhole development in the Imperial Valley. In Engineering Geology in Southern California, edited by R. Lung, Association of Engineering Geologists Special Publication p. 144-145.

Waters, M.R. 1980. Lake Cahuilla: late Quaternary lacustrine history of the Salton Trough, California. Master of Science Thesis, Department of Geosciences, University of Arizona, Tucson 74 p.

----- 1982. Late Quaternary lacustrine chronology and archaeology of ancient Lake Cahuilla, California. American Quaternary Association, 7th Biennial Meeting Abstracts and Program 1982:176.

----- 1983. Late Quaternary lacustrine chronology and archaeology of ancient Lake Cahuilla, California. Quaternary Research 19:373-387.

Watkins, R. 1990a. Pliocene channel deposits of oyster shells in the Salton Trough region, California. Palaeogeography, Palaeoclimatology, Palaeoecology 79:249-262.

----- 1990b. Paleoecology of a Pliocene rocky shoreline, Salton Trough region, California. Palaios 5:167-175.

Watson, A.O. 1977. Hydrogeology of Earthquake Valley. Undergraduate Research Reports, Department of Geology, San Diego State University, California 31(2):1-17.

Weaver, C.S., and D.P. Hill 1978. Earthquake swarms and local crustal spreading along major strike-slip faults in California. Pure and Applied Geophysics 117:51-64.

Webb, R.W. 1939. Evidence of the age of crystalline limestone in southern California. Journal of Geology 47:198-201.

Webb, S.D. 1984. Ten million years of mammal extinctions in North America. In Quaternary Extinctions, a Prehistoric Revolution, edited by P.S. Martin and R.G. Klein, University of Arizona Press, Tucson p. 189-210.

Webb, T.D. 1977. A geophysical survey of Kemp Ranch, Laguna Mountains, San Diego County, California. Undergraduate Research Reports, Department of Geology, San Diego State University, California 30(3):1-35.

Weber, E.M. 1979. Water quality control in the Colorado River basin. Geological Society of America Abstracts with Programs 11:536.

Weber, F.H., Jr. 1959. Geology and mineral resources of San Diego County, California. California Division of Mines and Geology, Scale 1:125,000.

----- 1963. Geology and mineral resources of San Diego County, California. California Division of Mines and Geology Report 3:1-309.

----- 1976. Preliminary map of the Elsinore and Chino Fault Zones in northeastern Riverside County, California. California Division of Mines and Geology Open-File Report 76-1.

Weber, G.E. 1962. Geology of a portion of the Indio Hills, Riverside County, California. Senior Thesis, Department of Geological Sciences, University of California, Riverside.

Weide, D.L. 1976. Regional environmental history of the Yuha Desert. In Background to Prehistory of the Yuha Desert Region, edited by P.J. Wilke, Ballena Press Anthropological Papers 5:9-20.

Weight, H.O. 1948. Nature's freaks on Salton shore. Desert Magazine 11(6):5-8.

----- 1952. Puzzel rocks of the Badlands. Desert Magazine 15(3):18-22.

----- 1956. Petrified palm in an ancient stream bed. Desert Magazine 19(7):13-16.

Weismeyer, A.L., Jr. 1968. Geology of the northern portions of the Seventeen Palms and Font's Point quadrangles, Imperial and San Diego Counties, California. Master of Science Thesis, University of Southern California, Los Angeles 63 p.

Weiss, R.B., T.S. Oldknow, and H.T. Meidav 1977. Geochemical-hydrological studies of the Imperial Valley, California. Transactions of the Geothermal Resources Council 1:30-307.

Weldon, R.J. 1984. Quaternary deformation due to the junction of the San Andreas and San Jacinto Faults, southern California. Geological Society of America Abstracts with Programs 16:689.

----- 1986. The late Cenozoic geology of Cajon Pass; implications for tectonics and sedimentation along the San Andreas Fault. Doctoral dissertation, California Institute of Technology, Pasadena 400p.

Weldon, R.J., and K.E. Sieh 1985. Holocene rate of slip and tentative recurrence interval for large earthquakes on the San Andreas Fault, Cajon Pass, southern California. Geological Society of America Bulletin 96:793-812.

Wells, D.L. 1987a. Geology of the eastern San Felipe Hills, Imperial Valley, California: implications for wrench faulting in the southern San Jacinto Fault Zone. Master of Science Thesis, San Diego State

University, California 140 p.

----- 1987b. Percent strain and slip rate estimates for a proposed buried extension of the Clark Fault, southern San Jacinto Fault Zone, Imperial Valley, California. Geological Society of America Abstracts with Programs 19(6):462.

Wells, D.L. and E.S. Feragan 1986. Wrench faulting, fault-propagation, and basal decollment in the San Felipe Hills, Imperial Valley, California. In Geology of the Imperial Valley, California, edited by P.D. Guptil, E.M. Gath and R.W. Ruff, South Coast Geological Society Annual Field Trip Guidebook, Santa Ana, California 14:84-95.

Wells, S.G., S. Connell, and J.J. Martin 1993. Geomorphic and soil stratigraphic evaluation of faulted alluvial sequence, eastern Coachella Valley, California. In Ashes, Faults and Basins, edited by R.E. Reynolds and J. Reynolds, San Bernardino County Museum Association Special Publication 93-1:39-49.

Welsch, B.W. 1959. Geochemical studies in the Julian, Banner mining district, San Diego County, California; determination of copper in the soil. Undergraduate Research Reports, Department of Geology, San Diego State University, California 3(3):1-29.

Wenk, H.R., and J. Pannetier 1990. Texture development in deformed granodiorites from the Santa Rosa mylonite zone, southern California. Journal of Structural Geology 12:177-184.

Werner, S.D., and M.S. Agbabian 1984. Soil/structure interaction effects at El Centro, California terminal substation building. Proceedings of the World Conference on Earthquake Engineering 8(3):1073-1080.

Werner, S.L., and C.R. White 1973. History and development of geothermal resources in Imperial Valley. Association of Engineering Geologists Annual Meeting, Program Abstracts 16:27.

Whistler, D.P., E.B. Lander, and M.A. Roeder 1995. First diverse record of small vertebrates from late Holocene sediments of Lake Cahuilla, Riverside County, California. Abstracts of Proceedings. 9th Annual Mojave Desert Quaternary Research Symposium, San Bernardino County Museum Quarterly 42(2):46.

White, D.E. 1955. Violent mud-volcano eruption of Lake City hot springs, northeastern California. Geological Society of America Bulletin 66:1109-1130.

----- 1963a. Geothermal brine well, mile-deep hole may tap ore-bearing magmatic water and rocks undergoing metamorphism. Science 139(3558):919-922.

----- 1963b. The Salton Sea geothermal brine, an ore-transporting fluid. Mining Engineering 15:60.

----- 1964. Deep geothermal brine near Salton Sea, California Bulletin Volcanologique 27:369-370.

White, D.E., and E.T. Anderson, and D.K. Grubbs 1963. Geothermal brine well - mile-deep drill hole may tap ore-bearing magmatic water and rocks undergoing metamorphism. Science 139:919-922.

White, J.A. 1965a. Kangaroo rats (family Heteromyidae) of the Vallecito Creek Pleistocene of California. Geological Society of America Bulletin 82:288-289.

----- 1965b. Late Cenozoic vertebrates of the Anza-Borrego Desert area, southern California. American Association for the Advancement of Science, Abstracts with Programs (unpaginated).

----- 1967. Late Cenozoic bats (sub-family Nyctophylinae) from the Anza-Borrego Desert of California. Miscellaneous Publications of the University of Kansas Museum of Natural History 51:275-282.

----- 1968. A new porcupine from the middle Pleistocene of the Anza-Borrego Desert of California; with notes on mastication in Coendu and Erethizon. Los Angeles County Museum of Natural History Contributions in Science 136:1-15.

----- 1970. Late Cenozoic Porcupines (Mammalia, Erethizontidae) of North America. American Museum Novitates 2421:1-15.

----- 1984. Late Cenozoic Leporidae (Mammalia; Lagomorpha) from the Anza-Borrego Desert, California. Carnegie Museum of Natural History Special Publication 9:41-57.

White, J.A., and T. Downs 1961. A new Geomys from the Vallecito Creek Pleistocene of California; with notes on variation in Recent and fossil species. Los Angeles County Museum of Natural History Contributions in Science 42:1-34.

----- 1965. Vertebrate microfossils from the Canebrake Formation of the Imperial Valley region, California. Society of Economic Paleontologists and Mineralogists, Pacific Section (unpaginated).

White, J.A., E.H. Lindsay, P. Remeika, B.W. Stout, T. Downs, and M. Cassiliano 1991. Society of Vertebrate Paleontology Field Trip Guide to the Anza-Borrego Desert. Society of Vertebrate Paleontology, Annual Meeting Guidebook 23 p.

Whitten, C.A. 1949. Horizontal earth movement in California. The Journal of Coast and Geodetic Survey 2:84-88.

----- 1955. Measurements of earth movements in California. California Division of Mines Bulletin 171:75-80.

----- 1956. Crustal movement in California and Nevada. Transactions of the American Geophysical Union 37:393-398.

----- 1973. Crustal movement in California and Nevada. In Report on geodetic measurements of crustal movement, 1906-71, National Oceanic and Atmospheric Administration-National Ocean

Survey National Geodetic Society 6 p.

Wiley, R.A. 1978. The geology adjacent to the Coyote Creek Fault on the south west side between Borrego Sink and San Felipe Creek, Borrego Valley, California. Undergraduate Research Reports, Department of Geology, San Diego State University, California 32(3):1-17.

Wiley, W.D. 1950. The salt in Salton Sea. Desert Magazine 15(10):26-28.

Wilkins, J., Jr. 1984. Editor's note: Picacho Mine update. Arizona Geological Society Digest 16:267-281.

Williams, A.E. 1988a. Delineation of a brine interface in the Salton Sea geothermal system, California. Transactions of the Geothermal Resources Council 12:151-157.

----- 1988b. Fluid density distribution in a stratified geothermal reservoir; Salton Sea geothermal system, California. Geological Society of America Abstracts with Programs 20(7):98.

Williams, A.E., and M.A. McKibben 1987. A brine interface in the Salton Sea geothermal system, California; mechanism for active ore formation. Geological Society of America Abstracts with Programs 19:890.

----- 1989. A brine interface in the Salton Sea geothermal system, California; fluid and isotopic characteristics. Geochimica et Cosmochimica Acta 53:1905-1920.

Williams, A.E., and C.S. Oakes 1986. Isotopic and chemical variations in hydrothermal brines from the Salton Sea geothermal field, California. International Symposium on Water-Rock Interaction, Extended Abstracts 5:633-636.

Williams, D.S. 1986. Saltwater upcoming potential in the Ocotillo-Coyote Wells basin, Imperial County, California. Master of Science Thesis, San Diego State University, California 128 p.

Williams, G.E., and H.A. Polach 1971. Radiocarbon dating of arid-zone calcareous paleosols. Geological Society of America Bulletin 82:3069-3086.

Williams, J.J. 1956. Geology of part of the Orocopia Mountains, Riverside County, California. Master's Thesis, University of California, Los Angeles 44 p.

Williams, P.L., and H.W. Magistrale 1988. Slip of the Superstition Hills Fault associated with the 1987 Superstition Hills, California, earthquake. EOS, Transactions of the American Geophysical Union 69:1448.

----- 1989. Slip of the Superstition Hills Fault associated with the 24 November 1987 Superstition Hills, California, earthquake. Seismological Society of America Bulletin 79:390-410.

Williams, P.L., L.R. Sykes, C. Nicholson, and L. Seeber 1990. Seismotectonics of the easternmost Transverse Ranges, California: relevance for seismic potential of the southern San Andreas Fault. Tectonics 9:185-204.

Willis, G.F., R.M. Tosdal, and S.L. Manske 1987. The Mesquite Mine, southeastern California; epithermal gold mineralization in a strike-slip fault system. Geological Society of America Abstracts with Programs 19(7):892.

Wilson, M.E., and S.H. Wood 1978. Salton Sea water-level records (1952-1977) and the southern California uplift. EOS, Transactions of the American Geophysical Union 59:242.

----- 1980. Tectonic tilt rates derived from lake-level measurements, Salton Sea, California. Science 207:183-186.

Wilt, M.J. 1975. An electrical survey of the Dunes geothermal anomaly and surrounding region, Imperial Valley, California. Master's Thesis, University of California, Riverside.

Wilt, M.J., and J. Combs 1975. Telluric mapping, telluric profiling, and self-potential surveys of the Dunes geothermal anomaly, Imperial Valley, California. United Nations Symposium on the Development and Uses of Geothermal Resources Abstracts 2.

Wilt, M.J., and N.E. Goldstein 1980. Resistivity studies at Cerro Prieto. Energy and Environment Division, Lawrence Berkeley Laboratory, California 10686:11-15.

Wingerd, B.J. 1979. Hydrogeochemical determination of geothermal potential in the Coyote Wells area, Imperial Valley, California and the role of the Elsinore Fault on groundwater movement. Undergraduate Research Reports, Department of Geology, San Diego State University, California 35(4):1-67.

Winker, C.D. 1987. Neogene stratigraphy of the Fish Creek-Vallecito section, southern California: implications for early history of the northern Gulf of California and Colorado delta. Doctoral Dissertation, University of Arizona, Tucson 494 p.

Winker, C.D., and S.M. Kidwell 1986a. Paleocurrent evidence for lateral displacement of the Pliocene Colorado River delta by the San Andreas Fault system, southeastern California. Geology 14:788-791.

----- 1986b. Planispastic paleogeographic model for the Neogene Colorado delta and northern Gulf of California. Geological Society of America Abstracts with Programs 18:199.

----- 1986c. Stratigraphic sequence of the Pliocene Colorado delta, Fish Creek-Vallecito section, western Salton Trough, southern California. Geological Society of America Abstracts with Programs 18:199.

Winterer, J.I. 1975. Biostratigraphy of the Bouse Formation: a Pliocene Gulf of California deposit in

California, Arizona and Nevada. Master of Science Thesis, California State University, Long Beach 132 p.

Wise, J. 1991. Structural geometries of the Santa Rosa Mountains detachment terrane along the western margin of the Salton Trough. Undergraduate Research Report, Department of Geology, San Diego State University, California, 64 p.

Wolski, W. 1989. Borrego. Glyph p. 17-21.

Wood, H.O. 1941. Seismic activity in the Imperial Valley, California. Seismological Society of America Bulletin 31:245-254.

----- 1942. Earthquakes and disturbances to leveling in the Imperial Valley, California. Seismological Society of America Bulletin 32:245-254.

Woodard, G.D. (ca. early 1960's). Neogene stratigraphy of the Vallecito badlands, western Colorado Desert. Manuscript on File, Anza-Borrego Desert State Park, California 2 p.

----- 1962. Stratigraphic succession of the west Colorado Desert, San Diego and Imperial Counties, southern California. Geological Society of America Bulletin 68:26-31.

----- 1963. The Cenozoic succession of the west Colorado Desert, San Diego and Imperial Counties, southern California. Doctoral Dissertation, University of California, Berkeley 173 p.

----- 1966. Neogene stratigraphy of the Vallecito Badlands western Colorado Desert. Society of Vertebrate Paleontology Field Trip, Manuscript on File, Anza-Borrego Desert State Park, California 6 p.

----- 1974. Redefinition of Cenozoic stratigraphic column in Split Mountain Gorge, Imperial Valley, California. American Association of Petroleum Geologists Bulletin 58:521-539.

Woodburne, M.O. 1987. Cenozoic mammals of North America, geochronology and biostratigraphy. University of California Press, Berkeley 336 p.

Woodring, W.P. 1931. Distribution and age of the marine Tertiary deposits of the Colorado Desert. Carnegie Institute of Washington Publication 418:1-25.

Wosser, T.D., D. Campi, M.A. Fovinci, and H.J. Degenkolb 1980. On the earthquake-induced failure of the Imperial County Services Building. In Imperial County, California, Earthquake, October 15, 1979, edited by D.J. Leeds, Earthquake Engineering Research Institute, Berkeley, California p. 159-172.

Wosser, T.D., D. Campi, M.A. Fovinci, and W.H. Smith 1982. Damage to engineered structures in California. In The Imperial Valley, California, Earthquake of October 15, 1979. U.S. Geological Survey Professional Paper 1254:273-288.

Woyski, M.S., and A.H. Howard 1987. A section through the Peninsular Ranges Batholith, Elsinore Mountains, southern California. In Centennial Field Guide, edited by M.L. Hill, Geological Society of America, Cordilleran Section 490(1):185-190.

Wright, F.F. 1962. A deformed layer within a Miocene turbidite, Split Mountain, California. Manuscript on File, Anza-Borrego Desert State Park, California 9 p.

Wright, L.B. 1946. Geology of the Santa Rosa Mountain area, Riverside County, California. California Journal of Mines and Geology 42(1):9-13.

Wyatt, F.K., and D.C. Agnew 1988. The 1987 Superstition Hills Earthquakes; strains and tilts at Pinon Flat Observatory. EOS, Transactions of the American Geophysical Union 69:1423-1424.

Wyss, M., J.N. Brune, and C.R. Allen 1969. Slippage on the Superstition Hills, Imperial-Banning-Mission Creek and Coyote Creek Faults associated with the Borrego Mountain Earthquake of 9 April 1968. EOS, Transactions of the American Geophysical Union 50:252.

Wyss, M., and T.C. Hanks 1972. Source parameters of the Borrego Mountain Earthquake. In The Borrego Mountain Earthquake of April 9, 1968, U.S. Geological Survey Professional Paper 787:24-30.

Yau, Y.-C. Microscopic studies of hydrothermally metamorphosed shales from the Salton Sea geothermal field, California. Doctoral Dissertation, University of Michigan, Ann Arbor 174 p.

Yau, Y.-C., and D.R. Peacor 1985. Wide-chain Ca-pyribole and actinolite intergrowths in primary euhedral crystals, Salton Sea geothermal field, California. EOS, Transactions of the American Geophysical Union 66:373.

Yau, Y.-C., D.R. Peacor, and E.J. Essene 1986. Occurrence of wide-chain Ca-pyriboles as primary crystals in the Salton Sea geothermal field, California, U.S.A. Contributions to Mineralogy and Petrology 94:127-134.

----- 1987. Authigenic anatase and titanite in shales from the Salton Sea geothermal field, California. Neues Jahrbuch fur Mineralogie Monatshefte 1987(10):441-452.

Yau, Y.-C., D.R. Peacor, E.J. Essene, and S.D. McDowell 1988. Microstructures, formation, mechanisms, and depth-zoning of phyllosilicates in geothermally altered shales, Salton Sea, California. Clays and Clay Minerals 36:1-10.

Yi, Z. 1987. On the shallow hydrothermal regime of the Salton Sea geothermal field, California. Master's Thesis, University of California, Riverside 87 p.

Youd, T.L. 1984. Liquefaction during the 1981 and previous earthquakes near Westmorland, California. U.S. Geological Survey Open-File Report 84-680:1-38.

Youd, T.L., M.J. Bennett, P.C. McLaughlin, P.C. Sarmiento,

and G.F. Weiczorek 1983. Liquefaction studies in the Imperial Valley, California, a natural laboratory. Geological Society of America Abstracts with Programs 15:373.

Youd, T.L., and R.O. Castle 1970. Borrego Mountain Earthquake of April 8, 1968. American Society of Civil Engineers Proceedings, Journal of the Soil Mechanics and Foundations Division 96:1201-1219.

Youd, T.L., and G.F. Weiczorek 1982. Liquefaction and secondary ground failure. In The Imperial Valley, California, Earthquake of October 15, 1979. U.S. Geological Survey Professional Paper 1254:223-246.

Youngs, F.O., W.G. Harper, J. Thorp, and M.R. Isaacson 1929. Soil survey of the Yuma-Wellton area, Arizona-California. U.S. Soil Conservation Service, Bureau of Chemistry and Soils 20:1-37.

Youngs, L.G. 1984. An annotated bibliography of geothermal information published or authored by staff of the California Division of Mines and Geology. California Division of Mines and Geology Special Publication 68.

----- 1988a. Aeromagnetic map of the San Diego/El Centro 0.5° by 4° quadrangle, California. California Division of Mines and Geology Open-File Report DMG OFR-88-09.

----- 1988b. Aeromagnetic map of the Salton Sea 1° by 2° quadrangle, California. California Division of Mines and Geology Open-File Report DMG OFR-88-15.

Younker, L.W., and P.W. Kasameyer 1976. Salton Sea geothermal field, source of the anomaly. EOS, Transactions of the American Geophysical Union 57:1017.

----- 1978a. A revised estimate of recoverable thermal energy in the Salton Sea geothermal resource area. University of California, Riverside, Lawrence Livermore Laboratory Report 52450:1-13.

----- 1978b. Application of the "leaky transform" model to resource estimation in the Salton Sea known geothermal field, California. Geological Society of America Abstracts with Programs 10:521.

Younker, L.W., P.W. Kasameyer, and J. Hanson 1980. Application of hydrothermal modeling to the selection of deep drilling sites in the Salton Sea geothermal field. EOS, Transactions of the American Geophysical Union 61:1149.

Younker, L.W., P.W. Kasameyer, and J.D. Tewhey 1982. Geological, geophysical, and thermal characteristics of the Salton Sea geothermal field, California. Journal of Volcanology and Geothermal Research 12:221-258.

Zakrzewski, R.J. 1972. Fossil microtines from late Cenozoic deposits in the Anza-Borrego Desert, California, with the description of a new subgenus of *Synaptomys*. Los Angeles County Museum of Natural History Contributions in Science 221:1-12.

Zernow, R.H. 1975. An evaluation of horizontal heat flow measurement for the detection of water movement in the Mt. Laguna area. Undergraduate Research Reports, Department of Geology, San Diego State University, California 26(2):1-22.

Zhang, Y., W. Thatcher, and R.A. Snay 1988. Coseismic slip in the 1940 and 1979 Imperial Valley earthquakes and its implications. EOS, Transactions of the American Geophysical Union 69:1433.

Zimmerle, W. 1971. Akzessorischer allanit im Rattlesnake-Granite (San Diego County, Sudkalifornien) und seine allegemeine petrogenetische Stellung. Abstracts of North American Geology p. 1746.

Zimmerman, R.P. 1981. Soil survey of Imperial County, California, Imperial Valley area. U.S. Department of Agriculture, Soil Conservation Service 112 p.

Zukin, J.G. 1986. Uranium and thorium series isotopes in the Salton Sea geothermal field, southeastern California: their applications in determining the rates of brine-rock interaction and radionuclide transport. Master's Thesis, University of California, Los Angeles.

Zukin, J.G., D.E. Hammond, T.L. Ku, R.A. Marton, and W.A. Elders 1985. Uranium and thorium radionuclides in the Salton Sea geothermal brines. EOS, Transactions of the American Geophysical Union 66:1144.

----- 1985. Uranium and thorium radionuclides in brines and reservoir rocks from two deep geothermal boreholes in the Salton Sea geothermal field, southeastern California. Geochimical et Cosmochimica Acta 51:2719-2731.